Regions of risk

THEMES IN RESOURCE MANAGEMENT
Editor, Professor Bruce Mitchell, University of Waterloo

Already published:
Adrian McDonald and David Kay: Water Resources: issues and strategies
L. Graham Smith: Impact Assessment and Sustainable Resource Management

Kenneth Hewitt

Regions of Risk

A geographical introduction to disasters

LONGMAN

Addison Wesley Longman Limited
Edinburgh Gate, Harlow
Essex CM20 2JE
England
and Associated Companies throughout the World

First published 1997

ISBN 0 582 21005 4

British Library Cataloguing-in-Publication Data
A catalogue record for this book is
available from the British Library.

Library of Congress Cataloging-in-Publication Data
A catalog entry for this title is
available from the Library of Congress.

Produced by Longman Singapore Publishers (Pte) Ltd.
Printed in Singapore

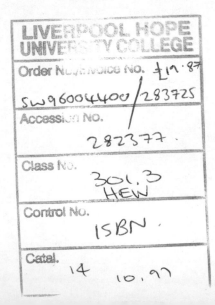

'. . . No amount of calamity which merely befell a man, descending from the clouds like lightning, or stealing from the darkness like pestilence, could alone provide the substance of [this] story . . . . The calamities . . . do not simply happen, nor are they sent; they proceed mainly from actions, and those the actions of men.'

A. C. Bradley (1906, 11)

Contents

List of figures   xi
List of tables   xiii
Foreword   xvii
Acknowledgements   xix

Introduction   **Danger and modernity**   1
Chronic and extreme dangers   4
The 'geographicalness' of risks   12

Part 1   **Approaches to risk and disaster**   19

Chapter 1   **Risk and damaging events**   21
Approaching danger   21
Perspectives on risks   24
Summary   39

Chapter 2   **The 'geographicalness' of disaster**   40
The space of danger and places of disaster   40
Dangerous geography   42
Where can one be safe?   50
Geographical calamities   53

Chapter 3   **Natural hazards**   55
Introduction   55
Domains and classes of natural hazard   63
Forms of endangerment   71
Floods and droughts   80

Contents

Chapter 4	**Technological hazards**	91
	The scope of technological risks	91
	Occasions of technological harm	93
	Technological hazards in context	100
	Transcendent hazards	107

Chapter 5	**Social hazards: violence and the disasters of war**	111
	The problem of violence	111
	Violent hazards	116
	Forms of damage and processes of harm	128

Chapter 6	**Vulnerability perspectives: the human ecology of endangerment**	141
	The idea and scope of vulnerability	141
	The anatomy of insecurity	143
	Interpreting vulnerability	151
	Vulnerability and disease	157
	Geographies of vulnerability	164

Chapter 7	**Active perspectives: responses to disaster and adjustments to risk**	169
	Introduction: questions of response	169
	The context and scope of action	170
	Organisation, practice and public policy	188
	Choices, values and powers	191

| Part 2 | **Communities at risk, places of disaster** | 195 |

Chapter 8	**'Unnatural' disasters: the case of earthquake hazards**	197
	Earthquake damage	202
	Earthquake risk	212
	The space of disaster events	219
	Earthquake country	224

Chapter 9	**Contexts of risk: mountain land hazards and vulnerabilities**	232
	Mountain habitats and risks	234
	Mountain hazards and disasters	236
	Mountain land vulnerabilities	253
	Disasters and development: two mountain land examples	255

Chapter 10	**Risk in the city**	266
	Urban problems	266
	Hazards and urban disasters	269
	Safe cities or defenceless spaces?	293

Chapter 11   **Place annihilation: air war and the vulnerability of cities**   296
              The disasters of air war   296
              The vulnerability of urban areas and civil life   305
              Place annihilation: an interpretation   319

Chapter 12   **The Holocaust: genocide and geographical calamity**   321
              The hazards: processes of the Holocaust   329
              Vulnerability and genocide   336
              Alternative adjustments to genocide   341

Concluding remarks   The perspective of ideas   349

Appendix   361

Bibliography   365
Index   383

List of figures

1.1 Elements of risk and their place in damaging events   31
3.1 Major disasters around the world, 1963–1992   60
3.2 Global distribution of major disasters around the world, 1963–1992   62
3.3 Riverine floods and low flows as threshold hazards: schematic diagrams to show the relations between flood height, extent and frequency   85
4.1 Airborne release of dangerous chemicals worldwide, reported 1900–1990   95
5.1 The global distribution of major armed conflicts in 1994   113
5.2 Deaths from armed violence, 1912–1990   118
5.3 Nuclear hazards from weapons, aging reactors and dumping in the former Soviet Union   126
5.4 Main areas of armed conflict, refugee movements and places of refuge in Africa during the late 1980s and early 1990s   134
5.5 Main areas of armed conflict, refugee movements and places of refuge in South Asia during the late 1980s and early 1990s   135
6.1 The HIV/AIDS pandemic. Estimates of victims to 1992 by major world regions   160
8.1 Global distribution of major damaging earthquakes, 1950–1979   203
8.2 The Lomo Prieta, San Francisco, earthquake of 17 October, 1985   216
8.3 The location of the 1985 earthquake in Mexico City and distances from the epicentre   221
8.4 Intervening conditions, vulnerability and major destruction in Mexico City from the 1985 earthquake   223
8.5 The relations between climatic dryness and earthquake disasters geography   228
9.1 The violent transformation of the people of Afghanistan   249
9.2 The global hard drugs traffic   252

9.3   The Karakoram Himalaya, showing the large cover of perennial
      snow and ice and extensive intermontane river basins, where
      settlement occurs   256
9.4   Locations of some of the large natural processes widely occurring
      in the upper Indus Basin that have been associated with disasters   259
9.5   Large earthquakes and volcanic eruptions affecting southwest
      British Columbia   263
9.6   Landslides and related events in southwest British Columbia   264
10.1 Disasters affecting urban centres in South America, 1974–1994,
      reported in mass media   270
10.2 Disasters affecting urban centres in South Asia, 1974–1994,
      reported by mass media   271
10.3 Location of short-term damaging events affecting the general
      public in southern Ontario, 1973–1984   274
10.4 Typhoons and the city of Osaka, Japan   281
10.5 Changes in the populations of the capitals of former North and
      South Vietnam between 1943 and 1979   291
11.1 The bombing of British cities, 1940–1945   302
11.2 The bombing of German cities: urban centres with civilian death
      tolls in excess of 700 from air raids in the Second World War   304
11.3 The bombing of Japanese cities: urban area razed in the attacks   306
11.4 The area of Tokyo burned out in the 1945 fire raids   308
11.5 Civilian deaths from air attacks on London shown by borough   310
11.6 Houses demolished by air attacks on Berlin during the
      Second World War, by district   311
11.7 Population change in Hamburg due to the disaster raids   313
11.8 Population change in Tokyo between 1940 and 1945 by city wards   315
11.9 Inner city devastation in Germany   318
12.1 Estimates of Jews from different national areas of Europe killed
      in the Holocaust   322
12.2 Estimates of Romany and Sinti in different national areas killed
      by the Nazis   323
12.3 Routes for the transport of Jews to the death camp at
      Auschwitz-Birkenau   332
12.4 The larger camps and main types of camp in the 'Holocaust
      system' of Germany and the occupied countries during the
      Second World War   335

List of tables

1   The main causes of untimely death and death tolls in selected countries, 1990   6

2   Natural, technological, epidemic disease and war disasters in the late twentieth century   7

3   International action on natural disasters   14

1.1   Hazards: broad classes of dangerous agents that may cause exceptional damage and disasters   26

1.2   Vulnerability: some of the basic forms in which it arises   27

1.3   Profiles of damage in three disasters   32

1.4   Distinctive social conditions of disaster and disaster response   35

1.5   Temporal sequences that may be involved in disasters, with reported durations and selected features of each phase   37

3.1   Classes of specific hazardous agent or condition, and the situations or events in which they are associated with actual disasters   56

3.2   Natural disasters 1947–1980   61

3.3   Atmospheric hazards: classes and examples of associated disasters with some estimates of losses   64

3.4   Hydrological hazards: classes and selected examples of associated disasters   66

3.5   Geological hazards: classes and examples of associated disasters   68

3.6   Biological hazards: classes and examples of associated disasters   69

3.7   Examples of fog-related disasters   73

3.8   Examples of snow-related disasters   75

3.9   Examples of flood catastrophes with loss of life, displacement and damage estimates where available   82

4.1   Numbers of technological disasters reported 1989–1993   94

4.2   Technological catastrophes: examples of large-scale devastation or casualties   97

4.3	Classes of technological hazard	101
5.1	Global incidence of armed conflicts, 1945–1992	112
5.2	Global distribution of armed conflicts, in regions with at least one major war between 1989 and 1992	112
5.3	Hazards of violence: dangerous agents and methods of indiscriminate warfare	117
5.4	'Dubious' agents of war and other armed violence	119
5.5	Incendiary attacks upon German cities in which firestorms were generated	120
5.6	Uses of dubious weapons: chemical, biological and nuclear weapons in the twentieth century	122
5.7	Estimated numbers of land mines in 1993, for the most severely affected countries	124
5.8	Forms of damage by violence: the processes of harm to which civilians, civil life and the habitat are subject	129
5.9	Traumatic experience of children in a war zone	132
5.10	Major examples of enforced displacement, 1900 to 1990	136
6.1	Forms and conditions of vulnerability	144
6.2	Active forms of vulnerability, with particular stress upon those that are socially imposed and enforced	154
6.3	The forms of geographical vulnerability: vulnerable places and people	165
6.4	Enforced geographical vulnerability or endangerment through geography	166
7.1	Alternative adjustments to natural hazards, or the theoretical range of adjustments to geophysical events	173
7.2	The range of adjustments to war hazards	174
7.3	Extending the notion of alternative adjustments to include actions to mitigate vulnerability and adverse contexts of risk	178
8.1	Earthquake disasters. Selected examples of catastrophic earthquakes, 1956–1995	198
8.2	Economic devastation in agricultural village communities, Maharashtra, India, 1993	205
8.3	Earthquake intensity: the Modified Mercalli Scale	207
8.4	Examples of earthquake disasters in which loss of strength and collapse of 'susceptible' soils or alluvial deposits resulted in major devastation	209
8.5	Selected examples of earthquake disasters, 1950–1995, in which a major part of the damage involved landsliding	210
8.6	Selected earthquakes with severe, sometimes mainly, tsunami damage	211
8.7	Earthquake disasters in which natural damming occurred, and where the bursting of these or artificial dams caused major damage	212

8.8	Selected earthquake disasters of 'moderate' magnitudes showing the large range of damage and loss of life	214
8.9	The social composition of earthquake deaths in agricultural village communities, Maharashtra, India, 1993	219
8.10	Geographical and damage attributes of the 243 most destructive earthquake disasters reported, 1950–1990	226
9.1	Examples of natural disasters in mountain lands associated with catastrophic Earth surface processes	238
9.2	Distribution of major natural disasters affecting mountain regions, 1953–1988	240
9.3	Examples of natural dam-break or outburst floods in mountain regions with disastrous consequences	241
9.4	Examples of the failure of artificial dams in mountainous terrain with catastrophic outbreak floods	245
9.5	Armed conflicts taking place in 1991 by major world regions, identifying those involving mountain lands	247
9.6	Armed conflicts involving mountains lands, 1900–1980	247
10.1	The degree and rates of urbanisation	267
10.2	Cities with more than 10 million residents in 1950, 1980 and projected for 2000	268
10.3	Disasters involving major damage in cities of Latin America and the Caribbean, 1974–1994	272
10.4	Short-term damaging events affecting the general public in southern Ontario, 1973–1984	273
10.5	HIV/AIDS and urbanisation	285
10.6	The HIV/AIDS epidemic in the United States	286
10.7	Examples of large evacuations from urban areas in the United States as a result of chemical accidents	288
10.8	A statistical profile of the civilian war victims in South Vietnam, 1954–1975	292
11.1	Estimates of the urban and civilian losses from raids on cities in Britain, Germany and Japan during the Second World War	297
11.2	The most lethal raids against British cities in the Second World War	303
11.3	Summary of civil losses in the twelve most lethal attacks on German cities	305
11.4	Distribution of civilian casualties in Hamburg	307
11.5	Summary of damage in the most destructive air attacks on Japanese cities in the Second World War	309
12.1	Processes of the Holocaust	328
12.2	The death camps: high and low estimates of the numbers killed	336
12.3	Adjustments to the Holocaust	343

Foreword

The Themes in Resource Management Series has several objectives. One is to identify and to examine substantive and enduring resource management and development problems. Attention will range from local to international scales, from developed to developing nations, from the public to the private sector, and from biophysical to political considerations.

A second objective is to assess the responses to these management and development problems in a variety of world regions. Several responses are of particular interest but especially *research* and *action programmes*. The former involves the different types of analysis which have been generated by natural resource problems. The series will assess the kinds of problems being defined by investigators, the nature and adequacy of evidence being assembled, the kinds of interpretations and arguments being presented, the contributions to improving theoretical understanding as well as resolving pressing problems, and the areas in which progress and frustration are being experienced. The latter response involves the policies, programmes and projects being conceived and implemented to tackle complex and difficult problems. The series is concerned with reviewing their adequacy and effectiveness.

A third objective is to explore the way in which resource analysis, management and development might be made more complementary to one another. Too often analysts and managers go their separate ways. A good part of the blame for this situation must lie with the analysts who too frequently ignore or neglect the concerns of managers, unduly emphasise method and technique, and exclude explicit consideration of the managerial implications of their research. It is hoped that this series will demonstrate that research and analysis can contribute both to the development of theory and to the resolution of important societal problems.

This book by Ken Hewitt is the eleventh and final book in the series. Over the past five or so years, numerous books have been published with a focus on "hazards", but Ken's is distinctive, and makes several significant contributions.

First, most of the other books represent the "dominant view" in hazards work by emphasising damaging agents and events. Ken's book provides a critique of the dominant view. He then offers a constructive alternative which can best be characterised as a *human ecology perspective of disaster*, with particular attention to the distribution of human vulnerability, intervening conditions and responses. Ken argues persuasively that rather than being exceptions, disasters arise out of normal life and living. In that regard, it is essential to be able to identify and consider the vulnerability of peoples, rather than concentrate only on the triggering agent of a disaster. Second, most other books dealing with hazards focus mainly upon "natural" or "technologically induced" hazards. Ken examines both of those, but also considers the hazards which occur from people's violence towards other people. As far as I am aware, no other geographers have examined the disasters and hazards generated by human violence towards other people in the manner in which he has. Third, Ken poses a profound challenge to hazards researchers and policy makers. The dominant view solution is inevitably technology-demanding and expensive. It may be that our societies can no longer afford the costs of solutions generated from the dominant perspective. It is therefore important that more attention should be given to other solutions which emerge from considering vulnerability of populations at risk. Such solutions are normally much less costly than those emerging from the dominant perspective.

Ken Hewitt brings much experience and insight to writing a book focused on human ecology, vulnerability and disasters. He has decades of field work related to these themes in the high mountain environments of Pakistan and other countries. In 1983, he edited *Interpretations of Calamity from the Viewpoint of Human Ecology*, in which he presented his initial critique of the dominant approach to hazards research. And in 1991, he received the annual Award for Scholarly Distinction in Geography from the Canadian Association of Geographers.

Bruce Mitchell
Waterloo, Ontario

December 1995

Acknowledgements

For background research on material used in the book I thank Dr Jayati Ghosh (industrial and Asian Disasters); Michelle Darvill (bombing of cities); Paul Blais and Tara Hewitt (general disaster inventories); Ian Gilbart and Tom Hammers (urban disasters in Asia and Latin America); George Yap (mountain hazards in Asia). For helpful comments and pointers on disease hazards, thanks to Dr Jodi Decker. Wilfrid Laurier University's Office of Research kindly provided a Book Preparation Grant. Original research results reported in Chapters 8–12 are from studies funded, in part, by the Social Science and Humanities Research Council (SSHRC), Ottawa, and by Wilfrid Laurier's Office of Research. The Department of Geography, especially Chairman Dr Alfred Hecht and my colleagues for many kindnesses. I thank Carol Allemang and Jo-Anne Horton for preparing the manuscript; Pam Schaus and Rachel Moser for fine and rapid preparation of the maps and diagrams. Throughout I received invaluable comments on the text by Dr Farida Hewitt and our daughters Nina and Sonia Hewitt. Last but not least, I acknowledge the fine support of Dr Bruce Mitchell, his painstaking and valuable commentaries on the manuscript.

Over the years my ideas have been influenced, in particular, by conversations with and the work of Drs Ian Burton, Jerry Hall, Ann Larimore, Kenneth Mitchell, George E.B. Morren Jr, Joseph Nipper, Gilbert F. White and Andrew P. Vayda. I thank them without implying any blame for what is lacking here.

sented at the World Conference on Natural Disaster Reduction, Yokohama, Japan; Fig. 5.1 from *Armed Conflicts Report 1995*, Institute of Peace and Conflict Studies, Conrad Grebel College, Ontario (Spring 1995); Table 5.7 from a table containing estimates of Landmines per country from Department of Humanitarian Affairs news, *DHA News* Sept./Dec. 1993, UNDHA; Table 5.9 from *Journal of Refugee Studies* p.7 and p.9 by permission of Oxford University Press; Fig. 4.1 (pp.104–6) and Table 10.7 (p.108) from *Living with Risk*, Hodder & Stoughton Ltd (Cutter, S.L.); Fig. 8.4a from *At Risk: natural hazards, people's vulnerability, and disasters* p.176, Routledge (Blaikie P.T., T. Cannon, L. Davis, & B. Wisner, 1994); Figs. 8.4b and c from *Geohazards: natural and man-made* p.101, Chapman and Hall (McCall, G.J., D.J.C. Larning & S.C. Scott eds, 1992); Fig. 9.1 from Afghanistan: the destination of the people, *Orbis* **31**:1 pp.39–56, p.54, Foreign Policy Research Institute, Philadelphia (Sliwinski, Winter 1989); Fig. 10.5 from *The price of war: Urbanisation in socialist Vietnam* 1954–86, pp.147 and 154, Routledge. (Thrift & Forbes, 1986); Table 10.6 from *The AIDS disaster* p.56, Yale University Press (Perrow, Guillen, 1990); Table 10.8 from *Victims and Survivors*, Quorum 1982 (Weisner, 1988) reprinted with permission of Greenwood Publishing Group Inc., Westport, Conn. © 1988.

For archival materials quoted in Chapters 2 and 11, the National Archives, Washington DC (Modern Military Branch); the Public Record Office, London, Ministry of Home Security files, and the Tom Harrisson Mass Observation archive at the University of Sussex. Detailed citations appear in the relevant papers by Hewitt cited in the text.

Introduction: danger and modernity

'Repetitive disaster events chip away . . . at conventional ways of ordering reality. The disconfirming, anomalous cases, refuse to disappear. Where single disasters do severely shake existing conceptions of order, these conceptions were already under serious and protracted stress.'

(Michael Barkun, 1974, 81)

The *Titanic*'s century?

A luxury liner glides swiftly across a darkened ocean. Lights blaze on the upper deck, where an orchestra plays, and first class passengers, some of them well-known personalities, are set to dine and dance the night away. Below decks is another world. More than 1500 persons prepare for sleep, many of them quite poor families in crowded berths, who dream of a better life in America. Here too, however, passengers and crew seem to have complete confidence in the safety of the ship. It is of the latest design. There are novel features said to make it 'unsinkable'. Its name, *Titanic*, speaks of colossal strength. They are entering ice-infested waters of the North Atlantic, south of the Grand Banks, Newfoundland, but reports of icebergs, from ships up ahead, go unheeded. The captain and his company, facing strong competition from other shipping lines, worry more about completing the voyage in a record time.

Just before midnight the vessel strikes an iceberg while travelling at about 41 km per hour. A huge gash in the hull causes it to sink two and a half hours later; 1522 people are drowned and 705 survive in lifeboats or rafts. Rich and poor perish, but deaths are disproportionate among those from the lower decks. The lifeboats prove insufficient for the number of passengers on board. Some cannot be launched and the crew seems to lack the training for such an emergency. There are stories of women and children being pushed aside or abandoned.

1

Introduction: danger and modernity

The date is 14 April 1912, early in a century that will bring many other and even larger calamities, but the sinking of the *Titanic* continues to fascinate and trouble later generations. Thanks largely to some costly safety systems put in place then, there has not been another such disaster in the North Atlantic.

Let us now move to 1995. The century is coming to an end. People are talking of the next millennium, about to start. Few places seem better placed to grow and prosper in it than the Japanese city of Kobe. Located on Osaka Bay, it has been a major centre in the development of international trade. This, in turn, has made Japan an economic superpower. The city is a favoured place for foreign corporations to base their activities. Most of Kobe's buildings have been erected since 1945, when a wartime incendiary raid devastated the city. Many of the main communications, port, municipal and industrial structures are less than 20 years old, built with the latest construction techniques. A rash of new building is underway in the ever-expanding industrial and container port areas.

It should be added Japan has a long history of earthquake disasters, but it is also a leader in the science of earthquakes, and in earthquake-resistant building design, engineering and planning. It has world-class institutions specialising in disaster research and training. Nevertheless, a great tragedy occurred in Kobe on 17 January 1995. An earthquake, centred about 20 km to the west, devastated the city, killing 4512 of its inhabitants (6336 in all). Some 122 500 buildings were destroyed. Arterial highways, the track of the famous 'Bullet Train' and most of the port works suffered major damage. The economic losses were put at around US$110 billion to US$150 billion – the highest ever actually quoted for a natural disaster. There were complaints of slow and ineffective emergency response, of official complacency and a history of failures to enforce building safety standards. Fires raged for days, unchecked, in crowded residential districts. A year later, almost 100 000 people remain in temporary, prefabricated housing. As many more are said to have moved to other parts of Japan in search of work and a new start.

Kobe is by no means alone in an apparent, remarkable lack of safe construction and emergency facilities in an earthquake-prone city. Japan is not the only wealthy country with a reputation for competence in dealing with certain hazards, yet these still cause major disasters. Recent earthquake, hurricane and flood disasters in the United States, for example, follow many decades of 'advanced' flood, storm and earthquake research. There have been vast outlays on protective measures against these dangers. Severe storm disasters in Britain afflict a nation that pioneered the study of the atmosphere and for long led the world in weather forecasting.

The sinking of the *Titanic* or the Kobe earthquake, like all great tragedies, were special and unique events, yet for those who study the problem of extreme risks there are compelling parallels with other disasters. We will find that many of the questions they raised are typical of the failures of technology and preparedness. They symbolise so many developments that are poorly constrained by concerns of public safety.

In many other ways, the modern world seems torn between conflicting visions of its achievements and its destiny. On one side are great improvements in material life, at least for some, and a promise of the same for all. On the other are expanding dangers from our own activities and evidence of human misery from uncontrolled natural forces or endangered habitats.

The main concerns of this book – dangers, disaster and the scope of human vulnerability to them – focus on one side of this dichotomy. They deal with daily reminders of the limitations and self-inflicted failures of modernity, but these, or how we view them, are not independent of the claims of technical and material 'progress'.

Destructive earthquakes in Japan or southern California, the AIDS epidemic in Canada, or several great oil spills on the coasts of Western Europe, are special cause for concern. They invite critical re-examination of the treatment of disaster in modern settings. Except in major wars, the human costs of disaster in wealthier countries tend to be less severe than in impoverished ones. Even so, they highlight the unfinished business in dealing with extreme risks where societies seem best equipped for the task.

Outside urban-industrial societies and wealthier enclaves, the dichotomy appears in terms of 'development' and disaster. Half a century of actual or promised modernisation is associated mainly with adoption or transfer of Western technical, educational, economic and administrative measures, and with economic globalisation. In nearly all countries, it has brought exponential growth in material production. With this, energy consumption, mechanisation, populations and urbanisation have surged. So too have poverty, disease, famine and social violence in most of the same countries.

The link between poverty and disaster is strongly drawn. We are told that perhaps a billion persons, a fifth of them children, suffer chronic hunger and malnutrition in a world where per capita, not merely total, food production exceeds that known in any other period. In poorer countries, natural disasters are singled out by huge death tolls, usually among the more impoverished people. Armed conflicts cause massive uprootings of defenceless folk and magnify disasters of famine and disease. An epidemic of human rights abuses afflicts civil populations and threatens the survival of minorities and indigenous peoples. A majority of the victims of disaster are not only relatively poor, but powerless, suggesting that lack of a political voice also places them at greater risk.

The emergence of a problem field

It is in this context that a modern field of hazards and disasters has taken shape. It is not concerned with all forms of danger and loss or with only some particular types of disaster. Rather, it has become engaged with those that seem to lie beyond, or to overwhelm, existing arrangements for public safety. Disasters, or dangers that threaten them, are obviously of this kind.

3

Since the Second World War, risk and disaster have taken their place among a number of related problem fields. They include 'the environment', 'development', (over)population, 'security' and cultural survival. Often they are discussed as though they are self-evident empirical realities, age-old, if suddenly more serious. In fact, each has been given new or redefined meanings over the past half century, and become a focus of popular and highly politicised concern. Notions like 'disaster' have been gradually, but in the end radically, redefined to suit modern, instrumental agendas that link science to professional and administrative, especially governmental, practices. If not exactly coherent disciplines, these fields have developed around 'communities' of researchers and agencies. Distinct from but drawing upon traditional scientific disciplines, the fields are commonly described as interdisciplinary.

The problem field that the present text seeks to introduce and critically review concerns dangers to human societies, especially the more extreme or novel forms. Those having a calamitous potential are the major interest. They may involve natural forces, large technological 'accidents' or the more severe consequences of social violence. However, we must also consider some pervasive, usually novel, dangers that societies seem ill-prepared for or unable to deal with. They may arise from technological innovations, dangerous consumer products, toxic industrial pollutants, drug-resistant diseases, global environmental deterioration or violent social change.

Although the phrase 'public safety' has not always had a happy history, it seems appropriate to our concerns. They mainly address dangers to substantial parts of communities, regions or states, even the global scene. Private risks or the internal safety measures of well-defined organisations are of interest, but to the extent that they affect or reflect public dangers.

Hazards and disasters have been seen mainly as an 'applied' field for science and scholarship. However, distinct questions of concept and method are raised and we must address them. Enquiry must satisfy the usual requirements of logic and critique, yet the work is never far from questions of practice and priority, of ethical and responsible conduct.

Before saying more about the scope and implications of a hazards and disasters field, some of the reasons for its existence must be presented. Since just about everything involves risk and may be harmful, the choice of dangers assumes a special significance. For that it is useful, initially, to recognise two broad domains of risk.

Chronic and extreme dangers

Routine risks

There are dangers that seem ever-present in the life, work and habitat of a given society. They might include endemic pests and blights that reduce crop yields in every year; the almost continuous toll of traffic accidents worldwide; a variety of common crimes and illnesses; or the childhood malnutrition that prevails in some regions. Household and occupational injuries, the risks of overeating, and substance abuse and cigarette smoking could be added. Such problems involve chronic or 'life-style' hazards. They are strongly reflected in gross statistics of a society's health, causes of death, life expectancy, nutritional status, disabled or unemployed workers, violent crimes and so forth. Some result in the largest overall, 'untimely' loss of life (Table 1).

Such risks tend to become integral or accepted, if feared, parts of everyday life. Their treatment is usually or mainly *routine*. They are not associated with marked or any interruptions in productive and administrative functions. They are, in some sense, absorbed within everyday activities, whether through public institutions or being left to households and personal decision. Permanent arrangements for dealing with them may, however, involve the larger areas of investment in risk reduction or management. Defence budgets and law enforcement in most countries, health, accident and unemployment insurance in some, direct a major share of public resources to particular dangers. This serves to define social priorities and shape the permanent risk environment. Specific examples would be the large and, for the public, largely 'hidden' costs of meeting standards for building materials and design in European cities, or the annual costs of winter snow removal and sanding of roads in Canadian cities to minimise traffic delays and accidents.

The practices associated with these chronic risks are as characteristic of a given culture as its food or festivals. They may be seen in hospital waiting rooms or funeral processions, a spot in the nightly news or the way word of premature death passes through a village. Increasingly, they include dangers associated with consumer goods and public services.

That they are common and widespread does not mean the dangers are shared equally by all. Some are accepted as 'occupational hazards', the necessary burden of a way of life or the tolerable price of a certain pleasure. Others are unwanted but inescapable predicaments – for the disabled, chronically sick or those obliged to cope with hazardous work places or not working at all. Risks may be forcibly imposed upon disadvantaged and exploited groups.

In sum, what links and distinguishes or perhaps contains these risks socially is their dispersed but widespread incidence, and routine treatment. Equally relevant to our concerns, in modern societies they are identified with prominent fields of study such as medicine, criminology, actuarial science, engineering safety and military history. To the extent that these chronic dangers are well-understood by established fields, and well-managed by responsible institutions, they are not the ones our field becomes involved with, though they are a key part of our working environment.

Extreme events

By contrast, there are threats and levels of damage that can overwhelm whole communities, or cripple aspects of everyday life. At least, they may bring widespread public anger and loss of confidence in existing safety measures. A period

Table 1 Chronic dangers: the main causes of untimely death and death tolls in selected countries, 1990. For comparison, numbers per 100 000 males and females are also given. The figures dwarf those for disasters, except for some wars and epidemics. However, the concentration in space and time of disaster losses can drastically impair the coping capacities of societies and magnify impacts upon survivors (after World Health Organisation, 1994)

	USA	Canada	Mexico	Japan	United Kingdom	Italy
Population (million)	248.7	26.6	81.1	123.6	55.5	57.7
All causes	2 148 463	191 973	421 736	820 305	641 799	543 708
M (per 100 000)	918.4	793.9	566.6	736.5	1123.1	1006.5
F (per 100 000)	812	652.7	432.0	602.8	1113.0	882.9
Circulatory systems	920 245	74 091	84 792	304 448	295 827	234 763
M	366.8	269.5	97.4	244.4	509.7	387.5
F	373.0	269.0	100.7	251.7	520.6	425.7
Malignant neoplasms (cancers)	505 382	52 426	44 125	217 413	161 230	145 036
M	221.3	220.4	46.1	216.4	300.4	304.2
F	186.0	174.7	51.2	139.3	262.2	201.7
Infectious and parasitic heart disease	30 424	1273	40 897	12 006	2785	2000
M	12.3	5.0	52.4	11.9	5.1	4.1
F	12.2	4.6	44.1	7.7	4.6	2.9
Pneumonia	77 415	6410	21 773	68 194	32 628	7649
M	29.8	24.1	27.6	64.1	43.1	13.0
F	32.4	24.1	24.0	47.4	69.9	13.5
Perinatal and at birth	17 482	1159	19 947	1756	2724	2806
M	8.2	4.9	30.3	1.5	5.3	5.4
F	5.9	3.6	21.4	1.1	3.9	3.7
Accidents	91 983	8849	39 240	32 122	14 008	22 901
M	51.1	44.6	71.7	36.8	30.3	48.2
F	23.6	22.3	21.2	15.9	18.8	31.7
Motor vehicle accidents	45 827	3645	13 926	14 398	5628	9123
M	26.1	14.4	26.1	17.2	14.3	250
F	11.1	8.2	6.9	6.5	5.5	7.1
Suicide	30 906	3379	1938	20 088	4643	4402
M	20.4	20.4	3.9	20.4	12.6	11.4
F	4.8	5.2	0.7	12.4	3.8	4.1
Homicide	24 614	554	14 455	744	413	1527
M	15.9	2.7	30.7	0.7	1.0	4.8
F	4.2	1.5	3.6	0.5	0.4	0.6

of acute distress ensues. If there is great material damage, outside assistance may be provided through governments, charities and international agencies. Sometimes the survivors, if any, are dispersed permanently and the society, or part of it, is brought to an end. Such are the features of disasters and catastrophes.

The news headlines, or 'Disasters' page of the yearbooks, are sad testimony to the continuing realities of extreme distress and devastation. They give us a preliminary overview of the landscape of calamities most often reported. Between 1989 and 1993, for example, the news media identified about 110 'technological' disasters, and almost 50 natural disasters in a given year (Table 2). If there are more of the former reported, natural disasters include more of the most catastrophic events, as do epidemics and wars.

The problems of uneven reportage and coverage in such information must be noted. Yet the major media, virtually alone, establish the commonly recognised, public events on the world scene. That in itself has a pervasive influence upon how disasters are perceived. In the so-called information society, people's view of the world beyond their immediate horizons depends increasingly on the mass media, as does the political importance given to problems. Disasters are among the items deemed most newsworthy.

While all the events reported involved or threatened the general public, many, especially technological 'accidents' such as air crashes, are fairly localised and dealt with by responsible agencies on the spot (see Chapter 4). They do not enter our concerns with the same force as disasters that bring huge, concentrated losses and overwhelm whole communities or even the resources of states. A profile of larger disasters or catastrophes over the past 10 years indicates the scope and scale of more extreme dangers in the late twentieth century (Appendix 1).

In addition to natural and technological disasters, roughly 32 major armed conflicts were documented in each of the past several years. They enter our concerns primarily because many, often the majority, of the victims are civilians, and because of massive impacts upon human settlements, habitats and means of livelihood. The more powerful industrial nations have enjoyed half a century without wars in their lands – albeit in the aftermath of the two most

Table 2 Extreme events: natural, technological, epidemic disease and war[1] disasters in the late twentieth century. A summary of reported incidence, 1989–1993 (after Encyclopaedia Britannica Yearbooks and Ploughshares Monitor, Waterloo, Ontario)

	1989	1990	1991	1992	1993	Total	Average
Natural disasters	46	52	54	32	51	235	47
Technological disasters	128	123	98	109	93	551	110
(New) epidemics	2	–	1	1	1	5	1
(New or continuing) wars*	32	31	30	30	35		32
						7	191

*Those that caused at least 1000 deaths.
[1] see Chapters 3–5 for details and Chapter 5 on why contemporary wars are treated as 'disasters' here.

destructive wars in world history, and with the unique dread of nuclear war. Elsewhere in the world, dozens of wars have taken place, some of long duration and unprecedented destructiveness. Meanwhile, there have been hundreds of clashes, acts of terror and other uses of armed violence short of full-scale wars, affecting ordinary citizens and minority groups. The huge refugee crises around the world, groups reflecting the peculiarly geographical calamity of enforced uprooting, mostly involve flight from violence (see Chapter 5). Some of the greatest calamities of this century have involved the systematic use of state violence to expel or to kill whole peoples.

There is no indication that, worldwide, the number of disasters and newly emerging dangers is declining. Various studies have found deaths or property losses, or both, to be increasing in the last half of this century. Particular places and regions appear to experience dangers and losses of unusual and growing severity. Some represent a persistence of old problems. Many are magnified or created by dramatic social changes associated, for example, with rapid urbanisation, environmental destruction, globalisation of economies, energy technologies, megaprojects and armed violence.

The problem of disaster

Many types and scales of disaster occur. What most have in common is concentrated harm, and often exceptional concentrations of death and injury. Damage is sufficient to disrupt significant parts of society's productive activity and administrative functions. Disaster overwhelms, at least for a time, the ability of established routines or responsible agencies to maintain public safety. Citizens cannot be supported and cared for in accustomed and acceptable ways, or only through extreme measures.

The shift from chronic or everyday risks to disaster brings out a different sense of the relations of risk to the social order. If most often defined by sheer quantities of casualties and damage, disaster also alters the conditions and scope of harm. Rather than being private and scattered, the tragedies are public and concentrated. The damages disrupt or destroy many different functions and institutions. They affect many different sorts of persons and domains at once. This makes them socially complex and uniquely perturbing.

The result can be a society-wide and systemic crisis, or at least, a collapse of public confidence. Such problems also burst out of, or cast doubt on, accepted areas of responsibility. There can be a breakdown or serious questioning of established social order and performance. Here too, is a basic difference from chronic risks.

It is possible, then, to distinguish two domains of risks, the 'chronic' and the 'extreme'. They tend to differ in incidence, social impact and treatment. However, the two are not separate in practice or, as will be argued throughout this text, in origin. The boundaries between them are uncertain and stretchy. Chronic or routine risks in one place or era become calamitous in another, and vice versa. Some dangers do not fit easily into either category, perhaps because of unresolved social conflict and ethical debate about what is acceptable or unavoidable. There are also aspects of modern life that further complicate this distinction, drawing us into some risks that are widespread but not routine.

Novel risks and shifting standards

Everyday life itself is far from static or merely repetitious. All societies are caught up in changes that may alter the scale and scope of common risks, or their awareness and tolerance of dangers. Developments that seem gradual or planned in themselves can, suddenly or imperceptibly, bring uncontrolled and disastrous outcomes. Such are problems of rapid urbanisation and motorised mobility, increasing or long-term loading of the environment with pollutants, deforestation, and soil depletion. Large-scale restructuring and relocation of industrial and commercial activity, or the arms trade, can undermine existing social and national security arrangements. The resulting novel dangers can have a singularly unsettling effect, seeming to threaten unnamed disaster if none actually occurs.

Extreme dangers can arise from novel developments, notably certain technological innovations or their deployment. Often presented as essential for the growth and improvement of countries and enterprises, these may introduce unprecedented risks. Meanwhile, nuclear power, supertankers and new biocides, for example, not only add new hazards and possible disasters; they also change the nature and course of everyday life. This is even clearer in the case of new, mass produced and widely distributed consumer products, when they prove to be dangerous or to lack adequate safety checks and safeguards.

Not least important for what separates 'routine' from extreme dangers are different standards and perceptions of acceptable and unacceptable risks. One group or society's 'normal', 'acceptable' risk may be intolerable to others. Novel risks and the globalising of concern and values as well as economy further complicate this. Some nations or cultures want to impose their own view of acceptable risks or standards upon others. Some do not hesitate to export a dangerous technology, product or waste product, or industry outlawed in their own country! These are ethical as well as material issues. They require awareness and understanding in terms of the geography of human differences.

Drawing these observations together, two broad concerns or constraints have shaped the preoccupations of hazard and disaster studies. The first involves questions of 'control' and the second of context and place. Both reflect how dangers appear in relation to modernity.

A concern with more extreme risks and disasters addresses problems that are seen to be *out of control*. This is also a construction placed on events as they relate to the expectations and practices of modern societies, where technical disciplines and official control are called upon to regulate all areas of national and public security. They involve the major uses of science and technology in the modern state and corporate enterprise. Public safety, among other things, is divided up, defined and 'disciplined' as a task of professional fields and administrative institutions. Our concerns stand out as evidence of the *limitations* of this social order.

Disaster also involves destruction or disintegration of the extensive, orderly patterns that bind together the large space and many places of modern material life. In terms of human geography, disasters appear as spatial *dis*organisation – areas and linkages where control is lost. As public crises, calamities burst the bounds of particular organised activities, installations, sites, land uses and territorial units. Residential areas become 'wetland' in a flood. Life-threatening industrial chemicals enter people's living space. Fighting forces destroy defenceless settlements and unarmed populations. Hence the sites of these events also appear *out of place*, violations of good order and cultural space.

Hazard reduction

A balanced view, and a full sense of our concerns, will hardly be achieved without recognising that much of the damage seen on the ground in natural and technological disasters is avoidable. Most or all of the deaths are 'unnecessary'. This is not merely so in principle, or as a matter of luck. The harm is avoidable with known practices, and is prevented in places and for persons where these are in use (see Chapter 7). Some great scourges of the past – notably certain lethal endemic and epidemic diseases – have been overcome entirely. There is every reason to think many other dangers could be reduced or prevented.

This is what gives to hazard and disaster studies a positive orientation. To help extend and improve safety and relief of suffering is surely the only acceptable purpose of research into extreme harm. It means we must identify, especially, the relations between the dangers to which societies are exposed and human practices that can and do increase, decrease or reallocate them.

Of course, the role of dangerous human agency and technological innovations raises special problems, the excesses of vicious political powers and ruthless groups even more so. Yet the despair or cynicism that these produce can be countered. There are peoples and political units, small and large, 'traditional' and modern, committed to maintaining public safety and improving social justice. We can point to a substantial decline in danger and damage from natural or human conditions in some places as a direct consequence of human effort. Admittedly, most are in wealthier states and enclaves with well-developed arrangements for personal and social security, but studies also reveal impressive arrangements for coping with danger and disaster in 'traditional' societies. A part, at least, of modern history has been the struggle to make public safety an integral and enduring feature of social life. It is mostly these human arrangements that divide the world into more and less secure people and places.

Scope of the hazards and disasters field

In sum, the common ground that defines this field mainly reflects the history of investigation. It has taken shape in response to late twentieth century contexts, rather than some fundamental logic of enquiry. However, this does not mean it is just a collection of *ad hoc* concerns, or of isolated hazards studies, only that contemporary history and intellectual constraints have largely decided our preoccupations. The interests are socially and, in some respects, narrowly constructed within the entire range of modern dangers and damages. Most of the relevant literature considers particular geophysical and technological dangers and particular technical considerations of, for example, building safety, transportation accidents, forecasting, emergency measures and post-disaster reconstruction. In the social sciences, the field developed around notions of collective stress and mass emergencies, social responses to disaster or risk analysis and civil defence. Among geographers it is identified mainly with 'hazards', with studies of floods and droughts, nuclear accidents and toxic chemicals.

Yet many common issues and conditions draw different forms of danger and occasions of disaster into common frameworks. Any actual society, modern or otherwise, is confronted with a range of dangers. By design or default, their relative importance and severity will be prescribed by the on-going pattern of life. The origins of disaster and the patterns of impact invariably reach back into the general contexts of material life, its environmental relations and geographical setting. In this sense 'risk', a term embracing possible as well as actual threats, past and future losses, seems to encompass the whole field, although it also embraces dangers we may not consider (see Chapter 1).

Disaster is generally taken to imply social breakdown and even chaos. Disaster studies may focus on single natural, technological or social hazards, but as events and predicaments, disasters are 'whole' crises for a society and a place. Any actual disaster affects society, its technology and its natural environment. These jointly define the fabric that is threatened and destroyed. This leads on to consideration of the diversity of different risk situations, and comparison of responses in different places and, in the same place, of different dangers.

In an introductory text we must draw upon work in both specialised and hazard-specific studies, and the broad view. However, it seems necessary to choose between a systematic review of individual dangers and a conceptual overview that emphasises the common ground of risk and disaster. The present text adopts the latter approach. There is no lack of accessible works on specific risks, notably natural and technological hazards. Several recent works by geographers cover them very adequately. Most of the research literature to which the reader can refer is also specialised. Here the case is developed for seeing this as a general problem field, with important considerations embracing many or all of the dangers discussed. I also want to suggest that this is an appropriate and special role for the geography of danger and disaster.

The 'geographicalness' of risks

As a geographer, I approach the field from a geographical perspective, anticipating that other geographers are my most likely audience. Nevertheless, I argue that geography is an intrinsic aspect of risks, and of central importance to all aspects of this field. Any given risk or disaster event is distinguished by its geographic location and setting. These are important keys to the origins of danger, the forms of damage and whom they most affect. They are critical for the appropriateness and deployment of organised response.

The scope of knowledge and concerns relevant to extreme, public dangers resembles the once dominant interests of geographical enquiry. We must take account of the interrelationships and distinctive mix of conditions that define human settlements and regions. Risks arise from or within the situational realities of particular places and their problems. Disaster causes an unravelling of those forms and associations variously termed 'people and place', 'land and life', society, and habitat. It is defined by the destruction of living space or ways of life. Great disasters invariably bring enforced displacement of resident communities, sometimes permanent exile. Refugee crises, the destruction of settlements and habitats, genocide and ecocide in extreme cases, describe some pervasive forms and painful consequences of disaster in the twentieth century. These are not only geographic aspects of disaster but *geographical calamities*.

Social hazards and disasters involve, in particular, the unique dangers of life and residence in areas subject to misrule and organised violence. Wars and civil strife often lead to flight or expulsion of populations, and annihilation of their places of settlement and sources of livelihood. These and many natural and technological disasters have their most severe impacts as destroyers of dwelling places and resident communities. They undermine the security of belonging to and living in a supportive habitat or larger cultural milieu. The security of home, neighbourhood, community or native land is threatened and destroyed. As such, they involve 'geography that matters' to those who are the main victims. These ideas will be developed in Chapter 2, where the 'geographicalness' of risks and calamity is a particular focus. Several of the case studies in Part 2 were selected to emphasise the significance of danger to people's geography, and the hazards of its destruction.

Geographers and hazards

A broad concern with danger and disaster problems can be found in the earliest work of geographers. Strabo's first century AD 'Geography', while intended to describe the known world, is a mine of information on earthquake disasters around the Mediterranean. He saw environmental stresses helping to explain the variety of peoples and places. Even earlier, Hippocrates' famous fourth century BC 'On airs, waters and places', formulated the classic Western view of health and disease, well-being and danger, as fundamentally dependent upon the geographical environment in which one lives. This prefigured environmental determinism, the prevailing general framework of geographical studies early in this century. It has been in disfavour for several decades, like the racial and imperial geography it tended to support, but an interest in the role of extreme weather events and other adverse environmental conditions has continued.

In the past half century, work by geographers developed from studies of particular natural hazards, especially floods (White, 1945), particular damaging events, such as the 'Atlantic Storm' of 6–7 March, 1962 (Burton et al. 1969) and the Managua, Nicaragua, earthquake of 1972 (Kates et al. 1973; Ebert, 1988). Experience, often unanticipated, in places subject to disaster has played an important role too (Waddel, 1975; Watts, 1983). In recent years, interests have broadened to include a variety of technological risks, some diseases such as AIDS, and social violence (Zeigler et al. 1983; Hewitt, 1983; Kates et al. (eds), 1985; Kirby (ed.), 1990a; Cutter 1993). Throughout, Gilbert White's work has been a major inspiration. Moreover, he placed questions of policy and practice at the heart of the field (see Chapter 7). This has always required consideration of issues going beyond a particular hazard or event, and attention to the geographical setting and the people involved. White recognised that hazards and disasters have broad social implications. He placed humanitarian concern at the heart of our professional responsibilities, and he wanted technical and managerial considerations to follow from, and be judged in terms of, that.

However, as with other late twentieth century problem fields, such as population control and environmental damage, the geography of disaster has been constructed as one of global and general human predicaments. But, also like them, it tends to be defined and treated, as Sachs (1990, 26) has said of 'development', according to 'how the rich nations feel'. Until very recently, the Cold War geography of two uncompromisingly different political worlds was superimposed upon that.

Disasters, and problems of responding to them, have come to be seen mainly in terms of a United Nations geography of 'worlds' – First, Second, Third, etc. – defined by certain economic parameters. This is clear in the agenda for the 1990s International Decade of Natural Disaster Reduction (IDNDR), which was committed primarily to deploying existing knowledge in the richer nations and advanced sciences, and training or technology transfer, for others (Table 3). This sense of geography has become so entrenched in the thought and language of these problem fields that it can seem almost impossible to speak of other and different geographies. At least they do place questions of geography at the heart of these problems.

Like a growing number of geographers, I would challenge the general validity of this picture of how the globe is divided up, and its automatic usefulness for our concerns. One is surprised that so many working in the international arena are comfortable with it. Meanwhile, transfers of knowledge and assistance between those 'worlds', and in terms of their stereotypical differences, can create more problems than they solve. The measures expressed in such gross and stereotyped divisions hardly recognise the actual

Table 3 International action on natural disasters

(a) Extracts from the United Nations resolution declaring the 1990s an 'International Decade for Natural Disaster Reduction' (Resolution 42/169, December, 1987)

- Recognizes the importance of reducing the impact of natural disasters for all people, and in particular for developing countries;
- Recognizes further that scientific and technical understanding of the causes and impact of natural disasters and of ways to reduce both human and property losses has progressed to such an extent that a concerted effort to assemble, disseminate and apply this knowledge through national, regional and world-wide programmes could have very positive effects in this regard, particularly for developing countries;
- Decides to designate the 1990s as a decade in which the international community, under the auspices of the United Nations, will pay special attention to fostering international co-operation in the field of natural disaster reduction;
- Decides that the objective of this decade is to reduce through concerted international action especially in developing countries, loss of life, property damage and social and economic disruption caused by natural disasters, and that its goals are:
  - To improve the capacity of each country to mitigate the effects of natural disasters expeditiously and effectively, paying special attention to assisting developing countries in the establishment, when needed, of early warning systems;
  - To devise appropriate guidelines and strategies for applying existing knowledge, taking into account the cultural and economic diversity among nations;
  - To foster scientific and engineering endeavours aimed at closing critical gaps in knowledge in order to reduce loss of life and property;
  - To disseminate existing and new information related to measures for the assessment, prediction, prevention and mitigation of natural disasters;
  - To develop measures for the assessment, prediction, prevention and mitigation of natural diasters through programmes of technical assistance and technology transfer, demonstration projects, and education and training, tailored to specific hazards and locations, and to evaluate the effectiveness of those programmes;
- Calls upon all Governments to participate during the decade for concerted international action for the reduction of natural disasters.

(b) From the 'Draft Strategy and Plan of Action for a Safer World', which came from a mid-term IDNDR World Conference in Yokohama, Japan, 23–27 May, 1994. The extracts show, in places, some much revised emphases (after UN/DHA, 1994c, 33–4)

- development of a global culture of prevention as an essential component of an integrated approach;
- adoption of a policy of self-reliance in each vulnerable country and community, including capacity-building as well as allocation and efficient use of resources;
- improved risk assessment, broader monitoring and communication of forecasts and warnings;

The Conference also affirmed the relationship between sustainable development, environmental protection, and disaster prevention, mitigation, preparedness and relief.

Table 3   contd.

However, even regional action may not reach far enough. The modern world is highly independent and in some cases international initiatives will be necessary. In particular, the Plan of Action adopted by the Conference hoped for:

- disaster prevention and mitigation will become an integrated component of development projects financed by multilateral institutions, including the regional development banks;
- the activities of the United Nations, Governments, NGOs, the private sector and other disaster mitigation actors will be more closely co-ordinated.

The General Assembly will be asked to consider adopting a resolution endorsing the Yokohama Strategy.

---

differences among settlements, regions, countries and environments. The sets of state units defined miss many of the complexities within, and decisive linkages between, places and peoples.

We will find that danger and damages in disaster are often more strongly differentiated within countries, communities and events than between them. They can arise through dangerous interactions of the different 'worlds', rather than what distinguishes them. Their treatment can hinge mainly on how countries in these seemingly different worlds interact, and whether they share or ignore common concerns.

Contested views and approaches

In fact, we begin our investigations at a time of re-examination of approaches to risk and disaster. This includes the International Decade, as the revised agenda from the mid-term Yokohama Conference suggests (Table 3,b). These debates will be addressed especially in the concluding chapter, but they have been an important factor in deciding the balance of concerns and emphases in the present text and need to be raised here.

In preparing this book, another aspect of the dichotomy discussed at the start has influenced my sense of purpose and direction. On the one hand, a remarkable number of new texts, compendia, conference reports and manuals have come out in the past four or five years, or revisions of older ones. The quality is often high. They deal well with many of the topics that I had expected should be covered here, and I will direct the reader to them.

What troubles me is the impression so many give, that hazards and disasters are much better understood, more cut-and-dried problems, than seems warranted. Authors tend to plunge as directly as possible into presenting technical knowledge developed by their own discipline, or of specific conditions, processes, events and organised actions. They present them as though they are unproblematic and clear understandings of the key issues. And even an intro-

15

ductory text like this one, whether for educational or professional use, must try to be clear, if not authoritative. It can hardly be written by someone who is confused or negative. However, overconfidence here seems more likely to confuse or mislead the reader than a certain scepticism, for, on the other hand, there is the dismaying incidence and scope of the disasters that keep occurring. They seem to outpace our knowledge and texts as much as the ability to reduce risks. The disasters we will consider reveal extraordinary and, in many places, growing vulnerability and loss for growing numbers of people. As discussed earlier, this is not confined to places notable for their impoverishment and harsh environments or corrupt administration. It applies, if in different degree, to countries and places where one could expect that our best understanding and practices are, or should be, available.

A spate of recent disasters overwhelms more than just existing arrangements to reduce or respond to crises. They challenge relevant understanding, cherished beliefs and expectations. The events might seem to be age-old problems of flood or storm; building collapses or transport accidents; malaria or tuberculosis epidemics in crowded, impoverished populations; or another war between old enemies. But they occur where it had been thought the problems were solved or unlikely to recur. Too often, it turns out that 'new' developments have fatally ignored, or have given a new potential to, old problems.

Meanwhile, reports of novel technological and disease 'scares' come across one's desk weekly, with claims that they are poorly researched, lacking even 'base line data'. It might be a new strain of drug-resistant disease, the side-effects of a popular technology or habit that could fill the hospitals to overflowing a few years down the line. Seemingly, new groups, social movements and weapons deployments exist against which there are few or no defences. Conference proceedings and compendia draw attention to these emerging dangers: sea level rise, the hole in the ozone layer, 'mad cow' disease, the Ebola outbreaks in Zaire, radon gas in homes, nuclear terrorism, contaminated blood – the list could go on and on. And this is to highlight the sorts of fears and crises that have caught the attention of the media and governments. New charities, lobby groups and even 'disciplines' spring up to address emerging or poorly recognised dangers, and to champion the needs of their victims. Often they demand new forms of risk assessment. They question the system of priorities or recommend radical changes to legislation and expenditures. No doubt this follows, in part, from the real pace of change and innovation in industry and employment, consumer fashion and geographical linkages. But it also creates doubts about established scientific and professional approaches.

In this book, the field of public safety and disaster will be examined as itself being problematic. This is already a familiar theme in the related problem fields of economic development, population control, health care and national security. There is more to such public and global concerns than just applying the latest knowledge, or well-crafted legislation, to problems that occur 'out there'. A critical and searching approach, rather than a merely prescriptive one, seems warranted by present conditions.

Suggested reading

Beck, U. (1986) *Risk society: towards a new modernity*, trans. 1992 Mark Ritter. Sage Publications, London.

Craig, D. and M. Egan (1979) *Extreme situations: literature and crisis from the Great War to the atom bomb*. Macmillan, New York.

Cutter, S. L. (1993) *Living with risk: the geography of technological hazards*, chapter 2. Edward Arnold, London.

Douglas, M. (1992) *Risk and blame: essays in cultural theory*. London.

Foster, H. D. (1980) *Disaster planning*. Springer-Verlag, New York.

Harrington, M. (1965) *The accidental century*. Penguin Books, New York.

Housner, G. W. (1989) An international decade for natural disaster reduction, 1990–2000. *Natural Hazards*, **2**, 45–75.

Kates, R. W. and I. Burton (eds) (1986) *Geography, resources and environment: themes from the work of Gilbert F. White*, 2 vols. University of Chicago Press, Chicago.

Kirby, A. (ed.) (1990a) *Nothing to fear: risks and hazards in American society*. University of Arizona Press, Tucson.

Quarantelli, E. L. (ed.) (1978) *Disasters: theory and research*. Sage, London.

Sorokin, P. A. (1941) *The crisis of our age*. Dutton, New York.

UN/Department of Humanitarian Affairs (DHA) (1994b) *Disasters around the world – a global and regional overview*. Yokohama, Japan.

Part 1

Approaches to risk and disaster

CHAPTER 1

Risk and damaging events

Anything can be a risk; it all depends on how one analyses the danger, considers the event.

Francois Ewald (1991, 199)

All kinds of intangible notions . . . are put forward as being the 'facts' of violence, but violence is an event not a condition. Violence is always an event, some say it is the most decisive kind of event possible.

Gil Elliot (1972, 15)

Approaching danger

Damages and risk

The understanding of hazards and disaster involves two broad areas of enquiry. We are concerned with actual damages, their incidence and distribution; and with how to explain them.

Damage is the empirical evidence of just how, where, and for whom danger is realised in harm. Death and injury, destroyed crops or buildings, and failed projects and emiseration are of foremost practical and humanitarian concern. Moreover, the processes or phenomena of damage tell us what needs to be addressed. Harm done is the unequivocal measure of the protection people did not have, but do require.

We pay attention to actual devastation and violence because they are more terrible, or at least, more final, than their possibility. When they occur, the community, civil society or technical plans have already failed. Efforts to respond to and reduce disaster pay, or should pay, most regard to its victims. In the places where disaster has occurred, our enquiries can be informed by the experience and concerns of survivors, those who have been 'at the sharp end' of these dangers. Safety measures, the limits of people's ability to cope, and the practices and performance of responsible agencies are truly tested in damaging events.

21

Second, however, there is the question of how people are placed in danger; the conditions that lead to disaster or may do so. What is it that promotes, limits or destroys public safety? In answering this I begin by rejecting a view of disasters, or most of them, as 'bolts from the blue' and so-called Acts of God; whether from unpredictable natural forces or human accident and failure. Earthquake and flood or oil spill and explosion may be the immediate or 'proximate' causes of disaster, and they may or may not be predicted. However, the severity and form of damages depend primarily upon the pre-existing state of society and its environmental relations. The argument supported here is that safety, and lack of it, are set up in the time before disaster happens. Security or danger are created and changed by human action. Thus, the assessment of danger goes well beyond actual damaging events. We find disasters being prepared by everyday life, in chronic areas of neglect and in disregarded implications of social change. This is not to say there are no surprises. Most people, even specialists in the conditions involved, often fail to anticipate the disasters that happen. But particular failures to anticipate future events are not the same as saying they are independent of preceding human activities and risk-taking choices.

The idea of 'risk' conveys a fuller sense of the field, in that it embraces exposure to dangers, adverse or undesirable prospects, and the conditions that contribute to danger. Thus, risk analysis considers, especially, potential and assessed dangers. The well-developed approach to insurable risks employs past damages to define profiles of danger attached to groups, activities and places having particular attributes. It provides a sense that risk resides in the fabric of everyday life or given projects. We do not have to wait for a disaster to say and to do something about it. If a certain material or design of building increases the fire hazard, replacing or changing it changes risk, hopefully with no fire.

For our purposes, this also directs attention to the human ecology and geography of conditions that promote or reduce safety. It suggests that risk is, in the broadest sense, continuously and socially constructed. It promotes an active and adaptive view of the responsibilities of human societies. That is how the notion of risk will be developed throughout the text. It is not, however, the only one in this field.

Assessing risks

Modern notions of risk developed especially from economic enterprise, or 'speculative risk', in which both favourable and adverse future results are at issue – benefits or profit, and losses. There is also a deep connection with gambling and games. Through the 'theory of games' it has been applied to some of the dangers of interest here, notably conflict situations. The problem of risk is also seen to apply to the chances of 'winning' and 'losing'. However, questions of public safety and social security, of insurable and disaster risks, have generally focused on adverse outcomes, their likelihood and possibilities for prevention or mitigation.

The adverse side of financial and property risks is helpfully discussed in texts on insurance. Risk estimation from empirical evidence is used especially in the 'actuarial' approach of the insurance industry. This employs tables or profiles of risk for persons, property or enterprises based upon their past performance. Unfortunately, the approach is poorly developed for the more extreme and novel dangers that concern us. The use of census-type, mass statistical indicators, such as probability of death, an 'accident' or a certain financial loss, is useful in exploratory work and to establish broad comparative overviews of risk. It helps clarify issues and debates over national priorities and broad regional or social differences, as seen in those between countries and by gender. However, there is a danger of greatly oversimplifying problems, the way in which they arise, and the concerns of those at risk.

There is a struggle between a narrow, essentially quantitative, technical view of risk and a broad social and cultural one. The narrow view seeks to estimate the probability of a certain measurable (adverse) outcome in a specific system or population. The purpose is to predict the frequency of, say, deaths and injuries, accident or monetary loss, over time or space, and in a way commensurable with other risks. Some regard this type of risk analysis, or even more rigorous formulations, as the only 'scientific' or sound approach. It is well-suited to technical work for well-defined practices and in planned projects, major areas of concern in which specialised technical knowledge is essential. However, the damaging events and human losses that concern us most are primarily threats beyond, or that break out of, technical and institutional frameworks. Where public and environmental safety are involved, we have to consider substantive issues not amenable to narrow and detached definition. Indeed, in such cases, problems of equity and responsibility, values and expectations loom larger and must be taken into account.

Technical risk assessment seeks to lay a grid over all eventualities in quantitative, standardised terms that are permanent and independent of the experience and the event. It may be in terms of probabilities of occurrence, costs, indemnity as percentages of all losses, or tabulated against other risks. And perhaps these are sensible methods for establishing impartial government policy, viable insurance schemes, and the resources needed for emergency preparedness. But in a human ecology of civil responsibility, danger and disaster involve more considerations than such grids of calculation. The sources and realities of danger for the people and places affected are of another order. For the person who loses the possessions of a lifetime, a loved one or a limb, damage is unique and irreparable. For the community whose town is destroyed by an earthquake or a bombing raid, the difference between before and after the event, between this lost place and surviving places round about, is profound and irreversible. Meanwhile, in many contexts around the world, there are few or no standard and representative data suited to formal risk assessment. What there is can be misleading or irrelevant. The researcher who wishes, or is required, to present the problem in that way must begin on the ground with direct observation and, often most effectively, by listening to the concerns and observing the ways in which those who live there cope with dangers.

Technical risk analysis has developed mainly with high-technology–high-risk dangers in mind, and in the hands of technical specialists. It has been applied in such areas as nuclear power, liquefied natural gas, oil pipelines, space weapons, genetic engineering or toxic chemical facilities. A special preoccupation has been the, as yet unrealised, potential for harm in new high-energy innovations and megaprojects. There has also been growing concern with the risks of 'global change' in, for example, future impacts of climate variability or sea level rise. These too are important issues. However, the vast majority of hazards and damages we consider involve dangers to or from more common-place technologies and everyday activities. They have no lack of precedents. They occur especially in places and for people who lack or have been denied modern, technical protection. To the time of writing, well-known if not age-old threats are the largest part of the actual landscape of human misery – losses in flood and storm; rail, road and marine transport 'accidents'; cholera epidemics and famines; and environmental or civil disasters brought about by 'conventional' warfare.

With these issues in mind, a broad, vernacular interpretation of risk is employed here. It seeks to take account of the varieties of experience and contexts, action and interactions around the world. A human ecology of risk is presented whose purpose is to describe and interpret the conditions that endanger or improve the security of communities.

Perspectives on risks

The conditions that endanger

For any form of danger or disaster, four broad sets of influences over risk or safety can be identified. Each contributes to danger and the form of damaging events, and describes major ingredients of a human ecology of disaster. They may be defined in terms of:

- hazards,
- vulnerability and adaptability,
- intervening conditions of danger, and
- human coping and adjustments.

You could call them the conditions or 'elements' of risk. Each involves a particular view of the realities of danger and response. In a disaster, they provide a perspective on the origin and nature of impact and loss, of survival and destruction. However, in most of the literature in this field they are found as distinct perspectives, often developed alone, with one or another taken to be more fundamental or useful.

Hazards

Often used to describe the whole field, a 'hazards' view emphasises phenomena, usually 'physical agents', in the natural or artificial environment that pose threats. Strictly speaking, something is a hazard to the extent that it threatens losses we wish to avoid. It is not the flood that creates risk, but the possibility of drowning or losing one's home. An earthquake or explosion that cannot adversely affect a society is not a hazard for it.

More careful discussions reflect these intrinsic meanings. For example:

> A hazard is a negative outcome which may take such forms as loss of life . . . risk is the probability that a particular negative outcome will occur . . .
>
> Ziegler *et al.* (1983, 17)

> The concept of hazards as external events impinging on unsuspecting people has been shed in favour of the interpretation that they emerge from interactions between people and environments.
>
> Mitchell *et al.* (1989, 107)

Yet, looking at the range of hazards work, these more sophisticated definitions apply to a small part. Most studies continue to treat hazards as objective conditions or agents in our environment – earthquakes or droughts, industrial explosion or oil spill, armed insurrection or economic blockade (Table 1.1). They are usually investigated as particular natural, technological or social processes. They may be considered as events like a tornado or forest fire, or conditions such as high temperatures or civil unrest. It is sometimes useful to look at specific harmful processes or substances, such as structural stresses, waterlogging, high-energy radiation, carbon monoxide or dioxin.

Because of our interest in the more extreme and novel dangers, hazards are also often defined in terms of *thresholds*. It may be the depth at which flood waters will start to do damage; the wind speeds or snow loads sufficient to damage buildings; or the radiation or toxic chemical dosage at which unacceptable harm may begin (see Chapter 3).

The investigation, modelling and attempts to predict such hazardous physical agents have been the main thrust of this field. Most work still tends to construct risk in terms of the attributes of dangerous conditions or processes. Hazard-based and hazard-specific work has been so pervasive that a *hazard perspective* prevails. It will be the focus of Chapters 3, 4 and 5. However, although hazards described as objective agents are necessary, they are not sufficient conditions for damage or to initiate a disaster. Other conditions of risk are also involved.

Table 1.1 Hazards: broad classes of dangerous agents that may cause exceptional damage and disasters

Natural hazards	
Atmospheric	e.g. hail, snow, tornadoes, hurricanes, blizzards
Hydrological	floods, sea-ice, glacier advances, drought
Geological	landslides, earthquakes, volcanic eruptions
Biological	epidemic diseases, blights, insect plagues, forest fires
Technological hazards	
Hazardous materials	physical (asbestos fibres), chemical (dioxin), inflammable (toluene), biotech (genetically engineered)
Destructive processes	structural failure, explosions, mass fires, ionising radiation
Devices, machines	explosives (e.g. TNT), aircraft, oil tankers, biocides
Installations	power plants, dams, LNG terminals, pipelines
Sector, organisation	petrochemicals, airlines, road transportation, mining
Violence and war hazards	
Weapons	conventional (firearms, bombs), 'dubious' (incendiary, chemical, biological, nuclear)
Release of dangerous natural forces	fire setting, triggering landslides, avalanches, floods, weather modification
Release of dangerous technological forces	targeting fossil fuel stores, nuclear facilities, chemical plants, dams and dykes
Armed forces and weapons systems	'special forces', strategic air power, guerilla forces
Strategies and tactics	economic blockade, environmental warfare, ethnic cleansing, sieges, counter-city bombing

Vulnerability and adaptability

This refers to attributes of persons, or activities and aspects of a community that can serve to increase damage from given dangers (Table 1.2). Decisions or actions relating to a hazard, whether deliberate 'risk taking' or preventive measures, are obviously relevant but will be treated separately. Meanwhile, much of human vulnerability arises, or is decided, with little or no regard to the particular hazards of, say, earthquake or toxic chemicals release. Rather it derives from the activities and circumstances of everyday life or its transforma-

Table 1.2 Vulnerability: some of the basic forms in which it arises

1. *Exposure to dangerous agents* and environments.
2. *Weaknesses*: predisposition of persons, buildings, communities or activities to greater harm.
3. *Lack of protection* against dangerous agents and for weaker persons and items.
4. *Disadvantage*: lack of the resources and attributes to affect risks or respond to danger.
5. *Lack of resilience*: limited or no capability to avoid, withstand or offset and recover from disaster.
6. *Powerlessness*: inability to influence safety conditions, or acquire means of protection and relief.

tions. How vulnerable people are may depend upon their age, gender and health status, and how society treats its members or different groups. Vulnerability also depends upon the quality and siting of buildings and land uses; public infrastructure and services; and ways of life and political authority. These are critical features of the exposure, safety and resilience of people in the face of dangers. A major concern in the modern and modernising world is how vulnerability can be affected more or less drastically by social change and patterns of development.

Like 'risk' and 'hazard', vulnerability describes a potential state of affairs. I am vulnerable to things that can happen to me, but they may not. I share a similar vulnerability with other persons similar to me in age, gender, life-style, wealth, or proximity to a dangerous facility. I am more vulnerable to certain dangers, less so to others, according to my characteristics and situation. My relation to some specific danger is not all that counts; the aspects of my life that may affect my ability to withstand, avoid, or cope with particular stresses are also important. In that regard, my personal attributes are much less important than my relations to and participation in collective life – the rights, protections and shared responsibilities I enjoy, or lack of them.

Vulnerability to natural and technological hazards has been treated in a variety of ways, often as simply exposure to hazards – 'being in the wrong place at the wrong time' (cf. Livermore, 1990). That is how it appears in notions of 'high-hazard locations' and 'harsh lands'. The geography of vulnerability is then defined by human occupancy of flood plains, drought-prone areas and war zones, for example. Vulnerability then becomes essentially part of a hazards perspective. However, its main significance as a distinctive element of risk arises with danger uniquely dependent upon the state and capacities of those at risk.

Vulnerability involves, perhaps above all, the general and active capacities of people – what enables them to avoid, resist or recover from harm. Whereas a hazards perspective tends to explain risk and disaster in terms of external agents and their impacts, vulnerability looks to the internal state of a society and what governs that. In relation to collective life, access or rights to services, information, support and protection may be crucial. Vulnerability in the modern world depends, perhaps most fundamentally, upon the legal, political and moral frameworks of civil society. In more traditional societies, it depends upon customary obligations and communal support structures. People vary greatly in such terms.

When Mary Douglas (1985, 27) defined 'hazards' as, ' . . . inability to cope', in a sense, she subsumed *hazards* under what is here being described as a vulnerability perspective. Danger from whatever may be 'out there' becomes primarily a question of whether societies have the means and ability to respond adequately. What is regarded as 'adequate', as Douglas's work emphasises, will itself be strongly dependent upon culture.

While 'vulnerability' suggests weakness and defencelessness, studies in this perspective actually challenge a common view of people as merely passive or unwitting victims of disaster. It is as much a question of the protection and active capacities people have, or are permitted, as the hazards they are exposed to and their innate weaknesses. We will find the geography of risk or disaster, in this regard, depends most critically upon *differential vulnerability* within and between societies. Moreover, danger and damages in disasters often relate to the most dynamic aspects of people's lives and history, social movements, environmental changes, and economic development. These are invariably reflected in changed, increased or decreased vulnerabilities. Likewise, social change tends to dramatically improve or undermine people's capacities to cope with, and adapt to, dangers.

Although not as widely employed as the hazards viewpoint, there is an emerging *vulnerability perspective*. The sense in which it is important will be developed in all chapters, but is the particular focus of Chapter 6.

Intervening conditions and contexts of danger

These draw attention to aspects of the habitat and society not directly related or tied to a given hazard, or human vulnerability to particular stresses. The emphasis is on circumstances that may intervene between the two. For instance, soil type, topography, vegetation cover and water tables can have a decisive influence upon the severity and impact of earthquakes on built structures. But these conditions themselves depend neither upon seismicity nor the built environment. They have their own, distinct geographic patterns. They intervene between seismic events and vulnerable structures, and to a different extent from place to place. Similar arguments apply to the way topography and tree cover can intervene to magnify or moderate impact and damage in severe snow storms, or topography and temperature inversion in the problems of photochemical smogs.

Equally, institutional and cultural phenomena may buffer or focus damage, without being tied to specific vulnerabilities or agents of damage. The form and development of settlements, occupations, and who is obliged to live where, also happen to be critical for storm or flood risks. International grain prices and local grain stocks can influence whether famine will be allowed to develop in a particular country. Food distribution policies and market prices can decide who, and how many people, will starve, even where sufficient food is actually available.

There is some artificiality in speaking separately of 'intervening' conditions. They could be called secondary hazards or contingent vulnerabilities. However, the issue is one of developing an adequate awareness of the circumstances that shape risk, loss and human response in different places. It is to make the case for looking more broadly at the *social and environmental context* of dangers. That avoids a common tendency to neglect influences not peculiar to a given hazard or aspects of society that are wreckable by it. Risk is recognised as part of the circumstances and settings of people's lives, and essentially a predicament integral to ways of life.

In modern societies, work and the responsibilities of most people tend to be specialised, even highly fragmented and alienated from each other. Yet our 'ecology' is pervasively mediated by and connected through artificial structures and public institutions. The energy and water supplies of cities, highway systems and police, mass media and the courts, taxation, education and national security are interrelated infrastructures or regulatory systems. Together they mediate and allocate danger. They not only stand between society and nature, but between the general public and particular institutions, and between different technologies. That surveys have found most North Americans and Europeans tend to rank socioeconomic and technological dangers higher or worse than natural hazards makes sense in terms of their experience. However, even where technology is less powerful, or people are in close and continuous contact with natural conditions, social arrangements are equally critical to danger and response.

The mediating conditions and contexts of risk are, in many ways, the key to its human ecology. They should show how particular vulnerabilities and exposure to hazards are prefigured by the social order and material life, and their relations to the habitat. Unfortunately, social change and economic development often occur without reference either to given hazards or changing vulnerabilities. Yet they are context-altering forces that drastically affect people's resilience and ability to recover from losses. This is the main concern of the case studies in Chapters 9 and 10.

Coping in disaster and adjustment to risk

Here we consider both what people do when confronted with dangers and crisis, and communal plans to reduce danger or prepare for emergencies. In an impending crisis or disaster, people's capacities to respond and what they

29

do, how well or quickly they recover, involve active responses. These in turn relate to who the victims are, their situation or responsibilities, what conditions were like before the disaster and contexts of recovery. Hence, the social geography of disaster responses, and what the testimony of survivors reveals about their concerns and losses, are important considerations.

Professional studies have given more attention to planned and official measures, especially by governments. These may be directed at hazardous processes and conditions, or at reducing vulnerability. They may be intended to adapt settlement and land use more favourably to intervening conditions. Organised emergency measures and humanitarian assistance in disaster are also major areas of study. These can reduce damage and speed recovery. Less often considered but perhaps more effective is how societies can improve public safety in a broader sense through general material uplift, social security measures, and broad legislation to protect people's rights.

The notion of 'a range of adjustments' to hazards, developed by Gilbert F. White and co-workers, was a major initiative by geographers and will be discussed at some length in Chapter 7. The idea had its origins in the 'multiple use' view of natural resources. It seeks to establish the set of possible accommodations to given hazards, and the best mix of them, that is, rather than undue reliance upon single solutions such as had prevailed, for instance, in the approach to flood hazards in North America through engineering control works.

The notion of alternative adjustments has also been directed mainly at planned responses by government. However, White emphasised the question of choice and the adjustments people at risk choose or would prefer. Much of the work which he set in motion considers the *choice of adjustments*, and the willingness of responsible agencies to pay attention to public preferences. The result has been to situate hazards problems in the realms of human decisions and public policy.

An important theme is the way the adjustments which people consider and make depend upon social organisation, and the *capacity* to choose adjustments – or lack of it. Choice is constrained by status and rights within a society. It is found to turn, for instance, on the way children and the elderly are treated, the position of women, or the situation of disadvantaged groups generally. Those least likely to have a voice in public safety and risky developments are so often the ones to suffer most in disasters. Freedom to choose seems to be as important as knowing or working out possible choices. And here questions of response are linked, especially, to those of vulnerability – which, in turn, is linked to the influence and resources of people at risk. Drawing these concerns together, we can begin to define an *active perspective* on risks, the focus of Chapter 7.

Four sets of factors enable us to define an initial profile of the conditions involved in danger (Fig. 1.1). At the same time, the risks which may arise from each imply, and are realised in, particular dangers and damages. Damage occurs when the relationships are unfavourable, disaster in the worst cases.

	HAZARDS	VULNERABILITY	INTERVENING CONDITIONS	PLANNED INTERVENTION
Elements	**Agents** Natural Technological Social	**of Persons** **Property** **Activity** **Land use** **Value system**	**Context** Habitat Built Environ. Social Cultural	Control works Forecasts Insurance E.M.O. Religious
Damaging Events	**Impacts** Casualties Damages Duration Areal extent Recurrence	Victims Losses Displacement	**Damage process** Primary Secondary Tertiary	Emergency measures External Aid Reconstruction

Fig. 1.1　Elements of risk and their place in damaging events

Damage: primary, secondary, tertiary

Each condition of risk has a bearing upon the forms of damage, whether they occur, and how severely they are realised in:

- loss of life, injury and impairment of persons;
- destruction of property, resources and heritage;
- disruption of activities, and denial of supplies and services;
- cultural, spiritual and ethical violations.

Each of these involves a variety of specific processes. They relate to the impacts of particular hazards, the vulnerabilities of particular persons, groups or activities and the concern or values of collective life (Table 1.3).

An important part of the case for a broader, human ecological perspective on risk and disaster stems from the observation that not all, and not always the main, losses come from the immediate impact of a given danger. Damage may be started by, say, earthquake shaking, falling snow, or industrial explosion. But other damages can follow on in *secondary* or *tertiary* processes. For example, the destruction of San Francisco in 1906 or Tokyo in 1923 was largely by fire, a *secondary* consequence of the earthquakes with which the disasters are generally identified. The presence of domestic fires and fuel stores in densely occupied, inflammable areas and windy conditions transformed earthquake damage into an even more devastating process, urban mass fires.

Secondary impacts generally involve secondary hazards. The Vaiont dam disaster of 1963 in northern Italy was initiated by a large landslide, but most of the 2500 lives lost and much of the property destruction were due to a flood wave. The waters were forced out of the reservoir by the landslide, over the dam and into the valley below, where property and population were exposed.

Tertiary damages may come from impairment of general social functions. Disease outbreaks can follow floods or typhoon destruction. There may be

31

delayed economic effects and forced migration, as survivors are unable to find work. Sometimes these can harm more people than the primary damages. We encounter examples where the worst of a disaster seems to have come *after* destruction by flood or earthquake, fire or warfare. It may even involve inappropriate, exploitative or unduly delayed actions by government and international agencies. A community may prove incapable of resolving conflict and blame arising from the disaster.

Relief measures are often critical to the course which a damaging event may take, perhaps heading off secondary and tertiary damages. Timely relief may spare people a range of other hurt, for example from exposure to the elements if their homes are destroyed, or starvation if their means of livelihood are lost. The extent and focus of rehabilitation and rebuilding may decide whether a

Table 1.3 Profiles of damage in three disasters

a) Hurricane Alicia, Texas, 17–18 August 1983 (after US National Academy of Sciences, 1984)

Total loss estimates

Human lives lost	17
Injured	3243
Temporarily evacuated	25 000
Made homeless	3000
Property damage	US$0.7–1.6 billion

Housing units damaged or destroyed:	Destroyed	Minor damage
Single-family homes	1209	12 472
Mobile homes	455	1034
Multi-family units (apartments, etc.)	633	2857

Losses in Galveston, Harris, Brazoria and Chambers counties

Types of loss	Value (millions of dollars)	
Residential	100	(9500 structures)
Commercial	9	(300 structures)
Industrial	4	
Public facilities	1	
Roads and highways	1	
Utilities	60	
Vehicles	19	(6250 vehicles)
Agriculture	51	
Marine	10	
Total	250	

Table 1.3 contd.
b) Earthquake, Mexico City, 19 September, 1985 (after Oliver-Smith and Hanson (eds), 1986)

Total human lives lost (officially 7 000)	20 000*
Total injured	40 000*
Total left homeless or with damaged homes	350 000*
Total receiving temporary shelter	30 000
Total persons unemployed	150 000
Total schoolchildren without classrooms	650 000
Total buildings destroyed (listings by category not available)	1091
Total buildings damaged:	
Residences	5025
Schools	3072
Offices	800
Services (public markets, recreation centers, theatres)	254
Commercial	244
Industrial	81
Hospitals (500 beds, 30% total)	24
Total vehicles destroyed:	
Private	1200
Public transport units	300
Total small industries affected:	
Retail damage	526
Total loss	800
Total insured damages	US$4 billion

c) Urban air raid: the 'firestorm' raid on Darmstadt, Germany, 12 September, 1944 (see Chapter 11)

Casualties (civilian)	Total	Women	Children
killed	10 550	41%	32%
injured	3 750 +		
Homeless (bombed out)	50 000		
Built-up area razed by fire	2km*		
Inner city destroyed	82%		

Buildings	Totally destroyed	Heavily damaged	Moderate to light damage
Residential: Buildings	4 563	891	12 687
Homes	17 332	2385	22 209
Commercial	871	243	869
Public	69	45	120
Industrial	5	27	38

* Estimates

community, or parts of it, suffers severe tertiary consequences, or fails to survive at all.

In fact, it is doubtful if many disasters, for some or all people and activities, conform to the stereotype of a single, massive blow, followed by a process of recovery. Perhaps relief goods do pour into an area, buildings pop up quickly and certain economic activities are resumed. The media or relief agencies are then likely to lose interest. Yet there may still be many, perhaps a majority, of the community suffering delayed harm or demoralisation. Follow-up studies that recognise this situation are few, but sufficient to give weight to such concerns. In the case of social violence, of course, there may be long-term, systematic undermining of groups targeted for oppression or successive blows against a place or population. Unless they are relieved, find effective means of resistance and protection, or are annihilated, the disaster has no real end.

What these points emphasise is that, although a particular hazard may initiate the damage process, its later course depends upon other conditions in and around the impact zone. The sense in which all four conditions of danger enter risk and damage is reinforced, as well as the need for human ecological understanding of context and community.

Events and places

To neglect any of the elements of risk, or the damages when they combine adversely, is to miss essential ingredients of risk. Nevertheless, as the previous section began to reveal, the most significant aspects of risk relate to how these elements can and do interact. That requires us to move from types or lists of factors to the features of damaging events and extreme situations. Then, the elements of risk appear and combine in specific processes of harm and constraints upon human response.

Most damages considered in this field occur in more or less well-defined events, especially crises and disasters. 'Routine' dangers may kill more people or cost more, but the large damaging event, concentrated in time or space, is clear evidence of out-of-control and out-of-place dangers. It is an *event*, in the sense described by Gil Elliot at the head of the chapter: a rift experienced in the routine or developmental.

A disaster is, also, always peculiar to and identified with a place, the inhabited area where destruction occurs. Disasters are remembered by and especially *in* the places where severe losses occurred. Sometimes the disaster and the place coincide almost completely, as with Bhopal, Aberfan, Love Canal or Nagasaki. Sometimes a particular place, perhaps where a disaster began, or the worst hit and most horrific location, comes to symbolise a much more widespread calamity. The 'San Francisco earthquake' (1906), Chernobyl (1986), or Auschwitz (1941–1945) were the most concentrated places of harm in, or epitomised, much more widespread calamities. Nor is this just a question of labelling an event with a place name. It draws attention to the essential context of harm, the setting of the event, the importance of geography.

34

In this sense, I would reject a common view of disaster as, essentially, about forces or conditions from 'outside' a human settlement, an industry or a society, or erupting accidentally within it. That is, when natural disasters are attributed essentially to the incidence and scale of earthquake or storm; technological or social disasters to 'unscheduled' or human failings, accidents or violations. These may trigger the damaging event, but its possibility has deeper roots. Responses are as much about whom and what is affected as the immediate source of harm.

Disasters

Concentrated, diverse damages and a collapse of the social fabric or its safety measures define disasters, or most of those that will concern us (Table 1.4).

Table 1.4 Distinctive social conditions of disaster and disaster response

1. Disaster overwhelms and destroys some or all:
   - facilities and organisations responsible for caring for the sick, injured, dead, impaired
   - services and professions that usually respond to damage and restore facilities, i.e. fire-fighting, repairing public utilities, broken appliances, household plumbing and wiring faults
   - essential life and work support systems or 'lifelines', i.e. roads, pipelines and power lines; stores and outlets for food, fuel, clothing, etc.

2. Disaster prevents or destroys some or all of the activities or functions of society.
   These may include:
   - economic production in fields, factories, workshops, etc.
   - service industries
   - consumer/retail trade
   - commercial business and services
   - schools, places of entertainment, civic institutions

3. Spatial disorganisation: isolating and fragmenting of communities or parts of communities from each other, and the larger administrative and economic space

4. Disaster not only fragments society, but breaks down or dissolves the separation of:
   - public and private domains
   - divisions of labour
   - age, gender and class barriers
   - institutional and professional responsibilities
   - levels and jurisdictions of governmental agencies

Disaster engages our attention because there is a fundamental shift in the scope and conditions of harm. Established responses break down, are destroyed or prove ineffective. Rather than private and scattered, the tragedies are public and concentrated. However, they may be concentrated in particular vulnerable sections of society.

Disaster damages are not only more concentrated or massive. They are socially complex and uniquely disturbing. Sometimes that is because particular types of person are victims – children, 'innocent' bystanders, defenceless groups. That brings a special anguish, blame or a sense of the violation of accepted values.

A disaster can disrupt or destroy many different sorts of functions and institutions all at once. It may bring society-wide or systemic crises. Hence, survival and recovery require the reintegration of a wide range of responsibilities which, in modern communities at least, tend to be carried out within particular institutions and distinct professions. Their collapse calls forth spontaneous, unofficial and non-specialist responses. That is one of the aspects that we refer to as *coping*, the actions of ordinary people or disrupted remains of institutions, in contrast to official and planned responses.

In the language of human geography, disasters are identified with landscapes of violence, and the disintegration of productive activities and public infrastructures. In human ecological terms, they destroy or undermine life support, the resources and established arrangements for producing and distributing supplies, and the relations with the habitat and surrounding communities. For survival, or at least to satisfy basic needs, an adaptive crisis response is thrust upon survivors – in due course, perhaps, with outside help.

The disaster itself is the core of our concern, but not necessarily of our work. For we are also, and in some ways more, concerned with why a disaster occurs and how losses might have been reduced. These are not only and from our perspective not mainly a function of the crisis itself. Rather, they appear as a result of pre-existing arrangements, or lack of them. In this regard, it is of note that disaster events are usually seen to involve 'before', 'during' and 'after' problems. Table 1.5 gives a more elaborate scheme.

Disaster work has focused mainly upon II, 'the disaster' and its particular phases. A danger of too much preoccupation with this, however, is the way it may minimise or disregard the preconditions and aftermath of disastrous damages. These seem to be the keys to the origin of disasters, and people's ability to cope with and to escape from a disaster-prone condition. We will repeatedly look at disaster experiences, but especially as they reflect the harm to and problems of everyday life.

Famine is perhaps the best-documented case, where it has been shown repeatedly that starvation primarily affects, uproots and kills the already hungry. But the principle will be shown to apply much more widely. Social and environmental change are found to prefigure disasters for some even as they may improve life for others. Defence expenditures can end up fuelling aggressive wars, and the civilian and environmental disasters they cause. In the case

Table 1.5 Temporal sequences or phases that may be involved in disasters, with reported durations and selected features of each phase (cf. Barton, 1969, 50; Turner, 1978, 85)

I Preconditions

Phase 1 *Everyday life* (years, decades, centuries)
'Lifestyle' risks, routine safety measures, social construction of vulnerability, planned developments and emergency preparedness.

Phase 2 *Premonitory developments* (weeks, months, years)
Turner's 'incubation period' – erosion of safety measures, heightened vulnerability, signs and problems misread or ignored.

II The disaster

Phase 3 *Triggering event or threshold* (seconds, hours, days)
Beginning of crisis; Barton's 'threat' period: impending or arriving flood, fire, explosion; danger seen clearly; may allow warnings, flight or evacuation and other pre-impact measures. May not, but merging with:

Phase 4 *Impact and collapse* (instant, seconds, days, months)
The disaster proper. Concentrated death, injury, devastation. Impaired or destroyed security arrangements. Individual and small group coping by isolated survivors. Followed by or merging with:

Phase 5 *Secondary and tertiary damages* (days, weeks)
Exposure of survivors, post-impact hazards, delayed deaths.

Phase 6 *Outside emergency aid* (weeks, months)
Rescue, relief, evacuation, shelter provision, clearing dangerous wreckage. Barton's 'organised response'. National and international humanitarian efforts.

III Recovery and reconstruction

Phase 7 *Clean-up and 'emergency communities'* (weeks, years)
Relief camps, emergency housing. Residents and outsiders clear wreckage, salvage items. Blame and reconstruction debates begin. Disaster reports, evaluations, commissions of enquiry.

Phase 8 *Reconstruction and restoration* (months, years)
Reintegration of damaged community with larger society. Re-establishment of 'everyday life', possibly similar to, possibly different from pre-disaster. Continuing private, and recurring communal grief. Disaster-related development and hazard-reducing measures.

---

of technological disasters, especially, Phase 2, or Turner's 'Incubation period' (1979), has been shown to be critical. This is when seemingly minor developments, conditions or evidence in the uncertain margins between jurisdictions can change the likelihood and form of disaster. Places outside the view or concerns of safety agencies, expertise or effective communication often prepare the ground for failure. Progressive changes in the character of everyday life may also create the conditions for disaster.

Disasters and catastrophes

Disaster is usually identified with sheer numbers or concentration of casualties and damage. In these terms, however, if you want an exact definition, there will be many events in grey areas. What constitutes a disaster, by any given measures, varies enormously from place to place – as we would expect if geography and the composition of human societies are important in them. The problem varies according to the size and capacities of the social unit affected, the things it values and its ability to cope or receive help.

Events called disasters occur at many different scales and show wide differences in scope. Those reported as transportation and other technological 'disasters' in the media, for instance, may be only marginally the sort of event described above. They have some impact on or casualties among the general public. They are briefly the subject of national or international debate, perhaps with some people agonising over their implications, but they are still dealt with largely or wholly in a routine way. Response remains within their controlling institutions and specialised measures, or by existing police, hospital and other facilities. A Chernobyl or Bhopal, or the 1976 Tangshan earthquake, is of a very different order. Then again, there are disasters whose casualties and costs appear modest but which affect whole communities that lack arrangements for coping, or whose losses are specially tragic, insupportable or final.

Quarantelli (1984) has proposed a way to address these issues by recognising two levels of calamity, which he terms 'disasters' and 'catastrophes'. Each has some of the extreme, crisis features discussed above, but a 'catastrophe' arises if:

- Most or all of the total residential community is impacted.
- The facilities and operational bases of almost all emergency organisations are themselves directly hit.
- Local officials often are unable to undertake their usual work roles, and this extends into the recovery period.
- Finally, most of the normal everyday community functions are sharply and simultaneously interrupted.

For individuals and local groups there may be no discernible difference between the two cases, which depend on community or society-wide consequences. How they apply will also vary widely according to the culture, size and organisation of societies. If a society or segment of it is small *and* weak, there may be more and smaller hazards that threaten catastrophe than if it is large and strong. Nevertheless, this sense of at least two levels of crisis or emergency highlights important differences among events that may be lumped together as disasters.

Summary

The radically impaired social context in catastrophes transforms the terms on which trauma and loss are, and perhaps can be, treated. They create a distinctive area of human experience and social concern, but it is not one of merely static or 'impact' phenomena. Disaster is a context of active or emerging responses. Survivors as well as emergency organisations do act and adapt. That too, however, depends upon pre-existing conditions, and the experience and concerns of those involved. And it depends upon where and how people live, or 'geography that matters'.

Suggested reading

Barton, A. H. (1969) *Communities in disaster: a sociological analysis of collective stress situations*. Doubleday, Garden City, New York.

Blaikie *et al.* (1994) *At risk: natural hazards, people's vulnerability, and disasters*. Routledge, London.

Burton, I., R. W. Kates and G. F. White (eds) (1978) *The environment as hazard*. Oxford University Press, New York.

Glickman, T. S. and M. Gough (eds) (1990) *Readings in risk: resources for the future*. Washington DC.

Mitchell, J. K., N. Devine and K. Jagger (1989) A contextual model of natural hazard. *Geographical Review*, **79** (4), 391–409.

Shrader-Frechette, K. S. (1991) *Risk and rationality: philosophical foundations for populist reforms*. University of California Press, Berkeley.

Waterstone, M. (ed.) (1992) *Risk and society: the interaction of science, technology, and public policy*. Kluwer Academic Publishers, Boston.

CHAPTER 2

The 'geographicalness' of disaster

> Geography does not denote an indifferent or detached conception: it concerns that which matters to me, or interests me, the most: my fears and cares, my well-being, my plans and ties.
>
> Eric Dardel (1952, 46)

> Of course, being out of Sarajevo, for people from Sarajevo, is also hell.
>
> Ademir Kenovic (1993, 14)

The space of danger and places of disaster

The geography of phenomena involves their whereabouts and distributions. It identifies their associations in spatial patterns, exchanges and interactions over the Earth. Human geography records, especially, the various, distinguishable habitats and cultural 'worlds', the places of shared existence around the globe. They are the subjects of the geographic or coexisting order of human life and environments. Three broad aspects appear important in the geography of danger and disaster.

The map of risk

First, there is a formal or mappable layout of risk-related conditions and damaging events. It was suggested in Chapter 1 that danger is a product of four sets of conditions – hazards, vulnerability, intervening conditions and context, and human responses. Each of these, or the particular phenomena they involve, has a distinct geographic arrangement. One cannot move from place to place, country to country – even within a village or urban neighbourhood – without some or all of them, and their mix, changing. With that, the degree and forms of risk change.

The variable overlap and interactions of these conditions create the map of risk. Damage and loss define the geographic space of their unfavourable overlap and its severity. Disasters mark the places where the interaction is most adverse.

If these dangerous conditions and relations, or suitable measures of them, can be identified, they provide a framework for technical analyses and practical response. Modern uses of geographical information are mainly in the realms of strategy and planning, monitoring, and administration. Work by geographers in this regard has been devoted especially to hazards mapping in such cases as flood plains or unstable slopes, and information systems for overlaying various ingredients of risk.

Spatial disorganisation

Second, hazards represent threats to the relations of society and habitat, to the practices and interactions that link human life over the Earth. Disaster is a disruption and unravelling of spatial or geographic order. Areas of death and destruction are more or less cut off from the organised functions of the larger society to which they belong. There is a breakdown or isolation of part or parts of the large space of political economy and civilised culture. That is obvious when roads or telephone lines are destroyed, when flood waters, deep snow or dangerous materials prevent movement within, or in and out of, an area. In the disaster zone itself there is a disintegration or ruin of living space. Its internal geography is a complicated map of worst hit, less severe, and unharmed people and property. Harm and survival, and human impairment and adaptive response, exhibit strong spatial variations.

As out-of-control or out-of-place problems for modern life, damaging events appear, especially, as geographic or spatial *dis*organisation. In the modern city and, increasingly, the modernised countryside, every aspect of life is linked to networks of control and supporting infrastructure. Disaster turns especially on loss of what are termed 'life lines' – the supporting structures that provide food, water and energy, or along which social and medical services are delivered. Destruction of these may threaten or harm people not touched directly by, say, storm or fire. Increasingly, rescue and relief in disaster zones, and their most costly aspects, are directed to restoring or replacing life lines, often superseding direct assistance to the victims.

The geography of fears and cares

A third aspect involves, in many ways, a different face of geography. It raises questions only hinted at thus far. This is the geography to which the quotations from Dardel and Kenovic speak – the geographicalness of experience and 'geography that matters'. It enters all levels of human organisation, from political decisions and legislation to popular concerns and humanitarian response. However, it is rooted primarily in shared identity with place and community values. The security of persons and ways of life, and our sense of safety, are most intimately bound up with the place of residence and the world to which we belong.

As with other fields there is a tendency to talk in terms of 'expert' and 'lay' views of risk and disaster. Lay knowledge is often regarded as 'soft', 'subjective', even uninformed. That may well be true compared with specialised knowledge of any particular threatening phenomenon, or a larger geographical context. However, there is a basic aspect of geography that reverses this sense of expertness. It concerns the knowledge acquired by 'being there', of knowing conditions on the ground and as a member of a community.

The geography of belonging in or being estranged from places turns upon first-hand involvement and socially acquired knowledge, upon speaking the language in every sense. The known context and shared expectations frame the experience of danger and disaster. Place and memory give it meaning. Geographic understanding in this regard begins with and should not lose sight of the people and places in danger. And if our concern is with the places of risk, it is their inhabitants who are most knowledgeable about them, at least as they affect the lives at risk. The experience and concerns of those who go through damaging events involve insights which we do not have. We were not there. In many cases we have never been exposed to such threats or degrees of danger in our own lives.

The geographicalness of danger, as of security, is closely bound up with what is happening in our surroundings, among those we trust or have meaningful dialogue with. Disaster is experienced as the upsetting or collapse of feelings of security. At worst, the home place is destroyed, one's people decimated, perhaps exiled never to return.

Although raised last, these considerations need special emphasis. Often neglected in the technical literature, in official and professional responses, they are an essential foundation for a geography of danger and disaster. Hence, *extreme experience*, especially the testimony of survivors of disaster, is the focus of the remainder of this chapter. That will serve to highlight the people and places whose plight is our foremost concern. It also shows how 'geography that matters' becomes more overtly recognised and expressed when severely harmed.

Dangerous geography

Geographic shock: of being lost, alone and 'strange'

Survivors of destructive disasters recall, among some of their first sensations, not knowing where they are, even in their own homes. They speak of being disoriented, lost and unable to recognise anything. A tornado survivor recalls:

> I've lived in this little town all my life. I mean, I was actually lost. I didn't know where I was at because there were no landmarks, nothing to go by . . .
> quoted in Wolfenstein (1957, 58)

A woman describes a friend after bombing raids on her home town, South-ampton, in southern England:

> . . . she could have cried when she saw all the shops down and everything. She didn't know where she was . . .
>
> quoted in Harrisson (1976, 173)

At Buffalo Creek, West Virginia, after the terrible flood disaster of 1972:

> "We find ourselves standing, not knowing which way to go or where to turn," said one survivor. "They should call this whole hollow the Bureau of Misplaced Persons," said another, "we're all just lost."
>
> Erikson (1978, 210)

To be lost has two related meanings: of not being able to find one's way, and of being *misplaced*. Both seem intertwined in survivors' recollections. 'Our' geography, geography that matters, is usually identified with knowing exactly where we are, and by a sense of belonging. Well-being is hardly possible unless one feels relatively secure and supported by the place where one lives. Disaster certainly destroys that. The known surroundings become unfamiliar. The members of a community are or feel forsaken.

A related feeling, in the aftermath of great damage, is of being alone or of loneliness. A child recalled going with her family to search for relatives in the burning streets of Hiroshima after the A-bomb:

> I had a terrible lonely feeling that everybody else in the world was dead and only we were alive. Ever since that time I haven't liked to go outside.
>
> quoted in Osada (1959, 140)

Another of the Buffalo Creek survivors links the loneliness and the lost place:

> My lonely feelings is my most difficult problem. I feel as if we were living in a different place, *even though we are still in our own home*. Nothing seems the same.
>
> quoted in Erikson *op. cit.* (211, my italics)

'Loneliness' has a special significance when thousands of other survivors, including many persons known to these individual victims, have gone through the same disaster. It reflects how much of a person, or of some communities, is tied up with a place. To lose that creates feelings similar to being abandoned by the people who are close to you. The lost place can come to signify and act as a constant reminder of the whole disaster.

A further way in which this is expressed is the sense of being 'a stranger' or of feeling 'strange' *in the home area*. A man returning to his home town, Hannover, in Germany, after it was destroyed by bombing said:

The 'geographicalness' of disaster

> I have the feeling that I am a stranger in my home town.
>
> quoted in Hewitt (1994a, 30)

A flood survivor echoes and amplifies these sentiments:

> We don't have a neighbourhood anymore. We're just strange people in a strange place. We feel our lives have been completely turned inside out by what has happened.
>
> quoted in Erikson *op. cit.* (211)

To be a stranger, in its usual meaning, is to *come from somewhere else* and from a different social context. It is an essentially geographic predicament. Only those who live in a place can try to make the stranger 'feel at home' – if they so choose.

A survivor of a tornado brought these various feelings and their geographicalness together in recalling how she felt when emerging from her ruined home:

> It seemed like a strange place, another world. I felt like I was all alone. It was the loneliest feeling in the world.
>
> quoted in Wolfenstein *op. cit.* (93)

Again, there is a special significance when people describe the sense of being 'lost' or 'a stranger' without having left the sites of their own homes. The geographic sense of this catastrophic experience appears here in a metaphor of uprooting. Their physical location has not changed, yet they are lost. What was familiar has become strange. Where people had felt at home, they now feel like aliens.

In the modern world, some even more problematic and sinister conditions arise for residents of many different environments, especially people still intimately connected with the land. This relates to the problem of 'hidden dangers' (Ehrlich and Birks (eds), 1990) and especially what Robert J. Lifton (1969) described at Hiroshima as 'invisible contamination'. We move to a city or buy a house because a job is available and it is what we had always wanted and can now afford. We do not see and are not aware of some deadly poison in the ground, or a distant factory whose chimneys quietly release dangerous pollutants in our direction. The habitat can even seem very healthy. Its waters look clean, its air pristine. And yet they are dangerous to our health. Perhaps as bad, or worse, are the unsettling psychological fears, the danger of unwarranted but debilitating paranoia over unseen threats, and the many reports of how such dangers have been covered up.

Human senses do not give warning of these risks. Their signs, at least initially, may be ambiguous or suggest other explanations. One thinks of such places as Love Canal, New York; Grassy Narrows, Ontario; or the Rocky Flats nuclear weapons plant, Colorado. Each has involved life-threatening health effects from invisible chemical or radioactive contamination.

How can one comprehend or express the scope of the 1986 Chernobyl nuclear disaster? Scientists could trace the fallout and rainout. But what of people on the ground who could neither see nor feel the radiation? The meaning, in terms of 'strangeness' and 'geographic loss' for those living close to the land that was contaminated, was expressed with great poignancy by a woman of the reindeer-herding Sami in Norway:

> It seems sometimes that things have become strange and make-believe. You see with your eyes the same mountains and lakes, the same herds, but you know there is something dangerous, something invisible, that can harm your children, that you can't see or touch or smell. Your hands keep doing the work, but your head worries about the future.
>
> quoted in Stephens (1987)

Of course, disaster survivors are also often uprooted physically, another major part of disaster experiences. A survivor of the 1985 Mexico City earthquake described how:

> . . . afterwards, when I had to face saving all I could from my house and leave it behind and, of course, change environment completely. Then, yes, I felt incapable of moving from one place to the other. I did not want to leave that spot . . . . It was dreadful, so lonely . . .
>
> quoted in Salgado (1988, 23)

These are not feelings that go away quickly, and may never do so for survivors of the most extreme losses. A survivor of the Holocaust (see Chapter 12) reported that:

> . . . after liberation, I suffered probably more from the loneliness and the isolation, more than during the Holocaust period . . . . I suppose it has to do with the fact that after, the life around you seems normal but *you* are abnormal . . .
>
> quoted in Langer (1991, 23)

As part of the geography of lived space, the significance of such sentiments goes back to the nature of *dwelling*. It resides, literally, in those practices and contexts with which people have had intimate and usually long attachment. Bachelard (1964) argued persuasively, in *The poetics of space*, that it derives especially from, and remains most deeply attached to, the critical learning and socialising time of childhood.

Wolfenstein's (1957) classic work on 'how disasters affect people' contains many other comments of this kind and a discussion of them in terms of the individual psychology of distress and, especially, the feeling of *abandonment*. Thus, it becomes identified with the deep-seated psychosomatic traumas of birth, and the fear or experience of being abandoned as a child.

Dardel discusses human relations to home and world as a more general, mature result of occupancy or residency, of coming to terms with the Earth and humanised landscapes as our larger 'home'. In these terms there is always a play of contrasts. Even within a house, there is a dialectic of the known and mysterious. The familiar landscapes can appear quite frightening in a certain light. Forests may seem friendly as well as threatening. City streets can be 'jungles' as well as a neighbourhood.

Because we deal with the stark feelings and transformations of extreme experience, this should not imply that the geography of safety and fear is ever a simple one. There are even those who, if not themselves hurt, enjoy or find disaster a thrilling or liberating experience – something embodied in Cottard, a character in Camus' novel *The Plague* (1947). People watching bombing raids from a distance, or unharmed by them, describe a peculiar sense of fascination and exhilaration.

'Poor possessions': ecologies of disaster

A further problem with these feelings of lost geography is pointed out by Erikson (1978, 211), who says, 'Once one has said that one feels "strange" and "out of place", one has almost exhausted the available vocabulary.' In part that is because to belong, to be tied to or, at least, to be fully 'at home' in a place is much like having a healthy body. You hardly think about it. To be safe and feel secure also means being able to get on with life. One of the best clues to places that matter most is that 'geography' and space are not mentioned, although they are presupposed. This has been referred to as the 'unselfconscious' or 'taken-for-granted' world.

That helps to explain why the geographicalness of disaster experience does not reside in spatial abstractions so much as in the threatened ties to things and persons, activities and happenings. Lived space and dwelling are found in or realised through the phenomena, rather than a separate plan or dimension of them. This is also another way in which everyday life becomes the reference frame of extreme experience. We will hardly understand one without the other. The meaningful vocabulary is not of spatial abstractions but the concrete terms of everyday life.

The gravest losses for most people in public disasters are still those affecting the home and home area. So often this is the space that decides, or comes to signify, the disaster. The home is the site of a host of other possessions and memories, irreplaceable when lost. Another victim of a flood disaster said:

> The whole thing is a nightmare actually. Our life-style has been disrupted, our home destroyed. We lost many things we loved, and we think about those things. Our neighbourhood was completely destroyed, a disaster area. There's just an open field there now.
>
> quoted in Erikson (1978, 196)

A woman in Frankfurt, Germany, after a bombing raid:

> The apartment house . . . was destroyed . . . and I lost my nice home. I am so depressed about it that I continuously cry. The only thing that keeps my mind

(off) the disaster is ... running around trying to find ... new pieces of ... furniture etc.

quoted in Hewitt (1994a, 11)

People will often risk life and limb to rescue their possessions, or stay and hope to protect their home, as fire, flood or bombs threaten or whirl around them. Not a few victims die in this process:

> We were in the cellar and noticed burning up above us, and ran out and tried to fight the fire. But we had no water and it always burned further down until the whole house collapsed . . . We saved about three pieces of furniture . . . [but] One is so distraught he left things which ordinarily he would have taken out . . . One's senses just stand still.

quoted in Hewitt (*ibid.*, 11)

Possessions, or some of them, are not just objects one owns, has worked hard for, or cherishes in themselves. They are part of the structure and evidence of continuity and *security*, symbols for family survival, even when damaged:

> . . . there was no street door, that had gone. There were no windows. All that lot was blown in . . . I thought to myself: 'Well, this ain't too bad now – they ain't knocked it down yet.'

quoted in Hostettler (ed.) (1990, 9)

The home itself is a comfort, an assurance that all is not lost. A woman of Kobe, Japan, talks of the air attacks of 1945:

> I was more or less used to the raids. But after [my] house was bombed I became more and more afraid

quoted in Hewitt *op. cit.* (11)

While drawing attention to place and the geography of experience, I do not want to put this ahead of, or separate it from, the *most* immediate and dire concerns of most disaster victims. Plenty of testimony shows that personal safety – in families, that of children and other dependents – comes first. The safety of neighbours and even strangers in distress may come ahead of one's own property. In these respects, there is great variety among and within disasters according to the severity of harm and impairment for individuals. Some people may be totally isolated and forced to attend to personal survival. The heroes, and many of the casualties of disaster, survive its initial impact. Then they throw themselves into rescuing and saving their loved ones. If there are some or many unimpaired survivors, they may devote themselves to rescuing others, even unknown to them. The same applies to the effort given to saving personal or communal property.

In all these cases, how people feel and react to danger and disaster is closely related to their pre-existing cares and responsibilities, to already established priorities and values. The geographicalness of their previous life is part of these priorities and, perhaps, indivisible from their other concerns.

Risks to family members assume special meaning in a public disaster, since women at home and children, the elderly and disabled may be exposed in great, perhaps the greater numbers. These, the principal 'carers' and their dependents in most societies, will tend to respond to danger and disaster in terms of their relations of love, dependence and social responsibilities. One of the reasons why women are unusually vulnerable in disasters is their special responsibility for dependents in most societies, and closeness to them.

In such terms, the loss of homes and personal possessions has a very broad significance. Not only is the survival of the 'home' second only to the safety of self and loved ones. The public disasters or catastrophes that concern us most are singled out by concentrated threat to and great losses of homes. The majority of victims are residents of the places destroyed. The numbers and proportions of the homeless commonly define the severity of disaster, and the needs of survivors.

Geography and security

When the home place is severely threatened, or has been destroyed, the sense of a safe and supportive context goes with it. The intimate connection of geography and security is revealed. Survivors at Buffalo Creek said, many months later:

> It's insecurity I believe. You're afraid when you walk out the door that you don't know what's going to happen next.. [and]
> . . . you don't feel secure around people you don't know.
>
> quoted in Erikson (1978, 240)

Charles Darwin described how feelings of security associated with the Earth itself are upset by natural hazards:

> A bad earthquake at once destroys our oldest associations: the earth, the very emblem of solidity, has moved beneath our feet like a thin crust over a fluid: – one second of time has created in the mind *a strange idea of insecurity*, which hours of reflection would not have produced.
>
> Darwin (1906, 289, my italics)

The catastrophic event severs the ties and expectations of on-going existence and destroys the safe haven. The physical fabric of the place is ruined, the nurturing habitat becomes a death-trap. People are forced to leave. The effect is one of being lost and deserted.

It may be added that in most disasters such feelings are followed more or less quickly by a tremendous reassertion of care and community. Disasters are often remembered as times of unusual 'togetherness' and unselfishness – at least in modern, urban communities where selfishness and alienation often seem the norm. We will return to these aspects when discussing active responses to risk and disaster in Chapter 7.

Inner space and outer world: the shock of remembrance

If the geography of security is set up in the time of residence and familiarity before disaster, the loss of it does not end with the destruction. The survivors must struggle with more than just losing those who are close to them, and the familiar surroundings. There can be an enduring pain of remembrance. It begins with finding that what has been destroyed 'out there', remains vividly 'inside'. Consider this woman's recollection after an air raid:

> . . . when I returned my house was burned down. I had the house key in my hand, but there was no door there.
>
> quoted in Hewitt (1994a, 19)

Her house no longer exists 'out there', but inside, and as the guide to action – where to go, to stop and take out her key – it remained a clear reality for her. Unless she was mad, why else was she standing there?

More than any other, the disasters of modern warfare have brought testimony to these realities of civil existence. From the city of Würzburg in Germany, after the annihilating firestorm raid of 16/17 March 1945, a survivor wrote to a friend:

> We persevere here, in search of the city. We find we still carry its [lost] dimensions and form within us. They grew inside through the long years when the homeland protected us. This inner portrait is stronger than the bad dream round about and surges from within to cover the ruins with its former life. Only when this inner vision is extinguished will Würzburg be definitively dead.
>
> Josef Dünninger, quoted in Hewitt (1994b, 270)

This is the shock of remembrance, a haunting realisation of so many survivors, like exiles and adults recalling childhood, that intact within them is the lost place. They fill in the ruined areas with that living mental place. Robert J. Lifton (1967, 29) quotes the words of a Japanese history professor after the bombing of Hiroshima:

> I climbed Hijiyama Hill and looked down. I saw that Hiroshima had disappeared . . . What I felt then and still feel now [1963] I just can't explain with words. Of course, I saw many dreadful scenes after that – but that experience, looking down and finding nothing left of Hiroshima – was so shocking that I simply can't express what I felt . . . Hiroshima didn't exist – that was mainly what I saw – Hiroshima just didn't exist.

The Japanese writer and survivor of Hiroshima, Ota Yoko, expressed clearly how the destroyed city is felt as a loss of significant landmarks. With them goes the sense of continuity with the community's past, and of orientation and

security in the present. She recalled, on coming to the site of Hiroshima Castle
' . . . toppled to earth and utterly flattened . . . ':

> Hiroshima was a flat city with no hills. Thanks to its white castle, Hiroshima
> became three-dimensional and preserved the flavour of the past. Hiroshima, too,
> had its history, and it saddened me to march forward over the corpses of the
> past . . . .
> Even in normal times, it was scary for me to cross Hiroshima's long, long bridge.
> But now the buildings of both banks that seemed to anchor the bridge were gone,
> as were the sections of the city that formed a distant backdrop. So the bridge
> seemed to be floating in the air, and I seemed about to be dragged off and down to
> the bottom of the river.
>
> quoted in Minear (ed.) (1990, 226–7)

On a still darker note, a survivor of the Holocaust tells of how the world of the
concentration camps and annihilations remained with her:

> I am not like you. You have one vision of life and I have two . . . it seems to me
> that Hitler chopped off part of the universe and created annihilation zones and
> torture and slaughter areas . . . We can't cancel out. It just won't go away . . . I
> talk to you and I am not only here, but . . . I see all that . . .
>
> quoted in Langer (1991, 54)

Without memory, of course, there is no disappointment and no achievement.
If humans depended primarily upon 'instinct', if we were biological machines
instead of socialised and adaptive participants in the world, disaster would
hardly exist for us. At most it would mark a sudden drop in population, frantic
activity, a special test of the inherited characteristics of survivors. The armchair
or laboratory scientist and bureaucrat may see things in such terms and think
them more 'realistic'. For those in crisis and calamity, however, memory is the
framework for experiencing and finding meaning or meaninglessness in disaster. Moreover, remembrance after disaster is intimately tied up with actual and
lost geographic space.

Where can one be safe?

Place and identity

A controversial issue involving the geography of risk is a widely reported desire
of survivors to remain in or return to the devastated home place. Often they
want to rebuild on the same spot in the same way or, perhaps, better than
before. That has been a frequent source of conflict with governments and
outside 'experts', who regard this as irrational and foolish, especially if, in
their assessments, the old place remains hazardous. You may even hear comments of the form: 'Why would anyone live in California (or Bangladesh or
Pompeii) after all the disasters, and threats of more?' It is not easy to see the

complexities of security and danger, real or imagined, for other people in other places.

Among places destroyed in a disaster, the town of Yungay in Peru is a remarkable example of partly successful opposition of survivors to official plans for their relocation. In the great earthquake of May 1979, an avalanche from the nearby Mount Huascaran, which turned into a catastrophic debris flow, buried most of the town and killed almost 90 per cent of its inhabitants. Despite their extreme loss and emiseration, the few survivors determined to resist enforced relocation. One of them is quoted as saying:

> We are the true sons of Yungay and one does not abandon an afflicted mother. We must defend our land.
>
> quoted in Oliver-Smith (1986, 212)

Another was overheard to say:

> This is where we want to be. We are accustomed to dying, to losing family. One dies everywhere. [Mount] Huascaran will keep on. We are strong-willed. We want to be here, to die where we were born.
>
> quoted in Bode (1989, 201)

The case of Yungay is known to us not just because of the scale of tragedy but also because the condition and actions of the people there, in the months and years following the disaster, were carefully studied. There were two major studies, by Anthony Oliver-Smith (1986) and Barbara Bode (1989). Through extended periods spent in the area, and a sensitive effort to understand the local situation and concerns of survivors, they have provided a rare degree of insight. One of Oliver-Smith's main conclusions (p. 18) was:

> The determination of the survivors to found and maintain a new Yungay, albeit one of vastly different appearance, near the avalanche which buried their old community, constitutes a refusal to allow part of their identity and their culture to die.

Bode (1989, 79) found the same feelings and determination in the nearby town of Huaraz, also slated for relocation:

> To survivors, it would not be Huaraz were it not 'in its place' . . . To be 'in its place' was right and good. It was hoped that the streets, too, could remain in their places and could retain their own names. "We would be very sad and confused if streets were called by other names," people said.

Bode also reports (p.201) an exchange with local residents concerning Yungay that highlights the complexities of these questions:

> . . . . with Cirila and Ruben one day, discussing the possible sites for a new Yungay, I found myself pointing to a place on the mountain side.
> "Is that strip of land there safe?" I asked.
> Serenely, Cirila answered, "The sierra is never safe."

> Ruben added, "There is no security anywhere. If you're in the United States, then the Russians or the Chinese come."
> "Where can one be safe?" Cirila challenged me.
> I had to admit I did not know.

Ironically, one of the sites chosen for relocating people in official reconstruction plans had been overrun by a mudflow 30 years earlier.

Studies of the plight of residents in the aftermath of the Buffalo Creek flood in the United States illustrate the same themes, but with an opposite result (Erikson, 1976; Gleser *et. al.* 1981). For a long time there was a preoccupation with imposing official measures, and defusing blame and grief. Survivors were removed and separated. They were obliged to wait for many months in temporary homes in unfamiliar places among strangers. In the end, the scale of their unresolved grief and demoralisation came to seem worse than the disaster. Their community became, in their own minds, beyond recovery. The prompt, strong relief and rehabilitation measures available in the United States increased this, not just the scale of the disaster. Survivors' concerns were met with official incomprehension and indifference. In due course, they filed a lawsuit claiming major psychic as well as physical damages. Several studies concluded the disaster was as much one of a total 'loss of community', aggravated if not caused by post-disaster measures, as of devastation in the flood.

In 1952, Argostollion, the main town on the Greek island of Kefallinia (Cephalonia), was largely destroyed by an earthquake. What remained was quickly demolished as unsafe by the army relief force – all except one house, whose owner told me it was saved only by his standing in front of the bulldozers! The Greek government also ruled that no houses or infrastructure could be replaced until plans were examined and approved. They had to meet strict earthquake-resistant standards. In itself this seems a wise decision, but it meant continuing and enormous hardship for the thousands of survivors evacuated, or living in tent camps and makeshift homes around their town. Most had also lost their means of livelihood in the disaster. Apart from the difficulties and costs of getting the necessary plans made, legal and bureaucratic hurdles took months and years to cross. It took a decade before the town began to recover its pre-disaster housing stock and facilities. By then, most residents had left in search of work and a future elsewhere. It was said the majority ended up and still live in Montreal, Canada. The new town was occupied largely by people from surrounding villages. The town that rose from the rubble looks very modern and smart and is, no doubt, much less vulnerable to earthquakes. Some old residents began to return in the 1970s, but the community, as well as the place that was there before the earthquake, was utterly lost.

It is true that the modern political economy, and the demands upon professionals who serve it, does not encourage a preoccupation with place, with people's feelings and concerns. We are not encouraged to emphasise the 'banalities' of everyday life, unless they can be reduced to some technical formula. Such age-old local concerns with family and neighbourhood tend to be displaced if the people are not. If dealt with at all, they are commonly subordinated to or redescribed in terms of technical abstractions or broad economic and political goals, the impersonal forces of population, environment, technology or economics. But what are the preoccupations of those at risk, including ourselves on our own ground? They are about the safety of family members and home; fear of losing personal possessions and dependents; neighbourhood safety and security of support; one's bit of land or job. For many folk, the plight of the less fortunate or disadvantaged is also a concern, which seems to demand reliable practices and fair treatment even more than further safety measures. These seem the primary issues in a public perspective on safety and crises, and basic to a human ecology of risk.

Geographical calamities

The concerns with place and attachment of communities to their long-time home areas may not seem 'modern'. Nevertheless, they do still loom large in the testimony of disaster survivors and perhaps involve the majority of people. At least, there is abundant evidence that people's attachment to place appears strongly in reactions to disaster and resistance to forcible removal.

These concerns define some particular forms or aspects of disaster that are essentially 'geographic'. The danger and loss comes from changes in people and place, the habitable Earth, or all of these. The forms of geographical disasters identified here and discussed later in the book are:

- *place annihilation*: meaning the destruction of the material reality, the landscapes and cultural foundations of life, especially in historic places of settlement;
- *enforced displacement* of resident populations: removal of the inhabitants of an area against their will;
- *ecocide*: destruction of the habitat or means of life in an area;
- *cultural annihilation*: the elimination of distinct societies. This involves, especially, an almost worldwide threat and atrocities against indigenous peoples or 'nationalities' – cultures fully interdependent with the land of their ancestors. Often this threatens or leads to:
- *genocide*: the deliberate, physical extermination of a people – their elimination from the 'map of man'.

These are all geographic disasters. They violate and destroy the basis of shared, settled existence in a land. They threaten and destroy the places of historic settlement or established communities. They eliminate them from the broader patterns and exchanges of human geography. The result is to produce or threaten to produce 'man-made' blanks on the map, or geographic 'extinctions'.

Natural disasters have, but rarely, annihilated both a people and all vestiges of a place. Even great epidemics have utterly destroyed only small communities. In general, survivors can rebuild, or others may join them to re-establish a decimated community in the old place. The gravest danger and worst examples of all these geographical calamities derive from human violence, especially in wars and other uses of armed force. Some cases, of which the Holocaust is the great and shocking modern example (see Chapter 12), involve all five of these processes of annihilation by and of a people's geography. We will pay particular attention to them in Chapter 5.

Suggested reading

Barkun, M. (1974) *Disaster and the Millenium*. Yale University Press, London.
Bowman, I. (1946) The strategy of territorial decisions. *Foreign Affairs*, **242**, 177–94.
Dardel, E. (1952) *Presses universitaires de France*. Paris.
Fried, M. (1963) *Grieving for a lost home*. 151–71 in Duhl, L.J. (ed.) (1963) *The urban condition*. Basic Books, New York.
Hewitt, K. (1983b) Place annihilation: area bombing and the fate of urban places. *Annals Tuan*, Association American Geographers, **73**: 257–84.
Lifton, R. J. (1971) *History and human survival*. Random House, New York.
Relph, E. (1976) *Place and placelessness*. Pion, London.
Tuan, Y. F. (1979) *Landscapes of fear*. Pantheon Books, New York.

CHAPTER 3

Natural hazards

Natural, as opposed to technological, hazards are those triggered by climatic and geological variability, which is at least partly beyond the control of human activity.

Palm (1990, 3)

Introduction

Hazards perspectives and the hazards paradigm

The study of hazards has had, as its primary focus, conditions and processes that are, or can be, direct causes of damage. As noted earlier, the term hazard, used strictly, has an interactive and an evaluative meaning. It depends upon the source of danger *and* the nature and concerns of human communities at risk. Nevertheless, in the language of most studies, 'the hazard' *is* fire or strong winds, toxic chemicals or nuclear weapons. The field is divided up mainly by work on natural, technological, biological or social hazards, and specialised studies of single hazardous agents within these domains (Table 3.1). Hazards are classified by particular dimensions or processes of nature, technology or society as dangers. However, danger is also realised in events composed of a mix of processes, even when only one of them is the main cause of damage. The latter part of Table 3.1 identifies 'compound' hazards and 'complex' disaster events. In the former, danger arises from the interaction of two or more of the main classes of natural, technological and social hazards. Disasters are 'complex' in the sense that they arise from damage by two of the different main classes of hazard. In reality, all disasters are 'complex', but in certain catastrophes the need to recognise the interaction of very different sources and forms of harm becomes paramount. Risk tends to be interpreted as a function of the properties of such agents.

Unfortunately, this has promoted a view of hazards as agents external to, or 'accidentally' erupting within, society. The geography of risk is then

55

Table 3.1 Hazards: classes of specific hazardous agents or conditions, and the situations or events in which they are associated with disasters.

Single condition/process (Agents)	Composite hazards (Events)
NATURAL HAZARDS	

Atmospheric

Temperature, fog, rain, (high) winds, lightning, hail, Snowfall, Freezing rain ('glaze')	Thunder/hailstorms, tornadoes Rain and wind storms Tropical cyclones Blizzards Glaze storms

Hydrological

Runoff (overland, stream) Snow on the ground Ground water Freeze–thaw Sea ice, icebergs	Floods: riverine, coastal (marine) Natural dams and outburst floods Glacier advance and 'surges' Ice-infested waters

Geological/Geomorphological

Seismicity, volcanoes Tsunami (seismic sea wave) Earth/rock materials: quickclay, quicksand Mass movements Radioactivity Geothermal heat	Earthquakes, volcanic eruptions Rockslides, rock avalanches, debris and mud flows Submarine slides Subsidence Domestic radon gas hazards

Biological and disease hazards

Viruses (e.g. measles, HIV, dengue) Bacteria (e.g. pneumonia) Protozoa (e.g. giardia, malaria) Fungal (e.g. pneumocystas) Algae Plants ('weeds') Insects ('pests') Animals ('pests')	Disease outbreaks/epidemics: bubonic plague, yellow fever, influenza pandemics Sexually transmitted diseases 'Red tide' (toxic algal blooms) Plant infestations, 'invasions' Insect plagues/infestations Locust/grasshopper plagues Rat infestations Shellfish poisonings

Table 3.1 contd.

TECHNOLOGICAL HAZARDS

Hazardous materials
radioactive materials
toxic substances (e.g. dioxin)
dangerous gases (e.g. carbon
  monoxide)
radioactive materials
mutagens
carcinogens

Contamination:
  buildings, soil, surface and
  ground water
Industrial pollution
Agricultural contaminants

Hazardous processes
radioactivity
fire

Release of dangerous materials:
  airborne (radionucleides, SO_2),
  waterborne (effluent, coolant)
Structure collapses
  collisions, explosions
'Accidents':
  transportation
  industrial plant
  mining
  medical, surgical

Hazardous devices
vehicles
power station
power line
explosives
birth control devices

SOCIAL VIOLENCE

Weapons
Firearms, incendiaries, nuclear,
  chemical, toxins, gas, biological

Bombardment (artillery, naval)
Air raids
Guerrilla warfare
CBT warfare
Environmental warfare
Sieges, Terrorism
Release of dangerous forces:
  oil spills/fires, chemical

Perpetrators
Armed forces
Governments
Terrorist groups

Methods
War, terror, subversion,
  sabotage, genocide

COMPOUND HAZARDS

Smog (fog + air pollution) (inversion + sunlight + pollution)
Artificial dam break ('accident' + flood wave)
Air raid 'firestorm' (bombing + mass fire + atmospheric storm)

Table 3.1 contd.

COMPLEX DISASTERS

Famines (drought + crop failure + food hoarding + poverty)
Refugee crises (famine + war)
Toxic floods (tailings dam break + toxic waste + flood)
'Dirty' nuclear tests and power explosions (nuclear explosion and
contamination + atmospheric circulation + rainout and fallout +
uprootings)

---

represented by the incidence of hazardous agents such as typhoons or insect plagues, and regions where they occur or are more frequent and severe. Such an approach involves what may be called a 'hazards paradigm': a view of risk as controlled by, or following from, the nature of geophysical agents, disease vectors, forms of violence and technological failures. The field has been dominated by this paradigm. However, a critical view of it is adopted here, as increasingly by geographers. A natural force is not dangerous in itself but becomes so in relation to human activities and values. To think otherwise results in too little attention being paid to other aspects of endangerment identified earlier – human vulnerabilities and intervening conditions, the past history of disasters and adjustments to them, and existing response capabilities or failure to invest in them.

However, the limitations of the hazards paradigm do not mean that knowledge of the properties of earthquake or flood, oil in the marine environment, or urban mass fires, is unimportant. On the contrary, it is quite essential to understanding in this field, most obviously where such agents are immediate causes of harm. Without the earthquake, and one of about that strength, location and other seismic attributes, there would have been no 1906 San Francisco disaster, nor such developments as the fire that did most of the damage in the city. Without the toxic gas leak there would have been no disaster at Bhopal, India, or not the particularly lethal and horrible one of December 1984. If the A-bomb had not been dropped, Hiroshima would not have been annihilated on 6 August 1945, along with more than 100 000 of its civilian inhabitants. Meanwhile, an introductory text can hardly do its job while ignoring the prevailing style of work. Hence, this and the following two chapters are devoted mainly to examining various classes of hazards, and damaging events associated with them. In doing so, however, an attempt is made to reconfigure the treatment of hazardous agents. What must be recognised is that they are *necessary* conditions of danger, but never *sufficient* ones. Since natural hazards have dominated work by geographers, it will be appropriate, firstly, to develop this approach in terms of them.

The scope of natural hazards

> . . . natural disasters have claimed about 3 million lives in the past two decades, adversely affected 800 million more people and resulted in immediate damages in excess of US$23 billion.
> UN General Assembly, *International Decade for Natural Disaster Reduction* resolution 42/169 (1987)

Natural hazards involve processes such as high winds, volcanic eruptions or forest fires that arise more or less spontaneously in the planetary environment. These would occur whether humans were present or not. Some are aggravated or brought on by human activity, for example, where the occurrence or severity of floods reflects deforestation of a watershed. Most are generated by large-scale planetary processes and are largely or wholly outside direct human control.

The importance of this subject is reflected in the continuing large tolls, and global incidence of disasters triggered by natural extremes. In recent years, an average of 48 natural disasters have been reported, worldwide, per year. Floods are the most frequently reported sources of disaster, followed by tropical cyclones, epidemics and earthquakes (Fig. 3.1). Drought events vie with or exceed the others in total number of persons affected while, in the course of this century, death tolls in drought-related famines exceed any of the others, except epidemics.

However, the significance accorded to these events will vary with the way their impacts are evaluated. They appear and are ranked somewhat differently by fatalities as compared to economic loss, total numbers of persons affected, or the relative importance of different types of event. Most assessments use national statistics, often the best or only ones available for global, comparative work. Yet there are large, sometimes larger, differences among groups within states, while a range of other conditions and factors influences where disasters occur and how severe they will be.

High human casualties are associated mainly with regions of dense, impoverished and poorly protected populations, notably in Asia (Table 3.2). Exceptional levels of economic loss, at least using standardised monetary values, are a feature of wealthier, urban-industrial areas. However, when considered in terms of overall national wealth, major economic losses are equally common in poorer countries. Overall statistics also tend to hide the great temporal variability in the incidence and impact of different hazards (Fig. 3.2). There are continuing and expanding losses from floods and tropical cyclones, reflecting the vulnerability to them of urban as well as rural societies. Earthquake damages are also increasing with rapid urbanisation, especially with poorly located and designed built environments (see Chapter 8). Drought-related losses are magnified by the increasing spread and productivity of commercial agriculture, when exposed to water shortage, and the marginalisation and impoverishment of populous rural societies in the tropical world.

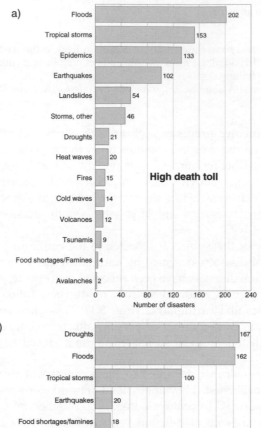

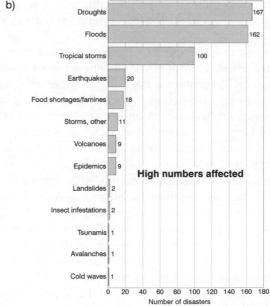

Fig. 3.1 Major disasters around the world, 1963–1992, classified by initiating natural agent and selected impacts in three categories:
a) High death toll (HDT), identifying all events with at least 100 killed.
b) High numbers of persons directly impacted (HAF), meaning at least 1 per cent of total national population.

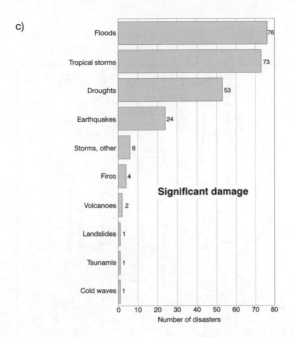

Fig 3.1 contd. c) substantial economic damage (HD), meaning losses of at least 1 per cent of annual gross national product (after UN/DHA, 1994b).

Table 3.2 Natural disasters 1947–1980, showing number of events, total estimated loss of life and average loss per disaster, by continental area (after Shah, 1983, 209)

Continental area	Lives lost (no.)	Disaster events (no.)	Average loss of life per event
North America	11 531	358	32
Central America and Caribbean	50 676	80	633
South America	49 265	75	657
Europe	26 694	119	224
Africa	25 540	34	751
Asia	1 054 090	437	2 412
Oceania	4 502	16	282
Total	1 222 298	1 119	1 092

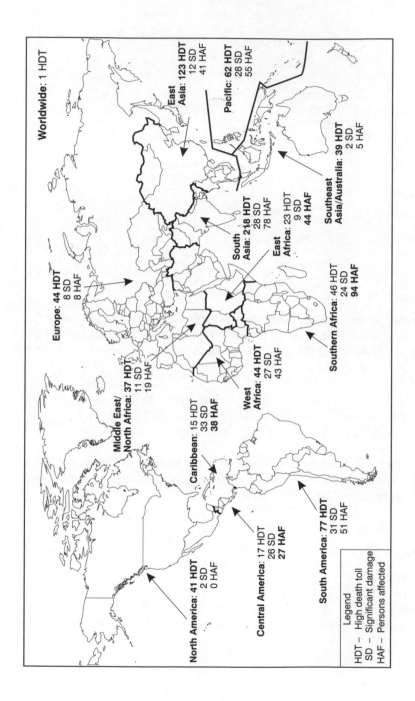

Worldwide: 1 HDT

East Asia: 123 **HDT**
12 SD
41 HAF

Pacific: 62 **HDT**
28 SD
55 HAF

Southeast Asia/Australia: 39 **HDT**
2 SD
5 HAF

South Asia: 218 **HDT**
28 SD
78 HAF

East Africa: 23 HDT
9 SD
44 **HAF**

Europe: 44 **HDT**
8 SD
8 HAF

Southern Africa: 46 HDT
24 SD
94 HAF

Middle East/ North Africa: 37 **HDT**
11 SD
19 HAF

West Africa: 44 **HDT**
27 SD
43 HAF

Caribbean: 15 HDT
33 SD
38 HAF

North America: **41 HDT**
2 SD
0 HAF

Central America: 17 HDT
26 SD
27 **HAF**

South America: 77 **HDT**
31 SD
51 HAF

Legend
HDT – High death toll
SD – Significant damage
HAF – Persons affected

Fig. 3.2 Global distribution of major disasters around the world, 1963–1992, classified by initiating natural agent, and by high death toll (HDT), high numbers of persons directly impacted (HAF), or substantial economic damage (HD) as defined for Fig. 3.1 (after UN/DHA, 1994b)

However, these are very general indicators of more complicated questions of place and context: of the spatial variability and relations among hazardous natural agents and endangered human activities. For instance, they do not reveal how certain segments of affected populations, most commonly women and children of the poorest families, can be disproportionately harmed in natural disasters. Small states or ethnic and other minorities are often at more severe risk, even from less severe or less frequent disasters, than large states or civil populations as a whole. Unfortunately, comprehensive, reliable data to compare with the usual profiles from national statistics are not available. We must deduce the prevalence of these problems from case studies and indirect evidence.

Domains and classes of natural hazard

The study of natural hazards has mainly accepted the division of the planetary environment into atmosphere, hydrosphere, lithosphere and biosphere. It is appropriate, therefore, to introduce these risks briefly in terms of natural domains. Even here, however, our foremost concern is to see how natural conditions relate to and affect human safety, rather than their properties, in and of themselves. It will soon become clear that dangers and disasters do not respect these divisions!

Atmospheric hazards

The atmospheric environment is the most widespread and frequent trigger of damaging events (Table 3.3). As an adaptive problem for humans and their technologies, atmospheric hazards arise from permanent and urgent resource needs, such as for oxygen, rain, warmth, cooling and drying, or sunshine. The price is more or less continuous exposure to the frequent, and sometimes large and dangerous, fluctuations of, especially, the weather layer of the atmosphere.

Modern and other highly mobile cultures are also at risk from the large geographical variations between different climates. Motor vehicles or electronic devices are expected to perform the same functions year round and almost worldwide, increasing the range of conditions that may cause them to fail. In cities, water is piped from wetter places, plant products are imported from other climates, and artificial energy is used to maintain climate-controlled spaces. Unfortunately, we are interested mainly in the cases where these climate-modifying and protective systems fail. Major disasters tend to be those in which people are suddenly exposed to unmediated heat or cold, high water, high winds or snow-covered ground. Even so, most, often the worst, weather disasters continue to affect rural populations and outdoor activities, where people's lives are more directly exposed to atmospheric conditions.

Damages can occur directly through the impact of, say, high winds, lightning discharges or extreme cold. However, atmospheric events lead to and strongly

Table 3.3   Atmospheric hazards: classes and examples of associated disasters with some estimates of losses

Class of event	Examples
Heatwaves	USA, 1901: 9508 killed Canada, 1936: 500 killed Southern USA, 1980: 1265 killed, US$20 billion damage Greece, 1987: 1000+ killed
Cold spell	USA, 1995 (mid-January): 'paralysed' east coast, mid-west cities, 140 killed
Lightning (thunderstorm)	Mülheim, Germany, 1988: aircraft struck and crashed, 21 killed
Hailstorms	Turkey, 1964: 12 000 people affected, US$2.2 million damage Fribourg, Switzerland, 1985: US$2 million damage Calgary, Canada, 1991: 8000 insurance claims, US$400 million damage
Wind storm	Britain, 1987: 20 killed, US$1.7 billion damage
Blizzards	New York, USA, 1888: 800+ killed Western Europe, 1956: 1000 killed Boston, Mass., USA, 1978: 29 killed, 10 000 homeless East coast, USA, 1993: US$1.6 million damage
Tornadoes	Mid-western USA, 1925: 689 killed, 13 000 injured, US$18 million damage Waco, Texas, USA, 1953: 114 killed, US$39 million damage Worcester, Mass., USA, 1953: 94 killed, 1306 injured, US$53 million damage Dhaka, East Pakistan (now Bangladesh), 1969: 50 killed, 4000 injured Mid-western USA, 1974: 315 killed, 5500 injured, US$0.5 billion damage Barrie, Ontario, Canada, 1984: 12 killed, 500 injured, 800 homeless, US$117 million damage
Tropical cyclones/ hurricanes/ typhoons	Haiphong, Vietnam, 1881: 300 000 killed Bombay, India, 1882: 100 000+ killed Bangladesh, 1979: 500 000+ killed, 1.1 million acres of rice destroyed, 1 million cattle drowned Hurricane Hazel: Caribbean, USA, Canada, 1954: 1000+ killed, 40 000 injured, 1.5 million people displaced Typhoon Vera: Japan, 1959: 4464 killed, 40 000 injured, 1.5 million people displaced Hurricane Flora: Caribbean, 1963: 7190 killed, US$400 million damage

Table 3.3 contd.

Class of event	Examples
	Hurricane Agnes: Atlantic coast, USA, 1972: 122 killed, 300 000 homeless, US$4.5 billion damages
	Typhoon Elsie: Philippines, 1989: 332 000 homeless
	Bangladesh, 1991: 139 000 killed, 9 million homeless, US$1.4 billion damage
	Hurricane Andrew: USA, Bahamas, 1992: 34 killed, US$15–17 billion damage

influence hydrological, geological and biological hazards. Nearly all floods, and most landslides, forest fires and insect infestations, are initiated by weather conditions. In that sense their separation can be misleading, both as an identification of the hazard and the terms in which human communities confront atmospheric dangers. We will return to this problem, and to more detailed assessments of selected atmospheric hazards.

Hydrological hazards

Hydrology considers all aspects of the movement and storage of moisture, including water in the atmosphere (hydrometeors like rain and snow), lithosphere (ground water) and biosphere (soil, natural irrigation and transpiration of moisture). Moisture-related risks can arise where human society interacts with each of these. However, the hydrosphere is treated mainly as extensive, continuous bodies of water dominated by the oceans, but including fresh water lakes and ponds, rivers, glaciers, and ground waters. Hydrological hazards are generally related to conditions and fluctuations in these (Table 3.4). Again, however, we should note that few if any hydrological hazards can be understood without reference to, if they are not initiated by, atmospheric or geological events.

There is a close association between the map of water bodies and that of human populations, agriculture, urban centres and most other settlements. Here too, risks arise mainly through the forms of human dependence on, and use of, water as a resource. That is firstly a question of large, necessary and direct demand for water for domestic, agricultural, municipal and industrial consumption, but also tapping the biological resources of seas, lakes and rivers. Perhaps more important in the rise of modern urban centres have been water-borne transportation, mercantile and naval strategy, waste disposal, coolant water, and amenities and recreation.

Hydrological hazards are identified, above all, with excessive amounts or inadequate supplies of water. Flood and drought studies outstrip all others in the field. Floods especially have dominated the development and, in many

Table 3.4   Hydrological hazards: classes and selected examples of associated disasters

---

Class of event
Floods: riverine, lake shore, sea coast
Droughts
Glaciers, including surges (catastrophic advances), ice dams and outburst floods
Avalanches (snow)
Ice-infested water (sea ice, icebergs)
Waterlogging

Disaster	*Examples*
Floods	Huanghe, China, 1887: 900 000 + killed
	Netherlands, 1953: 2000 drowned, 300 000 displaced, US$300 million damage
	Teheran, Iran, 1954: 2000 killed
	Mekong delta, Vietnam, 1964: 5000 killed
	Florence, Italy, 1966: 150 killed, huge damage to artwork
	Luzon, Philippines, 1976: 60 + killed, 630 000 homeless
	Mid-west, USA, 1993: 48 killed, 50 000 homes damaged/ destroyed, US$12 + billion damage
Droughts	'Dust-bowl', mid-west, USA, 1932–37: 50 million acres affected
	Laos, 1977: 3.5 million acres affected
	Ethiopia, 1983–84: 7.7 million acres affected (major famine)
	Western Canada, 1984: 10 000 farms affected, US$1 billion losses
	Italy, 1989: farmer losses US$1.5 billion, ski tourism losses US$1.3 billion

---

ways, the mindset of hazards investigations by geographers. In terms of losses and numbers of people affected, this emphasis seems justified. Later in the chapter, a review and comparison of droughts and riverine floods will be undertaken. However, in some parts of the world and for some activities, ice-infested waters, waterlogging or glacier behaviour pose major risks (see Chapter 9).

Economic growth and social change reveal great constraints and inflexibilities in water needs and related risks. In urban and heavily populated rural areas, growing dangers are associated with the ignoring of means to avoid and to prepare against hydrological hazards. Dangers arise increasingly from the pursuit of other concerns in places at risk from hydrological conditions. People intent upon being close to the city centre, or enterprises and municipalities wanting to exploit less expensive land, are willing to live or build in flood-prone areas. People utilising less desirable land in dry regions create residential areas where droughts quickly threaten their supplies of fresh water. These hazards can be modified and expanded by human activity. The land use of a watershed, notably deforestation, agricultural practices or urbanisation, can dramatically alter the scale and frequency of floods. In the tropical world especially, desertification – the spread and intensification of drought conditions and arid land processes – is attributed mainly to the activities of humans, or their increasing vulnerability to droughts and climate changes.

Geological hazards

These refer to natural processes involving tectonic, volcanic and geomorphological or Earth surface processes (Table 3.5). Danger may also arise from Earth materials creating chemical or radiation hazards. The main dangers are processes that suddenly alter, temporarily undermine or destroy the stability of the land surface including foundation and slope stability, physical and chemical erosion, tectonic movements, or volcanic eruption. The location and distribution of geological hazards vary with the relief and ruggedness of land and submarine surfaces, tectonic activity, and the composition, structure and weathering of rock and soil masses. However, in most cases, climatic conditions play an even larger role in whether dangerous events occur. Vegetation cover, soil and ground waters may also be decisive factors.

The solid Earth is the basis of human settlement and material security through, especially, the way it supports built structures. As artificial extensions of the solid Earth, artificial structures depend upon the stability of natural land surfaces. Any process reducing that stability or drastically changing surface conditions threatens the built environment. Most destruction in earthquakes, volcanic eruptions and landslides is to buildings. Most fatalities occur when the buildings that people are in collapse. Survivors are displaced by destruction of their homes and places of work. Major losses and disasters also occur as a direct result of erosion and deposition of Earth materials, but most are equally associated with and triggered by atmospheric and hydrological hazards. Earthquake risk is the focus of one of the case studies, and landslides are considered in the study of mountain lands (Chapters 8 and 9).

Biological hazards

These include dangers from and to the living world (Table 3.6). They arise both from human interdependence with the rest of life and from efforts to control or segregate parts of it solely for human use. We are endangered as hosts to, or potential 'resources' for, a wide range of organisms that cause diseases or are parasites in humans. Domesticated plants and animals, grown in artificially maintained environments, are at risk from so-called pests and weeds or infestations of blight, mould, disease and parasites. These can ruin field and garden crops or free-ranging and 'factory farm' animals, or consume or spoil

67

Table 3.5   Geological hazards: classes and examples of associated disasters

Class of event
Earthquakes (strong seismic motion)
Vulcanism (explosive eruptions, lava flows, pyroclastic flows, etc.)
Landslides or mass movements (rockslides, debris flows, mudflows, submarine slides, etc.)
Sandstorms and shifting sands

Disaster	Examples
Earthquakes	Shensi, China, 1556: 830,000 killed
(see chapter 8)	Sicily and Naples, 1693: 93 000 killed
	Lisbon, Portugal, 1755: 60 000 killed and 85% of city destroyed
	Messina, Italy, 1908: 150 000 killed (75 000–85 000 in Messina)
	Kansu, China, 1928: 180 000 killed
	Kanto Plain, Japan, 1923: 140 000 killed (mainly in Tokyo)
	Peru, 1970: 70 000 killed
	Tangshan, China, 1976: 245 000 killed (some say as many as 800 000)
Volcanic eruptions	Mont Pelée, Martinique, 1902: 30 000+ killed
	Mount Kelud, Indonesia, 1919: 5000 killed
	Mount Leamington, New Guinea, 1951: 3000–5000 killed
	Mount St Helens, USA, 1980: 34 killed, cost US$1 billion
	Nevado del Ruiz, Colombia, 1985: 23 000 killed
Landslides	Java, 1919: 5100 killed
	Khait, Tajikistan, 1949: 12 000 killed
	Nevado del Huascaran, Peru, 1970: 18 000 killed (earthquake-generated slide)
Sandstorm	Mauritania, 1995: 94 killed (passenger plane crash in sandstorm)

foodstuffs and natural fibres in storage. Many biological hazards have grown up with and are so closely adapted to human domesticates, artificial environments and economic activities that it seems misleading to call them 'natural' hazards. At least their geography, and that of the risks they pose or disasters they cause, is more intimately related to human activities, demography, health and vulnerability than to purely natural conditions. Certain insect pests or diseases are confined to particular climates or are more dangerous in certain regions, yet the infestations, epidemics and 'invasions' of dangerous exotic

Table 3.6 Biological hazards: classes and examples of associated disasters

---

Class of event
Communicable diseases of humans (viral, bacterial, protozoal, parasitic):
    epidemics (local, regional)
    pandemics (worldwide)
Epidemic diseases of domesticated plants (blights, etc.)
Epidemic diseases of domesticated animals (bacterial, viral, etc.)
Epidemic diseases of wildlife and plant resources (population replacement and
'plagues' of pests)
Plant and animal infestations (invasion of exotic pest species)
Biomass fires (forest, grassland)

Disaster	*Examples*
Epidemics in humans:	
bubonic plague	China, 1909: est. 1.5 million deaths
typhus	Europe, 1914–15: est. 3 million deaths
meningitis	Niger, 1923: est. 100 000 deaths
polio	Canada, 1953: est. 8000 affected
measles	Turkey, 1964–65: est. 100 524 deaths
yellow fever	Nigeria, 1969: est. 102 400 affected and 2000 deaths
equine encephalitis	Ecuador, 1969: est. 80 000 affected and 400 deaths
influenza	Japan, 1978: est. 2 million affected
hepatitis	India, 1986: est. 11 000 affected and 210 deaths
cholera	Somalia, 1986: est. 1307 deaths
Pandemics in humans:	
'Black Death'	1346–1351: Asia, 100 million
(bubonic and	Western Europe, 75 million
pneumonic plague)	
influenza	Worldwide, 1917: 20 million (some sources say up to 50 million)
	Worldwide, 1957–68: (low death toll but probably 50 million affected, more than in any to date)
cholera	Latin America, 1991–93: 901 000 affected, 8000 deaths
	Worldwide, 1993: 'tens of thousands' of deaths and 3 million affected
Epidemic crop diseases:	
potato blight	Ireland, 1845–47: 75% shortfall in the worst year,
(*Phytophora*	1846 (Solar, 1989, 114) (indirectly 1.5 million died of
infestans)	famine and 1.5 million were forced to emigrate)
Forest fires:	Maine and New Brunswick, 1825: 3 million acres, many towns destroyed, 160+ killed
	Wisconsin and Upper Michigan, 1871: 1.3 million acres, destroyed Peshtigo (1300 killed), 1500 killed

---

species are closely dependent upon the crops grown or animals raised, and upon social conditions and public health measures.

Again, our interest is mainly in sudden, large crises and losses. They include outbreaks of epidemic diseases, insect plagues and natural fires. However, biotic hazards are as fully given over to technocratic, official and corporate control as technological ones. Our field has given much less attention to these hazards than their importance would seem to require, since even the more calamitous risks are subject to extraordinary governmental and professional intervention. In part this seems justified by the successes of medical, veterinary, forestry and agricultural sciences in controlling many of these problems, and ridding the world of some of the most lethal diseases and pests. Yet, the AIDS pandemic, recent outbreaks of Legionnaire's disease and salmonella in North America or cholera in Peru, and the resurgence of tuberculosis and malaria in many parts of the world, show uncontrollable epidemics are not just a thing of the past. The development of strains resistant to existing treatments poses a particular threat of future, uncontrollable disasters. On-going and new examples of infestations by exotic pests and weeds, from 'killer bees' in the United States to purple loosestrife in Canada, are cause for similar concern with respect to agriculture and wilderness areas.

Meanwhile, there is striking evidence that public and habitat health are low on the agendas of many countries. It is clear that the burgeoning megacities of the world pose horrendous problems of public health and are continually threatened by epidemics, as are large, impoverished and malnourished rural populations. But countries like Canada, which had prided themselves that they maintained high-quality health care and environmental standards, in the wake of economic downturn and a preoccupation with government deficits seem unable or unwilling to resist pressures to cut these services severely.

As with rest of this field, the greatest risks come from the collapse of routine official and professional systems. Such developments in the social treatment and construction of biological risks, more than the chance appearance of drug-resistant or unusually virulent strains of bacteria, create the larger or more likely dangers. Of course, this reinforces the stress upon human ecology and human vulnerability, in contrast to a hazards paradigm preoccupied with, say, viruses or insect vectors.

Compound hazards and actual damaging events

Each of the domains of the natural environment has some distinctive relations to human well-being and gives rise to a variety of conditions that may threaten human societies. Each has served as the focus of particular hazards studies. However, all actual damaging events involve at least three, and usually all four, natural domains in influencing the forms and severity of damage, and constraining human responses. This follows from an obvious but basic observation: human life itself originates and takes place almost entirely at the interface of atmosphere, hydrosphere, biosphere and lithosphere. It is part of the continuous interactions and exchanges between them.

Meanwhile, few hazardous events involve just one force or form of danger. Usually, several different ones act together or follow from one another in secondary and tertiary damages, as described in Chapter 1. Hurricanes and typhoons involve high winds, torrential rains, lightning, waterspouts and, over land, hail and even tornadoes. Most deaths, and a large part of the damages from these 'atmospheric hazards', are due to flooding, a 'hydrosphere hazard'. Major, sometimes most, damage in earthquakes involves secondary ground or slope failure and landslides (see Chapter 8). Disasters, then, always involve *compound hazards*: the combined influences of air, water, land and living things. In confronting disaster, communities face these compounded risks and forms of assault. They rarely can or do respond by breaking down the problem according to the special properties of wind, flood or landslide, but in terms that reflect how they enter into human activities and concerns. And this is not yet to point to how most catastrophes are 'complex emergencies', in the sense of involving technological and social as well as natural hazards.

Forms of endangerment

Contexts and interfaces

> Everywhere geographical space is carved into matter or diluted into mobile or invisible substance.
>
> Dardel (1952, 9)

A geography and human ecology of natural hazards would seem to require an assessment of natural processes in terms of their relations with and significance for human society. The separation of society and environment in the hazards paradigm creates geographical and ecological fictions. Rather, it is a truism of environmental problems that society and nature are nowhere separate for a moment, but fully intertwined. Every society is constructed as a complicated 'negotiation' between artifice and nature, a two-way flow of materials, control and mutual adjustments. We have, therefore, to view natural hazards as aspects of, or breakdown in, the web of relations linking society and nature. We have to abandon the geographical fiction that danger and the incidence of disaster follow from the map of natural conditions or, indeed, of human populations and activities. Instead, they are seen to arise in the particular places where their interrelations are adverse.

The systematic, hazard-by-hazard account tends to preclude or not reveal this cultural and ecological view of hazards. However, the predominance of this approach and its importance for our understanding cannot be ignored. Therefore, selected hazards will be looked at to help to show further how danger arises at the interfaces of society and natural conditions, and how disaster depends upon the context in which that occurs.

Fog: a visibility and barrier hazard

This seemingly straightforward example will help clarify the relations of a natural agent to dangers. As an 'atmospheric hazard', fog comprises dense clouds of water droplets immediately above land or water surfaces, due mainly to weather conditions. Radiation fog occurs as a result of heat loss by radiative cooling of moist air, usually on clear, still nights. Advection fog occurs where warm, moist air moves over a cold surface, or cold air moves over a body of warm water. There is also 'hill fog', where moist air moving up-slope is cooled and moisture condenses out. Along humid coasts and in mountains, clouds, as normally described, can envelop land and water surfaces. Fog is generally a danger where these conditions recur more or less often. In such places, understanding of the conditions that will produce fogs is vitally important. However, that does not tell us what kind of danger fog poses, or why most fog-prone areas and the thousands of natural fog events in any given year involve little or no reported harm.

Fog hazards arise, almost exclusively, in relation to one set of activities: communications dependent on visual guidance. Fog endangers the movement of persons, vehicles and goods. More exactly, fogs become dangerous where visual sensing and signals are essential to safe activity or operations. In a practical or operational sense, fog is a *visibility hazard*.

However, if people are unable, or decide it is too risky, to move through a fog-bound area, fog acts as a *barrier hazard*. Though not an actual obstacle, it acts much like avalanches across roads, trees downed by a wind storm, or flood waters. Unlike tornadoes, floods or earthquakes, the fog itself does harm only indirectly, when movement of persons and goods is paralysed, seriously reduced or slowed down. People's lives may also be threatened where they go astray or get lost in treacherous environments: perhaps in the mountains, swamps or along a rugged coastline. Destructive damages occur when vehicles go off the highway, when trucks, trains, ships, aircraft and other mobile technologies crash into obstacles or each other.

In the modern world, dangers from fog have been transformed and magnified enormously by transport and communications systems. Required to work year round, day and night, and in almost all environments, these can turn fog into a severe or costly risk. It can hold up road, rail, marine and air transport. Where it does not, there is risk of disastrous collisions or crashes. This is to identify the main links or 'interface' problems between atmospheric and human conditions, whereby fog actually endangers the latter. If it seems rather obvious in this case, it clearly shows that the hazard arises from the relations of two 'systems' that, in themselves, operate on entirely different principles and reflect conflicting timetables.

Geographically, we see fog risks arising from spatial overlap between critical or highly valued activities depending upon visual guidance, and fog-generating atmospheric conditions. It is largely a hazard of busy coastal waters, airports, heavily used rail and highway routes and, more generally, of urban areas and commuter traffic. These dangers occur mostly in mid- to high-latitude coastal zones, in ice-infested waters, and in low-lying riverine areas. In certain tropical coastal zones, up-welling cold ocean currents also generate frequent fogs.

There are one or more fog-related air crashes almost every year (Table 3.7). The fog-prone Canary Islands provide a telling case, with four fog-related aviation disasters in the past 30 years, including the worst in peacetime. In 1977 two taxiing Jumbo jets collided on a foggy runway, killing 582 people. In most cases, fog dangers and disasters tend to be highly localised because of the specific nature and practicalities of the risk to transportation or outdoor amenities.

However, the most severe or lethal fog-related risks arise from the complex hazard of fog combined with air pollutants, or 'smog'. This too is almost wholly an urban hazard, increasingly dependent upon pollution from automobile

Table 3.7 Examples of fog-related disasters involving air traffic

Place	Date	Disaster
Tenerife, Canary Is.	1977	Two Boeing 747 jets collide on runway in fog. 582 people killed.
Lashkarak, Iran	1980	Boeing 727 crashed in dense fog. All 128 persons aboard killed.
Tenerife, Canary Is.	April 25, 1980	Boeing 727 crashed into cloud-enveloped volcano. 146 persons killed.
La Palma, Canary Is.	May 28, 1980	Spanish airforce airliner crashed into mountains in thick fog. 10 killed.
Madrid, Spain	December 7, 1983	Boeing 727 crashed into DC-9 on fog-shrouded runway. 90 killed, 30 injured.
Zaragoza, Spain	February 28, 1984	US military transport plane crashed into mountain in dense fog near airport. 18 killed.
Luzon, Philippines	June 26, 1987	Airliner crashed into fog-covered mountains near airport. All 50 aboard killed.
Mount-Crezzo, Italian Alps, Italy	October 15, 1987	Airliner crashed into mountains in fog and rain. All 37 persons killed.
Merignac, Bordeaux, France	December 21, 1987	Commuter plane crashed while approaching airport in fog. All 16 aboard killed.
Lake Constance, Austria	February 23, 1989	Aircraft plunged into lake in dense fog, killing 11 persons aboard.

exhausts but historically associated with fossil fuel burning in households and industries. The worst fog-related disaster on record was the 'killer smog' of April 1952 in London, England, to which some 12 500 deaths were attributed. It was the worst of a series of such smogs and was to lead to Britain's smokeless fuels legislation. However, this broaches the more common forms of hazard, which are compounded of several processes, as well as involving natural and technological agents.

Snow: towards a sense of the complex hazard

About half the world's land area is subject to snowfall. Snow lies on the ground for part of the year in most mid- to high-latitude areas. Snow-prone areas extend into lower latitudes in the continental interiors and, at high elevations, even into the equatorial mountain lands. However, snow-related disasters and costs are remarkably concentrated in space and context. Like fog, snow is overwhelmingly a danger to transportation and human mobility. The problems affect urban transportation especially, and arterial road and rail transport between cities. This is reflected in the incidence of the disasters (Table 3.8).

Geographically these disasters are distinguished less by being associated with snow-prone areas than with major urban regions or centres – to a lesser but growing extent, certain winter recreational areas – *and by the exceptional number and share reported in North America*. This is the one natural disaster in which loss of life is highest for North America. Thompson (1982), who compiled an inventory of 'large-area disasters, 1947–1981', meaning those with damages reported over two or more $10°$ latitude and longitude squares, found eleven involving snowstorms or blizzards. Nine of them occurred in North America. In nearly all cases, the main damages involved what the media usually call 'paralysing' of cities: primarily interference with, and closing of, urban and intercity transportation. In a study of 13 snowstorms in southern Ontario, Hewitt and Burton (1973, 119) found major disruption reported for road transportation in every case, and closures of many urban functions and damage to fuel, power and overhead communication lines in most.

Falling snow can be a visibility hazard, sometimes a severe one for motorists. Even a brief snowfall, if heavy and coinciding with, say, rush hour in a city, may slow traffic and result in multiple vehicle collisions and other problems. Loss of traction or control of vehicles on snowy and icy roads compounds these dangers. However, in most disastrous snowstorms, the problem for human mobility is direct physical impediment, or a barrier-like hazard. With increasing snow depth, more types of vehicle and persons are unable to pass. In his pioneering work, Rooney (1967) found a fairly good correlation between the occurrence and severity of losses and the depth of snow recorded in the given storm and, to a large extent, in the snowfall year. Others have reinforced his findings, if showing the source of risks to be more complex. Higher snow-falls are also associated with added visibility, traction and other problems. There are large differences between the impacts of heavy, wet snow and

Table 3.8   Examples of snow-related disasters. It will be seen that nearly all involve heavily urbanised areas

Place	Date	Disaster features	Deaths
Iran, Teheran	Dec. 5, 1974	Airport terminal roof collapsed under snow load.	17
USA, Mid-west	Jan. 25–6, 1978	Blizzard stranded 8000 motorists, closed airports, factories, highways. Damage estimated in US$100s of millions.	100+
Canada, Cranbrook, B.C.	Feb. 11, 1978	Aircraft swerved to avoid snowblower on runway, crashed and broke up.	41
Turkey, nr. Ankara	Jan. 4, 1979	Crowded passenger trains collided in blizzard. Frozen switching gear. 190 to hospital.	56
USA, New York, New Jersey	Feb. 19, 1979	Snowstorm paralysed urban areas, hundreds stranded on snowy highways, etc.	13
USA, Eastern	March 2, 1980	Snowstorm from Pennsylvania to Florida stranded hundreds of motorists.	36
USA, Northeast	Feb. 11–12, 1983	Blizzard paralysed nearly every city, incl. New York, Washington DC, and Baltimore.	11
USA, Eastern and Mid-west	Feb. 28, 1984	Major snowstorm paralysed cities from St Louis, Missouri, to Buffalo, NY. Closed highways, airports, schools, offices, factories for two days.	29
USA, Eastern and Rockies	March, 1984	Three more snowstorms paralyse cities, highways.	?
India, Kashmir	Nov. 17, 1985	Snowstorm trapped 300 road workers, soldiers, drivers on mountain highway.	60+

Table 3.8 contd.

Place	Date	Disaster features	Deaths
USA, Eastern	Jan 22, 1987	Storms and blizzards, Maine to Florida, paralysed traffic. Shut down government offices in capital.	37
USA	Dec. 12–16, 1987	Blizzard closed airports, highways, schools, downed power lines.	73
Turkey, Southeast	Feb. 1–3, 1992	Exceptional snowfalls (3–5 m), isolated 7000 villages, cut off power for days, followed by avalanches, which caused the deaths.	210
Canada, Maritimes	Dec. 31, 1993	Blizzard brought most of region to standstill, disrupting road, rail, ferry and air transport.	?
Canada, Montreal, Quebec	Jan. 5, 1994	Heavy snowfall, high winds, low temperatures closed airports, bus terminals, bridges, tunnels. Brought 100-car+ pile-ups on highways. Prolonged power failures. 6000 homes without heat or light.	6
India, Kashmir	Jan. 23, 1995	Snowstorms and avalanches buried 100s of vehicles on Jammu–Srinagar highway.	200+

light, dry snow; between new snow and old, compacted snow. Hence, attendant air temperatures or time of year complicate the hazard.

A particular complication of the barrier hazard is the behaviour of snow in windy conditions. Not only during snowstorms but also at times when none is falling, the wind moves the snow around, building drifts that may cross highways and other routes. Drifting also compacts the snow, making a denser barrier. This important secondary hazard is influenced by wind, topography, vegetation and the arrangement of the built environment, as well as by snow itself. Meanwhile, in major urban snowstorm disasters, multiple vehicle pile-ups, abandoned vehicles or parked vehicles buried in snow can decide the scale and duration of the crisis.

The most disastrous events or snow catastrophes tend to occur with blizzards. While associated with snow, these are accompanied by high winds and low temperatures. The snow affecting people, buildings and traffic is wind-driven, kept in the air, moved over the ground, and built into drifts in great concentrations. As well as exceptionally poor visibility, loss of traction and snow barriers, the blizzard threatens severe exposure dangers for humans, and domestic and wild animals. Deaths and injuries in North American snow disasters involve hypothermia and frostbite for stranded motorists. A common tertiary type of harm is death from carbon monoxide poisoning when using the engine to keep warm without proper ventilation or with snow blocking the exhaust pipe.

More 'traditional' forms of snow danger still occur, though they are less prominent than those described so far. They include persons and farm animals caught outside or stranded by snow, and collapse of structures due to weight of snow. In the great Alberta, Canada, blizzard of 15–16 December, 1965, more than 1000 cattle died of exposure and starvation, and there was also massive wildlife death. Scattered damage and collapse of huts and barns is common in rural areas and heavy snowfalls. More rarely, major disasters result from the collapse of roofs in large indoor facilities, such as sports complexes (Morrison *et al.*, 1960).

Snowfall-related hazards arise from snow avalanches in mountains, mainly affecting arterial transport routes and winter recreation, or from floods when heavy snow cover melts quickly. However, sufficient has been said to show how this is a complex hazard and that related disasters involve compound rather than single forms of damage.

Forms of impact and dangerous conditions

These examples begin to suggest the range of ways in which natural hazards act as dangers at the interface of physical processes and human activity or well-being. For introductory purposes it will be sufficient to recognise six broad forms of endangerment or damage process:

- *Destructive physical forces*: mechanical stresses or physical impacts, dangers commonly conjured up by hailstorm, landslide, earthquake or flood. In addition to snow, destructive loading may also occur in ice or 'glaze' storms, from the deposition of volcanic ash, and from icing of ships' hulls and shoreline structures in cold weather.
- *Combustion, electrical discharges and explosive forces*: fires may originate from vulcanism, lightning or inflammable gases. Explosive forces in nature are associated mainly with volcanic events, severe electrical discharges with lightning in storms.

- *Physical constraints or indirect hazards*: harm to humans and their property resulting indirectly from the limiting or preventing of certain activities and by reducing margins of safety.

  Examples already identified are:

  - visibility hazards,
  - barriers to movement,
  - loss of control or use of a technology, for example where surfaces are slippery.

  To these may be added:

  - severe ambient conditions: extreme cold, heat or dryness may prevent activities, reduce the performance of a technology, threaten the lives of people, plants and animals, or exaggerate the impact of other hazards.

- *Deprivation or denial hazards*: cutting off essential life support, for instance, respiration (drowning), water (thirst and plant wilting in droughts), food (starvation), or essential supplies for economic activity. These may be secondary effects of other hazardous processes.
- *(Natural) pollution, spoilage and contamination hazards*: include smoke from forest fires, toxic clouds from volcanic eruptions, and build-up of sterile minerals in soils (e.g. salinisation). Food grains and other stored goods, clothes and furnishings may be ruined by flood waters. Heatwaves or severe cold may result in the spoiling of perishable goods.
- *Biological assaults*: epidemics, infestations, and biomass fires involve unique forms and pathways of danger introduced as biological hazards.

Particular natural phenomena may give rise, predominantly, to just one of these forms of endangerment or damaging process. Most disasters involve multiple dangers.

Hazard dimensions

The approach suggested above may be helped by, but contrasted with, a more common one defining observed dimensions of natural agents associated with risk and its variability. Some relate to the damaging processes discussed above, or to rather specific features such as wind shear or toxicity. Dimensions suited to generalising and comparing hazards are:

Spatial dimensions

The *areal extent* or reach of dangerous forces or damages is important to the scale and scope of disasters. Some natural agents, like hurricanes, affect relatively large areas and some, like tornadoes and many mass movements, very local areas. In other cases, such as flood hazards, events may range from extremely local to those affecting extensive river basins or coastal zones. The significance of area affected is not very meaningful, however, without measures of intensity. A catastrophic rockslide or lava flow may affect a fairly limited area but cause total destruction of everything there. A snowstorm can cover a large area, but snowfalls may be relatively light over most of it. Thus the relations between the areal extent and intensity of the natural event, and vulnerable populations and wealth exposed to it, are crucial and complex. Great disasters generally occur when areal extent and exposure are large as in great droughts, or hazard intensity is locally very great in a populous place, say strong earthquakes in urban areas.

Temporal dimensions

- *Rate of onset*, the suddenness with which dangerous forces affect a community, can be critical to damage, at least with respect to saving lives and movable property. Some hazards, like droughts and some plant and animal infestations, develop slowly. In others, like tornadoes or rapid landslides, the event comes and goes with extreme suddenness. Other general classes, such as floods and volcanic eruptions, include a range of cases from sudden flash floods or explosive eruptions to slowly rising water levels or lava.
- *Duration* of the dangerous forces or event. Duration itself may multiply the problems of human response. Sudden events of great intensity over a short time often do the most severe and unmanageable damage. But, like areal extent, duration has little meaning without measures of intensity and the exposure of vulnerable people. Drought is the main geophysical hazard that is relatively long-lasting and becomes progressively more severe with time. Then again, landslides or forest fires can leave a legacy of destruction that, in a sense, extends the hazard long after its immediate impact phase.
- *Frequency*. Estimates of risk are commonly given as the likelihood of occurrence of dangerous forces, estimated from evidence of their recurrence frequency in records of, say, flood flows, rainfall or earthquakes. Most events we consider are regionally rare and appear almost random in their occurrence over time.

Compound parameters

Since any one of the above measures may be offset or intensified by the others, it is probably more useful to look for combined spatial, temporal and intensity or magnitude and frequency dimensions. The 'worst cases' will count high on two or more of these dimensions. Physical constraints mean that many hazards cannot be high for all of them, and those that may be – such as the major meteorite impacts now thought to have caused mass extinctions and geological change – must be extremely rare and improbable events, otherwise our survival would be precarious indeed.

79

Thresholds

In some ways, the most characteristic and influential hazard dimensions define and measure *thresholds* of impact or damage. For any given person, structure, activity or community, there will be a level of impact from a natural hazard below which damage does not occur or is routinely managed. It might be very weak earthquakes, or the lower ranges of wind speed, river flow or height, soil movement or numbers of pests. However, when these values or numbers pass beyond certain levels, the consequences may become increasingly severe or calamitous. This suggests the existence of thresholds, or a series of thresholds, at which damage can be initiated and above which it becomes more serious. Important thresholds apply to impact variables like mechanical force or the concentration of a dangerous substance. Along an inhabited river valley or coastline, an important factor in the initiation and amount of flood damages will be the reach of the flood waters.

Generally, danger from any given physical agent involves 'bundles' of thresholds that apply differently in different places and times, according to land uses, and technological and societal characteristics. Even more relevant to this argument, they apply in any *one* place and time. In earthquake damage zones, totally demolished homes are found beside others with little or no damage. In this and other damaging events, there will be a spectrum of damage between the least and worst. This is not only, and not mainly, due to spatial differences in the strength of the damage-initiating agent, seismic shaking, but to differing exposure, strengths and tolerances in the human realm. In other words, thresholds of damage are as much expressions of the vulnerability of human items as of dangerous physical processes.

Floods and droughts

We will conclude with a review of these hazards, which have played so large a role in the development of the whole field. By looking at them together it will be shown how what appear to be related aspects of hydrology turn out to be very different in the ways they arise as dangers for, and interact with, human societies. However, this does not mean they only affect different groups of people. There are societies where risks from flood and drought do involve the same persons, notably in some agricultural societies.

Flood hazards

Flood hazards involve harmful inundation of occupied land. They appear as water 'in the wrong place', or 'at the wrong time'. They involve the most frequently reported and costly of natural disasters worldwide (see Fig. 3.1).

Virtually any place on land where water can be introduced faster than it drains away may be flooded. However, flood disasters generally occur around permanent water bodies or water carriers, mainly in high water levels along coastlines and river flood plains. Here, we will consider only riverine floods (Table 3.9).

Rivers usually establish well-defined channels, which contain most flows passing along them. Rarer, higher flows spread out over more or less well-defined flood plains. Human settlement tends to stay out of channels and the more frequently inundated parts of flood plains. Communities often encroach where inundation is rarer, and this leads to most losses from disasters. In regions with ephemeral streams and long periods of seasonal or secular low flow, temporary but well-established uses of dry channel areas are found. People are then endangered by unseasonal or premature river rise.

Various damage forms identify different aspects of danger and human vulnerability:

- *Drowning and injury* to persons, their domestic animals and crops, possibly to wildlife resources upon which they depend.
- *Physical impact*: loading and abrasional damage of entrained debris.
- *Contamination and deterioration* of materials, household objects and equipment.
- *Barrier hazards*: the disruption of pedestrian, road, rail and air traffic, although access by water may be increased temporarily and used for relief purposes.
- *Denial* of life-supporting supplies, services, access to means of livelihood or work places. Inundated or destroyed homes result in a housing crisis.
- *Secondary and tertiary damage* includes outbreaks of waterborne disease and spread of unwanted pests. Snake bites or insect plagues are secondary hazards in some tropical areas.

There has been considerable research into the relation of hydrological parameters to flood risk, with flood height usually a critical threshold parameter. Often defined as both a flood level on the ground and the height that statistical estimates suggest will be exceeded within a given time frame, it leads to such notions as '25-year' or '100-year' floods. The fact that floods tend to be short-lived, threshold hazards suggests a close temporal and spatial relation between their incidence as hydrological events and the scope of human responses to them (Fig. 3.3). Flood severity also depends upon the rate of rise to flood peak, especially where forecasts are absent.

The nature of the damage and the role of various flood parameters depend equally upon human exposure and vulnerability. In areas with rural populations and an important agrarian sector, losses of human life and domestic animals or ruined croplands may be the greatest dangers. In urban-industrial areas, damaged materials, severed communications and lifelines, or lost amenities, may loom larger. Land uses and land use changes in flood plain areas are often, and increasingly, seen to be more directly responsible for the scale of flood risks or changing risks than hydrological conditions.

Table 3.9 Examples of late twentieth century flood catastrophes with loss of life, displacement and damage estimates where available

Place	Date	Disaster features	Deaths
SW. France	July, 1977	Torrential rains and worst floods this century. Millions of dollars damage, 90% tobacco and cereal crops destroyed, large livestock losses.	26
Mozambique, Zambezi R.	March 27, 1978	Century's worst floods. Millions of dollars damage. 200 000 made homeless.	45+
N. Mexico, SW. USA	Feb., 1980	Nine days of rainstorms. Floods and huge mud-slides. Damage estimated at US$0.5 billion.	36
India	July–Aug., 1980	Monsoon flood damaged over 20 000 sq. km. Damage estimated at US$131 million+.	600+
India, W. Bengal	Aug.–Sept., 1980	Monsoon floods, landslides.	1500
India, Orissa	mid-Sept., 1980	Torrential rains and flood waters burst dam. Homes of 300 000 people inundated.	200
China, Szechuan Prov.	July 12–14, 1981	Monsoon rains caused Yangtze R. to burst banks. Injured over 28 000; 1.5 million homeless. Damage est. US$1.1 billion.	1300+
China	October, 1981	More floods, also triggering landslides.	240
China, Kwangtung Prov.	mid-May, 1982	Torrential rains, heavy floods; 450 000 flooded out. 46 000 homes collapsed.	430
India, Orissa	early Sept., 1982	Monsoon floods; 8 million made homeless; 5 million receiving air-dropped supplies in inundated areas.	1000+

Table 3.9 contd.

Place	Date	Disaster features	Deaths
N/C. Germany	February, 1983	Severe winter storms caused major floods.	?
S/C. Germany, N. France	May, 1983	Floods on all major rivers. Drownings in R. Rhine around Köln and in France.	18
Bangladesh	May–July, 1984	Monsoon floods affect 30 million +, making almost 1 million homeless.	300+
Philippines	late June, 1985	Torrential monsoon rains and floods, make 100 000 homeless.	60+
Bangladesh	August, 1987	Weeks of floods affected 20 million persons.	1000+
South Africa, Natal Prov.	Sept., 1987	Torrential rains generated large floods making 50 000 homeless; US$500 million damage.	200+
Sri Lanka	early June, 1989	Monsoon floods and landslides, injured 700+, made 125 000 homeless.	300+
NE. Brazil	late Dec., 1989	Torrential rains and floods made 200 000 homeless.	35+
Malawi	March 10, 1991	Torrential rains, floods, collapse of mountainside, left 150 000 persons homeless.	500+
Afghanistan	June, 1991	Flash floods from rainstorms.	5000+
Brazil, Argentina, Paraguay	May–June, 1992	Continuous heavy rain for two weeks causcd floods on major rivers, inundated hundreds of towns and villages; 220 000 evacuated.	28+

Table 3.9 contd.

Place	Date	Disaster features	Deaths
Pakistan	Sept., 1992	Late monsoon storms, floods, landslides in Indus basin, Himalayan headwaters, but worst damage/loss of life from flood wave in Indus plains after sudden opening of Mangla Dam floodgates by panic-stricken manager.	2000 +
Nepal, India, Bangladesh	July, 1993	Monsoon storms and floods made 4.5 million + homeless. Damage US$12.6 billion. Deaths 3000 in Nepal, 1000 India.	4000
USA, Missouri–Mississippi Rivers	July, 1993	Eight states declared 'disaster areas'. Some 40 000 homes and businesses damaged or destroyed, 150 000 persons homeless and 100 000 evacuated. Losses US$10 billion +.	33
China, southern provinces	June, 1994	Extensive, destructive floods damaged homes, industries, agriculture. 400 000 homeless. Losses $6 billion.	1400
NW. Europe	Feb., 1995	Extensive flooding N. France, Belgium, Germany. Worst in Netherlands since 1954, with 250 000 evacuated and US$1.5 billion to save dikes.	40 +

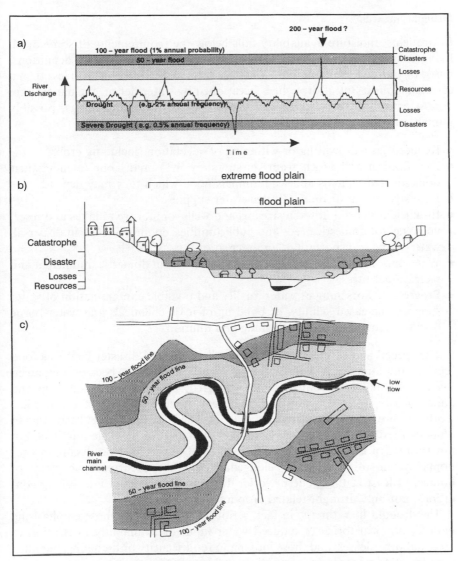

Fig. 3.3 Riverine floods and low flows as threshold hazards: schematic diagrams to show the relations between flood height, extent and frequency, and how these relate to disaster through patterns of land use:

  a) temporal thresholds: graph of a hypothetical stream height record, indicating levels at which flows are useful or manageable. The bands indicate likely return periods for floods of given heights, and relations to damaging events.

  b) river height thresholds: cross-section of river valley indicating the height of hypothetical river flow events.

  c) spatial risk thresholds: plan of river valley area relating stream flow heights to topography and more or less vulnerable land uses.

Drought hazards

In droughts, moisture availability falls below the requirements of some or all living communities in an area. Droughts are often given official definitions, based on exceeding a minimum number of days without rain or with less than a certain amount. But, if drought is occasioned by water deficit, the dangers are almost entirely in *deprivation and denial hazards* for living things, possibly human settlements and industries, involving:

- Reduced growth, wilting or withering of vegetation, including crops.
- Dehydration and death from thirst, rarely in humans but often in their domesticated animals and wild animals upon which they may depend.
- Excessive overdraft on controlled water supplies.
- Insufficient natural flows in streams and wells, or storage in lakes and reservoirs (i.e. for domestic needs and public utilities, cooling or effluent dispersal, water-borne transportation, and recreation).
- Water rationing, which may include restrictions on domestic, municipal and recreational uses.
- Progressive worsening of water quality and possible contamination of water supplies, increased salinity, or build up of toxic chemicals and water-borne diseases due to inadequate flushing and dilution.

The special and severest association of drought and disaster concerns food security and famines. A majority of the worst famines have been accompanied by drought and exacerbated, if not triggered, by drought-related crop and animal losses. However, not even the great modern famines of the tropical world or any since the early years of the last century have been due to *absolute* food shortage. People have died of starvation in large numbers, but not because of an overall insufficiency of food in the food-producing or economic system to which they belonged. Hence, the occurrence or extent of famine, at least in the modern world, depends upon human intervening conditions, not only drought-related crop loss.

The drought literature is, in fact, seriously split between those emphasising natural, physical forms of reduced water input or availability in relation to crop and other losses, and those looking to food security or humanly magnified desertification hazards for answers (see Chapter 6).

Spatially and environmentally, patterns of drought, especially severe and potentially disastrous ones, depend upon atmospheric conditions affecting wide territories. Although possible in most regions, these tend to be more prevalent or severe in semi-arid subtropical areas, and certain continental interiors with recurring runs of more and less dry years. Examples include northeast Brazil, northwest India and Pakistan, the western Great Plains of North America and, most notably, the Sahel of West Africa.

Landforms, soils and vegetation cover, by influencing the moisture-holding or -conserving properties of sites, act as intervening variables to moderate or hasten the impact of dry spells. In this way, the severity or the development of droughts involves fairly complex spatial patterns of sites in the same region. Some have more, or earlier, severe impacts, and others become serious only after prolonged dry periods.

Comparing flood and drought

Hydrologically, flood and drought may be discussed in terms of the same generative processes or conditions in different parts of their range – rainfall and snowfall or their absence; high and low stream flows; and lake and ground water levels. Since they are largely regulated by atmospheric heat and moisture conditions, they are hardly separate from atmospheric hazards. However, the direct *human* connections with flood or drought tend to be via rivers and shorelines, wells and soil moisture, and plant growth or failure. Meanwhile, this may be the most misleading instance of hazards being reduced to physical processes, not to say an abstract, statistical view of hydrological data sets. With respect to the human ecology of dangers, it would be difficult to find two types of hazard that create risks of such different forms, patterns and implications.

The flood *hazard* may be about water. It is not about water *supply*, whereas that, or its insufficiency, is the essence of the drought hazard. Drought risks are integral to the needs and patterns of water consumption, and arise directly from them. Flood involves water supply only indirectly, to the extent that it puts people and property in the path of excessive moisture.

Thus flood and drought tend to affect human societies in different parts of their activities and in substantially different ways. Their apparent link as hydrological extremes is largely incidental. Drought is, above all, a physiological crisis or killer for agricultural crops and animals, increasingly an amenity and economic crisis for urban-industrial communities. Floods tend to have their primary impacts in killing people by drowning, in property damage and in paralysing land transportation. As an agricultural risk, floods often have the compensating effect of abundant residual water to help crops and herds recover.

Disastrous floods generally develop quickly and are of short duration. Droughts develop slowly, even imperceptibly, but may last weeks, months or even years. Smith (1992, 246) refers to them as 'creeping' hazards.

The size and duration of a flood is often known ahead of time, or soon after its onset. Floods are mostly events that arrive, peak and end rather quickly. A drought is not recognised or declared until a dry spell has gone on for a long time, though each preceding day without rain has actually contributed to its severity. Only a more or less long-delayed wet spell brings it, perhaps quite suddenly, to an end. Droughts are 'escalation' hazards, of relatively long duration, extending in area and intensifying in effects the longer they last, but the drawn-out escalation of drought impacts also provides opportunities for varied and geographically extensive human responses to modify risk while damage is in progress. That is one major reason why there is, or need be, no inevitable relation between drought and famine.

Flood and drought often have directly opposite spatial patterns of development and incidence. Floods are linear or patchy in spatial extent, mostly following watercourses or coastal zones, and reflecting their topography. Severe drought events invariably embrace extensive regions, partly because of the climatic conditions giving rise to them, partly because local water shortages can usually be readily offset. Areas most prone to drought, or that feel its effects earlier, such as well-drained parts of farmland, are often the opposite of those most prone to floods. Poorly drained and flood plain areas, for instance, tend to feel the greatest flood effects.

These differences also mean that responses to flood and drought are, or perhaps should be, very different. A conspicuous exception seems to be the dam construction option, which can be used both to store water for dry periods and to hold back flood waters. Yet, as Gilbert White pointed out long ago, these are not necessarily the most environmentally sound or secure ways to respond to water resource risks. They are not always compatible goals, coming into conflict just when the problem is most serious – as when the pressure to keep reservoirs full to avert water shortages conflicts with the risk of inadequate storage if a flood situation develops.

A tragic example occurred in Pakistan in September, 1992. Normally at this time of year, it is essential to top up reservoirs on the Indus streams with the last of the glacier melt water and monsoonal runoff from the Himalaya to supply irrigation, industrial and municipal needs through the winter dry season. However, unseasonally late monsoon rains created huge floods. One of the major reservoirs, Mangla Dam, was approaching its capacity at that time and the local manager, fearing it might be overtopped by the unexpected floods, opened the floodgates and allowed even more water to flow out than was coming in from above. The resulting 'man-made' flood wave caused the worst damage and loss of life in the whole episode. This is an extreme but not atypical example of how flood and drought needs can come into conflict in the operation of storage reservoirs.

The undue preoccupation of modern states with such hydraulic control works has also encouraged a hydrological hazards paradigm in the treatment of both these hazards. Thus, the profound differences in the ways flood and drought enter human circumstances, their differing geographic patterns or temporal features, may be ignored. In that too, perhaps, their importance in our studies may have helped to reinforce the hazards paradigm.

Suggested reading

In part the choice of approach here is justified by the many recent texts that systematically review hazardous agents from a hazards perspective. There is also a huge research literature in that mode, which the reader can readily refer to for particular hazards and most disasters that are of special interest. General studies of hazards and the development of a hazards perspective include:

Abbott, P. L. (1996) *Natural disasters.* Wm. C. Brown, Dubuque, Iowa.

Alexander, D. (1993) *Natural disasters.* Chapman and Hall, New York.

Ebert, C. H. V. (1993) *Disasters: Violence of Nature, Threats by Man*, 2nd edn. Chapman and Hall, New York.

Jones, D. K. C. (ed.) (1993) Environmental hazards: the challenge of change, *Geography*, 161–98.

McCall, G. J. H., D. J. C. Laming and S. C. Scott (eds) (1992) *Geohazards: natural and man-made.* Kendall/Hall, Dubuque, Iowa.

Smith, K. (1992) *Environmental hazards: assessing risk and reducing disaster.* Routledge, London.

White, G. F. (ed.) (1974) *Natural hazards: local, national, global.* Oxford University Press, Toronto.

CHAPTER 4

Technological hazards

Those same features of the most complex human communities which indicate their ecological advantages also suggest an unusual degree of inherent ecological risk. Such communities are complex and delicately balanced, and depend utterly upon artificiality . . . . Their very technical perfection may destroy them in time . . .

Philip Wagner (1960, 23)

Often, society reexamines the application of new technology only after it is too late, after the device is thoroughly integrated into social institutions, after the device had produced a series of undesirable second-, and third-order consequences, or worse, after the device has caused a disaster and a body count . . .

Michael R. Reich (1991, 6)

The scope of technological risks

'Technology' has been dangerous since the first spear went astray, the first building collapsed under its own weight, and camp fires set a forest ablaze. Such age-old tools and techniques may still take a larger toll in lost production and human injury than recent innovations. A majority of the world's people still use 'traditional' methods in most activities – cooking fires and homemade implements; hand- and animal-powered devices; or processes driven by sun, wind and water power. However, these are not widely monitored or reported dangers. They rarely give rise to major public emergencies. Meanwhile, the idea of 'technological hazards' has arisen in relation to recent, mainly industrial technologies. Those with scientific and engineering applications of 'high' and novel forms are most prominent. If natural hazards are commonly seen as external threats and limits to modernity, technological hazards appear as its internal failures.

Although geographers came to them later, the hazards of modern technology have had major significance for this field. Britain, for example, often described

91

as the first industrial nation, has had dozens of official enquiries and reports into mining, marine and other technological disasters (see Turner, 1978). Prince's (1920) classic study of the 1917 Halifax Harbour munitions ship explosion in Canada, or Killian's (1956) of the Houston, Texas, fireworks explosion, are also among those that helped to lay the foundations of modern approaches to risk and disaster.

If technological risk analysis and policy may seem new, harm from innovative industrial technologies is not. Consider disasters such as the 1879 collapse of the Firth of Tay steel railway bridge in Scotland, plunging an express train into the river; the collapse of the South Fork Dam above Johnstown, Pennsylvania, in 1889 and, especially, the *Titanic* sinking in 1912. In their day they symbolised technological failure and provided subjects for social debate and soul searching about modernity. For some they exemplify the 'price of progress', for others, the dark side of technological prowess or dependency, the lack of wisdom in competitive innovation at any price. Such attitudes are not irrelevant to how, and to how seriously, the risks are taken.

The sinking of the *Titanic* might seem exceptional at any time, but many of yesterday's unprecedented disasters are today's routine accidents. Annually, as many as 20 aviation disasters occur with casualties worse than, say, the notorious *Hindenburg* airship disaster at Lakehurst Field, New Jersey, in 1937, when 36 people died. However, part of the problem is how we seem constantly to rush ahead into new dilemmas. Any significant technology is subject nowadays to constant innovation, increasing scale of production and expanding geographical extent of its use. Hence, its risks are never static, sometimes improving dramatically, sometimes worsening or presenting novel dangers. The same applies to the rapid obsolescence and abandonment of technologies. Some may cease to pose serious risks. Others leave remnants or wastes that are severe and poorly handled dangers. Every field, from rail transport and municipal water supply to pesticides and fuel pipelines, has its story of innovation and safety crises. During the latter, when the problem is perceived to go beyond, and call into question, routine controls, our field becomes involved. Public enquiries and questions of public policy are raised, as well as the performance of the technical professions directly concerned.

In general, the hazards of urban-industrial societies have dominated this work, especially problems emerging in North America and other fully industrialised countries. They include uniquely dangerous and unforgiving technologies, if the extent of the risks remains controversial. Topics such as nuclear power, large dams, toxic chemicals or genetic engineering receive much attention. Some have involved especially grim disasters. Witness the dark meaning associated with geographical names like Chernobyl in the Ukraine, Bhopal in India and Minimata in Japan. There is growing concern over widely disseminated but dangerous innovations in such products as medical drugs, food additives and birth control devices. Trade names like *Thalidomide* or the *Dalkon Shield* are identified with especially horrible dangers from new products. Great oil spills and oil fires exemplify major environmental disasters associated with the dominant energy and transportation technologies.

However, these dangers are by no means confined to urban-industrial settings. Virtually every modern product and process is disseminated to most countries and social settings. Of 25 nations with operating nuclear power stations, at least eight are commonly placed in the so-called Third World and six in the 'Second'. In almost every country and environment, we find large dams, thermal generating stations, commercial oil fields, oil terminals and large-scale storage of highly inflammable fuels. The smallest of countries may have chemical and explosives plants, metallurgical and electronics industries, commercial airports, and high-rise structures.

If concentrated in and around major cities, these technologies or their damaging impacts are not confined there. Ancient fishing communities sit next to oil refineries and LNG terminals. The fallout from nuclear tests has contaminated the habitats of indigenous desert, alpine and sub-arctic communities. The Grassy Narrows calamity in Canada illustrates how toxic chemicals from industrial plants – specifically methyl mercury from pulp and paper mills – invisibly concentrated through the food chain, can endanger and destroy the lives of hunting and fishing communities.

Occasions of technological harm

Fire, crashes, explosions: recurring technological emergencies

A sense of the overall landscape of technological risk is gained from disasters described in the mass media. In the five-year period 1989 to 1993, taken to represent the situation in the late twentieth century, about 110 serious technological disasters were reported in an average year (Table 4.1). Of these, 79 per cent were in transportation, mainly mass passenger transport, to a lesser degree cargo transport. Almost a quarter were aviation disasters.

Apparently, few of these technological disasters called for major national or international emergency assistance. This is in marked contrast to natural disasters, although only half as many of them were recognised. Events generally involved tens to several hundred deaths, and occasionally some thousands of people were, at least temporarily, evacuated. The general public was affected and moderate to large numbers of them were briefly placed in danger, but these appear, otherwise, as recurrent emergencies whose treatment is almost routine. Although these events are more newsworthy than chronic hazards, responses remained largely within established agencies, professional safety teams, even firms. This may be why they are as likely to be called 'accidents' as disasters.

The presence of these events in the same lists as highly disruptive disasters and catastrophes reflects *social constructions* of danger and technological concerns of modern societies. A majority of the events have two features in common. First, as public disasters they involve relatively well-to-do, visible classes and consumer groups. Often the victims come from people with collective, if

93

Table 4.1 Numbers of technology-related disasters reported 1989–1993, by broad classes of technology (after Encyclopedia Britannica Yearbooks)

Rank		1989	1990	1991	1992	1993	Totals
	Transportation:						392
1	Aviation	44	25	18	38	20	145
2	Road (mass transit)[1]	17	24	15	22	20	98
3	Marine	22	24	15	13	17	91
5	Rail	17	13	27	12	15	58
4	Fires and explosions	15	13	27	12	15	82
6	Mining	9	11	4	8	1	33
7	Building/structure collapse	2	7	3	5	3	20
8	Mass poisoning (food or drink)	1	3	3	5	4	16
9	Toxic chemical release	1	3	1	1	2	8
All technological		128	123	98	109	93	548
All natural (in same source)		46	52	54	32	51	235
Epidemics[2]		2	–	1	1	1	6

[1] Mostly involving bus traffic.
[2] Newly identified – excludes AIDS, TB, malaria, etc, which continued growing through this whole period.

not individual, large disposable incomes or credit ratings. Airline passengers and tourists; tenants of urban housing, offices and factory buildings; the people at major sports, religious or entertainment events, can be vocal and politically active social groups.

Second, the events directly affect the economic life or credibility of modern enterprises. Disasters for air and shipping lines, mines and factories, or in major public meeting places, involve prominent corporate and governmental institutions. They cause lost productivity and usually reflect failures of administrative responsibility and control. They threaten public confidence. Concern over their cumulative impacts upon the environment and public health also draws them into broader risk debates.

We might expect that not all types of disaster are equally likely to be reported. Air or train crashes will tend to be reported everywhere. Unless very large or destructive, oil spills and toxic chemical leaks may escape media notice or be successfully covered up. A careful search by Cutter (1993, 107), for example, showed about four 'acutely toxic chemical accidents' a year in the 1980s.

The distribution of disasters by technology and by country, if partly an artefact of reporting, leaves no doubt of the global incidence of these problems. In a survey confined to airborne chemical releases that might endanger the general public and the environment, Cutter (*ibid.*) found that over one-third of 339 events had occurred in Asian and Latin American countries (Fig. 4.1). This is considered just the tip of the iceberg on dangerous chemical releases.

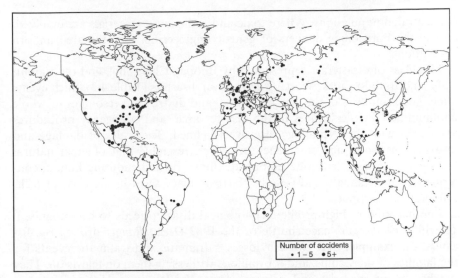

Fig. 4.1 Airborne release of dangerous chemicals worldwide, reported 1900–1990. They include chemical plant, vehicular transportation and pipeline accidents. However, they do not include chemical spills and other events not yielding immediate airborne vapours and toxic clouds. There were 339 events in total, 60% of which occurred in the 1970s and 1980s, partly a reflection of reporting, mainly of the massive growth of the chemical industry (from Cutter, 1993, 104–6)

She quotes other evidence to suggest, in the 1980s alone, almost 11 000 chemical accidents. While more occurred in the major industrial nations, the worst ones tended to occur in less wealthy countries due to lax controls and safety measures. However, most technological disasters are locally concentrated around urban centres, and along commercial and transportation networks. They follow the artificial geography of modern, material life rather than the map of broad cultural and environmental differences or human populations. To be sure, some people are much more vulnerable to them than others, often those marginal to modern life, but exposure arises mostly from involvement in or proximity to cities, industrial activity and its resource hinterlands. The tendency – sometimes justified – to explain hazards elsewhere as due to lack of modern technology and technical organisation should not distract attention from this.

Technological catastrophes

Major technological disasters or catastrophes seem to occur much less frequently than natural ones or, as we will see, than social calamities. Yet they occur. And when they do, events of singular destructiveness and far-reaching implications may be involved. They may threaten distant as well as locally

95

concentrated populations. Where toxic and radioactive materials are involved, and in the burgeoning dangers of genetic engineering, they threaten unborn generations.

This class of disasters is represented by Bhopal, Chernobyl, and the Vaiont, Italy, and Buffalo Creek, West Virginia, dam disasters (Table 4.2). Such events affect large numbers of the general public, and destroy the resources of whole communities. Public services in the disaster area, and emergency procedures associated with the technology, are overwhelmed. The scales of damage and disarray, responses and emergency measures, resemble that of great natural disasters. If less sudden in development, the scope and enduring human consequences at Minimata and Grassy Narrows place them in the class of technological catastrophes.

Another kind of high-profile technological disaster needs to be recognised. The 1986 *Challenger* space shuttle or the 1989 *Ocean Ranger* drilling rig disasters, for example, were singular losses – dramatic and traumatic events for the families of the victims, and for millions witnessing them on television. They sparked major enquiries and policy reviews. Yet they were disasters of, and largely contained within, the technological system that failed. If they enter our concerns, it is in a different sense from disasters that primarily affect the general public or the environment.

'Living with risk': consumer and personalised products in mass society

Recent study of technological risks has been less prone to disregard or separate chronic, if not 'routine', risks than occurs with natural hazards research. Recurring problems of consumer and waste products, 'life-style' and built environment hazards have received a great deal of public attention. Consumer products in a mass market economy may go to tens of millions of customers, and into every walk of life, yet they are produced by and are disseminated through centrally controlled industrial technologies. They involve constant innovation and ever-expanding markets, and hence, multiple opportunities for exposure to new hazards. They raise product safety issues in terms of the whole life cycle, from ingredients of production to final disposal, or those parts which are not adequately controlled. Products are the subject of manipulation through mass advertising and the construction of imagery, taste and value, and of government intervention to regulate risks or calm fears. There are environmental dangers in the use and disposal of many items. External to both the manufacture and uses of products, the wastes are issues of public policy and an ecology of risks.

The sense in which this causes *everyday life* to bring large-scale and, for many, unacceptable public dangers has been gaining ground. It accords with the notion of technological hazards as 'living with risk' (Burton, *et al.* 1982; Cutter, 1993). They are not only or mainly a problem of external agents and accidents that may intrude upon everyday life. Widespread danger may lie in mundane necessities. Staple foodstuffs and the water we drink have become

Table 4.2 Technological catastrophes: examples of large-scale devastation or
casualties, and the more deadly threats to the general public from
technology-related accidents and failures

Place	Date	Disaster features	Deaths
Atlantic Ocean	14 April, 1912	*Titanic* luxury liner struck iceberg on maiden voyage.	1517
Oppau, Germany	21 Sept., 1921	Huge explosion from chemical mixing at BASF chemical plant, dynamite + ammonium nitrate.	561
Honkeiko, Manchuria	26 April, 1942	Coal dust explosion. Most lethal mine disaster.	1572
Texas City, USA	16 April, 1947	French ship carrying nitrate fertiliser exploded. Blast and fires caused devastation.	576
Cali, Colombia	7 Aug., 1956	Army trucks carrying dynamite for Public Works Dept. exploded near town centre.	1150
Windscale, UK	10 Oct., 1957	Plutonium reactor fires, partial core meltdown released radioactive iodine. Danger covered up.	?
Vaiont, near Belluno, Italy	9 Oct., 1963	Catastrophic rockslide into reservoir. Water forced over dam. Flood wave below caused deaths, most destruction.	2600
Aberfan, Wales	21 Oct., 1966	Mudslide from coal mine tailings pile, after heavy rain, engulfed school, killing 114 children.	144
Torrey Canyon, off Cornwall, UK	Feb., 1967	Oil spill. 100 000 tons crude oil into sea, coastal pollution on massive scale.	–
Flixborough, UK	1 June, 1974	Explosion at plant making artificial fibre components.	28
Amoco Cadiz, off Brittany, France	March, 1978	Oil spill. 230 000 tons crude oil, massive loss of marine life, coastal pollution. US$640 000 fine (1992).	–
IXTOC I Oil Platform, Gulf of Mexico	March, 1979	Oil spill, 460 000 tons over many months.	–
Morvi, Gujarat, India	11 Aug., 1979	Macchu II, masonry and earthfill dam failed after overflow in heavy rains and floods. approx. 150 000 people affected.	2000+
Cubatao, Brazil	25 Feb., 1984	Gasoline pipeline leak into squatter settlement, fire and explosion. Deaths by fire and asphyxiation.	500+

Table 4.2   contd.

Place	Date	Disaster features	Deaths
Mexico City, Mexico	19 Nov., 1984	Gas truck exploded in liquefied gas storage depot in San Junico suburb. Fires over 20 blocks. 4350+ persons severely burned.	450+
Bhopal, India[2]	3 Dec., 1984	Toxic gas leak from Union Carbide fertiliser plant. Adverse, continuing health effects on 200 000 persons. Worst chemical industry catastrophe to date.	4037+
Chernobyl, Ukraine	28 April, 1986	Explosion and fire at nuclear power plant. Fallout and rainout radionucleides over much of W. Europe. Permanent evacuation of 135 000 and 70 million at risk of radiation- related disease. US$10s billions losses.	31+
Alaska, USA	24 March, 1989	Oil spill. *Exxon Valdez* ran aground. 10 million gals. crude oil spilled. Massive coastal pollution, wildlife killed (400 000 birds; thousands of marine mammals).	–
Guadalajara, Mexico	22 April, 1992	Series of explosions of gasoline and volatile gases leaked into sewers from state petroleum installation. Many children among dead and injured.	200+
Kozlu, Turkey	3 March, 1992	Coal mine, methane gas explosion. Worst mine disaster in nation's history.	265
Baltic Sea, Estonia	28 Sept., 1994	Sinking of car ferry *Estonia*. High seas and equipment failures. Worst post-war European maritime disaster.	859
Taegu, South Korea	28 April, 1995	Gas explosion from leaking main at construction site. Dead and injured (200+) mainly commuters passing site in morning rush hour. More than half children.	100+

Table 4.2   contd.

Place	Date	Disaster features	Deaths
Seoul, South Korea	29 June, 1995	Sampoong Mega-Department Store collapse, blamed on faulty construction. 1000–2000 persons in store, 900+ injured. Victims included workers and customers. 25+ charged with corruption.	421

[1] cf. Table 9.4 for other dam failures.
[2] see chapter 10, p. 287.

some of the most contentious and politically volatile issues in the heartlands of modernity, but there are large differences in exposure as a function of wealth, habits, gender and other social variables. Hence, the issue is again not solely one of 'safe products' for majorities. We must consider a range of personal and interpersonal vulnerabilities (see Chapter 6).

Industrially produced consumer hazards are not confined to North America, Western Europe and other industrialised nations or wealthier enclaves, if most commonly debated there. Women in remote Himalayan and Andean villages apply new, strong pesticides to their kitchen gardens, where residues wash into irrigation channels and accumulate in ground water. They adopt and have sometimes been forced to adopt harmful, industrially produced birth control devices.

In the poorly or unregulated realms of international product marketing, a great many items are being sold, even 'dumped', in dozens of so-called Third World countries, without regard to, or information on, the risks they bring. A growing problem is hazardous items, including antibiotics and pesticides, subject to stringent controls and even banned in some countries – often the country of origin – but which are still manufactured and marketed elsewhere without such constraints. Health risks identified with particular industrial processes or wastes are also being exported from countries that regulate them to those that do not. This cynical exploitation of the global system by competitive business interests reshapes and magnifies the geography of everyday technological hazards.

'Global change': trend and tip-of-the-iceberg hazards

A special way in which out-of-control technological risks arise is through cumulative impacts and trends. Examples include acid rain, eutrophication of lakes, waterlogging and salinisation of irrigated lands, poisoned ground water aquifers, chemicals that destroy the ozone layer, and 'greenhouse gases' from fossil fuel consumption, which may drive climate change and sea level rise. They are, largely, indirect consequences of technology. The threats develop through regional and global environmental change. In this sense, they are more accurately defined as 'compound' risks. They arise from the interaction of technological, natural and social processes, sometimes particular human activities such as overzealous use of irrigation water, pesticides or fertilisers.

Recently, global change has become a major focus of international scientific research, but it moves uneasily between general environmental and health concerns, and our own field. The most intractable modern risks are not neatly confined within the compartments of particular disciplines and hazardous processes, or the responsibilities of particular government agencies or the disasters field. A further difficulty with some of these risks is the possibility of, as yet unrealised, calamities from more or less unprecedented events. Here there is plenty of room for debate, or misconceptions and obfuscation where powerful interests seem threatened. At the same time, fears about distant, if great, calamities may draw attention and resources away from existing risks and more immediate damages, perhaps from the same technologies.

Technological hazards in context

Dangerous agents or organisations?

The 'hazards paradigm' also reigns supreme in technological risks. As Steve Rayner (1992, 85) expresses it:

> Almost without exception, attempts to understand human behaviour related to technological risk begin with an event, an activity, or a statement of the probability and consequences of an activity assumed to be the stimulus for human response.

Danger is identified most often with specific agents, devices, components, processes, products, by-products and practices that may be dangerous (Table 4.3). Safety assessments may examine larger units of technology such as hydroelectric dams, strip mines or waste disposal systems. They may look at the safety record and prospects of large organisations such as the nuclear industry or, perhaps, a field such as genetic engineering. But the danger is generally seen as a technology or agent that may go out of control.

There is also some uncertainty or oversimplification in these classes of technological risk. Shall we identify the hazards of smoking with the activity? Tobacco as a substance or a carcinogen? The cigarette as a particular artefact? The tobacco industry as a huge organisation producing and promoting the product? Media heroes who encouraged the habit? Smoking as a life-style or social hazard? Or cancers, heart disease and emphysema that may result? At least they are all facets of the same risk, drawing attention to and influenced by technology in rather different ways. Each has a measure of independence in

Table 4.3   Classes of technological hazard

Class	Examples
Hazardous materials (substances, processes)	coal dust, PCBs, paints, leaded gasoline, drugs, tobacco, mutagens, carcinogens
Destructive processes	radiation, fire, structural failure, ionising radiation
Devices, artefacts, machines	spray cans, explosives, power tools, vehicles, aircraft, x-ray machines, trains, hand guns
Installations, plant	power plants, suspension bridges, dams, strip mines, refineries, LNG terminals, power lines, overpasses, high-rise buildings, pipelines
Occupations, practices	mining, construction, crop spraying, flying, automation
Technosystems or organisations	agribusiness, petrochemical industries, public utilities, airlines
Sectors	industry, transportation, military

different contexts and events, or relates to particular professions and jurisdictions. Yet some or all of these aspects are intertwined as parts of the same risk.

Another difficulty relates to Lewis Mumford's (1966, 4) warning about the 'tendency to identify tools and machines with technology: to substitute the part for the whole'. Technologies are also about human organisation, about technique, training and discipline, and established practices. These are as important as the so-called hardware and software. A factory, laboratory, public utility, airline or power company – any distinct organisation that manufactures, utilises, markets or controls technologies – is a technical system. Each of its main tasks is technically defined, disciplined and controlled. To the extent that technological hazards arise from technology itself, the organised relations of persons and machines, processes and practices, are critical to risk. As a technology, a system may be very sophisticated and able to handle a wide range of problems. However, the demands made of personnel, the ordering of activities in space and time, and centralised control, tend to be exacting and often rigid.

The designers, engineers, chemists, machinists may, alone, understand or be able to operate particular, specialised units. Yet theirs are not the only human interventions that make a technology safer or more dangerous. They may be subordinate to the enterprise behind and managerial system overseeing technical performance. These are subject to pressures from other levels and conditions external to the organisation that can influence safety decisions more than technical matters. It might be commodity prices, competitors' behaviour, adverse legislation, interest rates, investment decisions, or a state of national emergency. There are government pressures, corporate cultures and foreign debt to take into account.

Danger and involvement

Technological risks arise in two main types of situation: dangers or losses for deliberate users, including organisations operating or producing the technology, and those that affect people and places beyond them. The first involves work place, physical plant or corporate safety, user competence and training. The second defines the dangers to the general public and the surrounding habitat, the problems most likely to be a concern of our field. Of course, the design, workmanship, management and safety standards applying directly to a technology are always relevant to public safety issues, but the way in which individuals and communities become involved with a technology is equally decisive.

In practice, the contexts of human involvement with a technology are varied and complex. There are facilities and installations whose purpose is to cater to private users and clients. Passenger planes and trains are examples, as are hospitals, amusement parks, sports stadiums and university laboratories. These may be further distinguished according to the extent to which the user is an active participant who can directly influence the safety of the system, or a passive one. The motorist has a more or less large measure of responsibility for highway safety. The airline passenger or hospital patient usually has very little influence on safe operation. In either case there is a distinction between the safety of the equipment, technical organisation and its practices, and the involvement of 'lay' users. But public risk is always a matter of the relations among them.

The risks from widely disseminated consumer products, small pieces of 'technology' intended to pass into the public domain, create yet other, distinct contexts of risk. On the one hand, the consumer who purchases the item has a measure of choice as to what to buy and how to use it. On the other, the safety of the product depends mainly upon the large technological organisation that manufactures and distributes it – the technical expertise, quality standards and product safety checks encoded, as it were, in each ordinary item. How well these work depends, in turn, upon the vulnerability of particular or 'targeted' users, and a host of factors that may intervene between product purchase and final use. Our field tends to become involved when opportunities for abuse or widespread breakdown of safe use exist, and have disastrous results.

Some great hazards seem to have been largely overcome. In almost 90 years there has been no repeat of the *Titanic* disaster on one of the most heavily used shipping routes in the world. To a great extent this must reflect the sea and iceberg monitoring, and the shipping and other safety regulations put in place after 1912. As little as a century ago, one of the greatest hazards for urban living was the development of uncontrollable mass fires in congested districts. Fire remains a serious hazard, but gradually fire regulations, insurance and related urban planning made the extensive mass fire a thing of the past in industrial cities. At least, that seemed so until weapons of mass destruction were directed at them in war (see Chapter 11).

A problem with assessing risk over time is how to balance a remarkable safety record of a technology against its rare but, perhaps, unacceptable disaster. An *Exxon Valdez* could be set against tens of thousands of kilometres of oil transport without a major spill. Then again, such an event creates mass scepticism after thousands of hours of televised advertising that said it could not happen. A Three Mile Island nuclear safety failure undermines the credibility of an entire industry. But then, with such unforgiving technologies, *once may be too often.*

In any dangerous technology, the past safety record is only as good as continued vigilance, the ability to foresee and forestall changing conditions of risk. It is an unending predicament of high-risk technologies and a world in which dangers are continually redefined by innovation and changing social conditions.

Meanwhile, the dominant hazards paradigm has directed attention to the technology itself: how it endangers or fails. Yet enquiries into technological disasters rarely point the finger at the damaging agents themselves. Less and less is that tempting convenience, 'human error', plausible. Rather, the source of disaster is found in failures of organisations, in information and communication problems, in gaps in professional and organisational set-ups, and in narrow attitudes and judgment. Cumulative institutional and management problems often lie behind failures to recognise, head off or contain threats, or to make products safe.

The business 'climate' and decisions, or the responses of higher level management to them, commonly play a large role in the erosion of safety measures. This was found to have occurred before the 1974 Flixborough chemical plant explosion in Britain, and the 1984 Bhopal toxic gas leak. The Aberfan coal tip disaster of 1966 in Wales was blamed mainly on shortcomings or failures of communication and vigilance among the various agencies responsible for mine and public safety. The Chernobyl disaster has been traced, ultimately, to the set-up of the Soviet nuclear agency, and to attitudes and practices among its personnel. The key question there was how the staff at the plant could have conducted a safety system test that actually precipitated the catastrophe.

All these disasters were 'technically avoidable'. The necessary safety procedures and know-how – at Aberfan, specific warnings of the danger – were available. They were simply neglected or ignored by key actors, sometimes but not always cynically, in the run-up to conditions that would trigger the disasters themselves. This is no cause to be complacent about dangerous substances and processes themselves, but it situates the problem mainly in the realms of technical organisation, its social construction and contexts.

Safety organisation

The context of risk may also be one where two or more types of technology or technical organisation interact. They might be different transport systems and industries, or a regulatory agency and an industry. Technological disasters are often attributed to failure to deal with problems of interacting organisations and overlapping jurisdictions. Danger and safety measures at a railway crossing, for example, involve several more or less independent technical systems coming together – trains, trucks, buses and private motorists – as well as 'ordinary' pedestrians. Safety measures and the degree or form of danger vary markedly for these different users. Docklands, airports, coalfields and amusement parks suggest even more complex interactions and opportunities for technologies to cause trouble.

These distinctions have important ethical implications. Voluntary or participatory risk and involuntary, especially 'innocent', victims describe essential distinctions for public policy and legal protection. They identify certain practical and human ecological features in the links among technological hazards, human vulnerability and intervening conditions.

These observations can be fleshed out by looking at some findings of a reassessment of enquiries into 'man-made disasters' in Britain. Barry Turner (1978) began by comparing three disasters. One was the Aberfan disaster, already mentioned. In it, 144 people were buried in a mudslide from a coal tip, 116 of them children who were in school at the time. The second occurred at Douglas, on the Isle of Man, in 1975, when an amusement centre caught fire. Of some 3000 tourists caught inside, 50 perished. The third, in 1968, involved a loaded road transport vehicle being struck by an express train at a railway crossing in Hixon, Staffordshire. Eight passengers in the train and three railwaymen were killed. It appears that the heavy transporter could not get across in the time allowed at a new automatic crossing. These are typical of the kind of 'recurring emergencies' that we find most often in reports of technological disasters.

Ill-structured problems

Turner (1978, 75) sought to define common features among these events, or 'factors which may combine to produce a disaster'. He saw dangers arising primarily in the areas of organisation and communication. Relations among different institutions, and their interactions with the general public and at particular sites, were especially critical. Here is where 'things fall apart'. Critical failures and omissions in safety management are then most likely to end up in disaster. Turner became quite critical of disaster studies that focus only on the event itself, or attribute it solely to the immediate damaging agent or trigger. Rather, he stresses the background to and places where disasters are 'incubated' (see Table 1.5).

A key insight, with respect to formal protection of the public by safety management, is that problems leading to these disasters tended to be 'ill-structured', the antithesis of how we think of a technology, especially a modern one. But Turner shows disasters being 'incubated' in fuzzy areas within, between and outside rather rigid organisations. By contrast, their safety measures are set up to deal with 'well-structured' problems, and clear areas of responsibility.

He even points out that this is a limitation of the disaster enquiries themselves. Having uncovered the ill-structured sources of the known disaster, they invariably assume a clear problem or prescribe a well-structured approach next time.

Turner identifies the origin of such ill-structured problems, and their intractable nature for technical organisations, in areas that seem central to our concerns. He stresses information and communication problems between and beyond responsible organisations, especially government agencies. He finds that centralised and hierarchical command structures, and routine operations, can underestimate or misidentify the rarer or odd, but potentially disastrous, risks. At Aberfan, worried individuals thought the danger from the waste tip was much diminished when, in response to complaints, the tipping there was stopped. He calls this 'solving the wrong problem'. However, two other notions are of special relevance for the human ecology and geography.

'Strangers' and 'sites'

Turner called 'strangers' anyone not belonging to the responsible organisation. They might come from other organisations but, more importantly, might include members of the general public. Of special concern in problems of public safety are people who live near, or have occasions to visit, sites where safety is the concern of, and supposedly guaranteed by, the given agency. (It should be said that these persons, whether living in a coal mining community or using a railway crossing, might well consider themselves the very opposite of 'strangers'. Rather, they are in their home places, or going about their everyday business. The remote offices of, say, the National Coal Board or the railway safety agency might well seem the strangers in the places of risk.)

Turner found that many technical organisations are poorly able to cope with, understand or adequately inform his strangers. Problems arose most often where the latter were involuntarily or casually associated with the activity or dangerous situation, and the safety organisation had limited control over them.

A critical aspect of communication turned out to be between responsible organisations and anyone outside it. Members of the general public are again most relevant to our work, and communities like Aberfan, situated in the vicinity of dangerous facilities. Turner shows their concerns being dealt with more as a public relations issue, if not ignored or dismissed altogether. Those who had to be answered, he found, were often 'fobbed off' with comforting assurances that were beside the point. Often, there seems to be greater worry about 'alarming the public' or undermining the credibility of the organisation than effectively informing people at risk. However, breakdown of internal communications was more likely to involve the information for or about 'strangers' *and* 'sites'.

The question of 'site' addresses a universal problem, and one of central interest to the geographer:

Technological hazards

> Any problem which is concerned to create a site (planning, town planning, hospital planning, level-crossing planning) or which uses sites, becomes involved at that site in a multiplicity of systems, some designed, some unpredictable . . . except on the smallest and most accessible of sites, any problem involving a site is potentially an ill-structured one . . . Human ingenuity is endlessly resourceful in finding ways of manipulating objects in a concrete situation in a manner unforeseen by designers of one abstract aspect of the situation.
>
> Turner (1978, 69)

To travel by ship or airline, visit a weapons system facility or be a patient in a hospital is to experience highly structured but routinised systems with a high level of commitment and competence in controlling the 'strangers' that enter them. But even they can fall apart in situations where established control is lost. Meanwhile, the 'sites' where members of the general public encounter a great many technologies typically include ill-structured and unlikely circumstances. As when entering a technically reliable elevator in an anonymous high-rise building in a crime-ridden neighbourhood, modern people, or some of them, repeatedly find themselves in 'social no-man's-lands which it becomes perilous to enter' (Turner, 1978, 70).

No doubt, a double jeopardy emerges from Turner's analysis. Professionals and organisations surely must be right more often than wrong in the safety judgments for which they are responsible. They may well 'waste' time and scarce resources looking into ill-founded worries and complaints. Yet routine technical preoccupations and organisational inertia, and the sheer volume of actual safety inspections and decisions, can undermine or blur judgement. It becomes more likely they will fail to identify, above all, what Turner calls the 'remote' problems. And these are the sort that lead to disaster. Persons on the ground may well be the first to recognise them, even if they cannot express the problem in technical language.

Context and everyday life

Turner makes clearer than most the sense in which technological risks are embedded in society. They are not simply about how machines go wrong or how their operators make errors, but they arise from the techniques of living in and through artefacts and artifice. The dangers, or how we choose to see and deal with them, are even more clearly socially constructed than for 'natural' hazards. Technologies arise within the social order and must be designed, legitimised, used and maintained by its members.

A difficulty of Turner's analysis, however, is with the places and persons where he sees his 'ill-structured problems' arising, and disaster being 'incubated'. These are often the contexts and the persons who, in general, make up the bulk of the interests of geography, and other fields not uniquely concerned with disasters. They may well lie outside the organisation and direct control of the technological system or agency responsible for safety. But they are not only 'inside' communities and the activities of everyday life; they are the overwhelming majority of people and places whose safety is the central issue here.

Improvement in the awareness and performance of technical organisations may be important and even, as Turner seems to think, the only reliable answer to these problems. Yet this marginalising and stereotyping of those outside such organisations, let alone civil society, seems to put the cart before the horse. It tends to direct all attention to making the organisation more efficient and extending its control, rather than addressing the needs and concerns of populations and communities at risk. Moreover, in 'technological societies', a grave danger is that the only organisation that may seem capable of handling risks through expert and centralised management will be, or will want to become, totalitarian. The dangers in such systems may exceed, and in experience to date have far outweighed, the security the systems may seem to offer. Such has long been the fear and danger of urban-industrial societies, if best known from fictional works such as those of George Orwell or Karel Capek, or movies like Fritz Lang's *Metropolis* (1926).

Transcendent hazards

Networks of risk

Technological risks are bound to be pervasive in an urban-industrial society. As the quotation from Philip Wagner at the head of the chapter suggests, technology has become our primary environment. Just one example is the way modern life is almost as fully enmeshed in electricity and electrical devices as the air we breathe. After motorised transport, and sometimes before it, electrification is taken as a first definitive step in modernisation. The chequered history of danger and deaths, concern and regulation in the electricity industry goes back to the days of its inception more than a century ago. The past decade has seen an upsurge of yet other concerns. There are the possible if contested health hazards of electromagnetic fields surrounding high-voltage transmission, and radiation hazards from electronic equipment in offices and households. There are the risks in electricity generation. Thermal power, the dominant use of coal and civil use of atomic energy, and the proliferation of large hydroelectric dams are among the foremost technologies identified with large health and environmental risks.

In comparing, say, the risks of nuclear power with those of fossil fuels, it has seemed desirable to look at every phase of these energy systems. Different hazards, and an accumulation of dangers, are observed from the original extraction of their fuel stocks in mines to the disposal of waste products. And these individual forms of risk do not act or develop independently but affect and are affected by the overall system and regulation of the industry.

Many forms of damage and hazardous process may arise from a single technology. Chemical industry hazards, for example, include some of the fires and explosions listed under recurring emergencies. Some involve catastrophic forces, leaks and explosions. Hazardous cargoes spread these dangers beyond the site of manufacture. The products of an industry may go to thousands of farms and hospitals or millions of households. And all of these, through industrial and consumer wastes and spills in transport, add to long-term loading of the environment and the dangers of global change.

'Megamachine' dangers

Modern life is testimony to the incredible scope of technologies in two ways. There is an ever-growing integration of modern commercial, industrial and defence technologies in vast, complex systems drawing together many components, often all of the different sectors and systems described above. Along with that, there may be tens of millions of individuals linked through use of a given product or device, or as neighbours of particular types of installation, like petrochemical plants or open-pit mines.

It matters little, in these contexts and especially in cities, whether we look at natural, technological or social hazards. Each is most likely to be expressed in or experienced as a breakdown of artificial networks and built environments. Snowstorms or smogs, toxic chemical fires or terrorist activity, affect most people mainly through the networks and expectations of urban services or life support. Droughts or chemical spills are both likely to be felt in contaminated municipal water, water rationing and related inconvenience.

This is a special but not a uniquely modern predicament. Mumford (1966) argued that civilisation has always involved the kinds of technologies that organise large numbers of people, tasks, supplies, devices and uses. He called it 'the megamachine'. Once, and in many places still, it employed human and animal muscle power, or natural flows of energy, rather than fossil fuels and power tools. This does not mean an absence of large-scale, systemic order. Ancient civilisations such as the irrigation cultures of the Near East or dynastic China were huge centralised and collaborative enterprises. They made possible or supervised vast systems of production and storage of material wealth, and control over nature and society. Their hazards were linked equally to the overall functioning of their megamachine. This is conveyed in a discussion of early irrigation economies of Asia:

> The slightest carelessness in the digging of a ditch or the buttressing of a dam, the least bit of negligence or selfish behaviour on the part of an individual or a group of men in the maintenance of the common hydraulic wealth becomes, under such circumstances, the source of social evils and far-reaching social calamity.
>
> Metchnikoff, quoted in Walter Benjamin (1968, 123)

This quotation identifies technological risks in many parts of the world today. In all cases, safety is affected by the two tendencies towards centralised control and standardised technologies. Dangers become contingent upon the practices of society as a whole and tempt those responsible for safety to look for totalising systems.

Geographies of technological hazard

The incidence of 'recurrent emergencies', looked at earlier, shows a global pattern of risk having little respect for national, climatic or cultural boundaries (see Table 4.1). There may be more aviation disasters in the United States than in India or China, but the problem is significant in all three countries. Indonesia and Peru each reported more air crashes than all of Europe. India heads the list of major traffic and railroad disasters. Such general and mediated statistics are hardly an adequate basis for understanding these distributions, but they do establish the worldwide presence of modern technological hazards. They challenge some common stereotypes. The tendency of cultural geography texts to present, say, India or Indonesia through images of peasant agriculture using manual and animal power, and hence of suffering primarily from natural hazards, is seen to be misleading if not offensive.

This evidence from the more commonly reported disasters is just the tip of the iceberg. There are many new and spreading risks from technology transfer and globalisation of economic activity and industrialised production. And this is not to ignore large parts of the human population marginal to or only casually involved in the modern 'world system', who may be the least protected from its dangers. They may be affected through dangers carried by wind and water into remote environments, as with acid rain in northern Canadian lakes, or radioactive fallout in northern Finland after Chernobyl. But economically and politically marginal communities may be in close proximity to dangerous modern technologies. For example, at Grassy Narrows in relation to nearby pulp and paper mills, or for the Navajo in the uranium mining areas of New Mexico. The resource hinterlands of urban-industrial North America and much of the rest of the world often create exceptional risks for such peoples.

Hazard spaces

However, the media and many technical works identify disasters with national units. This reinforces the illusion that the geography of risk is essentially one of state geography. Rather, the web of metropolitan, commercial exchange and global economic functions provides the base map for plotting these events and their variations. Thus, one can envisage technological hazard in terms of what Zeigler *et al.* (1983, 14) called distinctive 'hazard zone geometries' – occurring as 'point', 'linear' and 'areal' hazards. Reality complicates these, of course. An oil spill may originate along the linear geometry of a shipping route or pipeline, but where it occurs, it may become an 'areal' hazard. Conversely, chemical spills into a waterway from a single plant can spread as a linear hazard, as with the toxic spill from Sanoz and its course down the Rhine in 1990. The path of the Chernobyl radioactive cloud created a pattern of risk much more complex than areal risk around the plant, although the worst fallout was there. Meanwhile, in most parts of cities, residents are subject to or lie within the reach of hazards that have all three patterns.

The public impact and geographic patterns can also be governed by external, intervening conditions other than the artificial networks. Damage may be due to a 'natural' process triggered by technological failure. The outburst flood waves from the failure of artificial dams are not readily distinguishable from natural dam failures. Some of the most widespread and severe 'technological hazards' are realised in, or compounded by, interaction with natural processes. In the Chernobyl example, they appear in the attendant weather patterns that carried the radioactive cloud over Europe. Other examples include photochemical smogs or concentration of dangerous pollutants through natural food chains. In a case like the *Exxon Valdez* oil spill in Prince William Sound, a disaster is identified with a particular industrial product but magnified or singled out by the threat in a particular habitat.

Suggested reading

Erikson, K. (1994) *A new species of trouble: explorations in disaster, trauma, and community*. W. W. Norton, New York.

Gould, P. (1991) *Fire in the rain: the democratic consequences of Chernobyl*. Johns Hopkins University Press, Baltimore, Md.

Kates, R. W., C. Hohenemser and J. X. Kasperson (1985) *Perilous progress: managing the hazards of technology*. Westview Press, Boulder, Colo.

Mumford, L. (1963) *Technics and civilisation*. Harcourt Brace and World Inc., New York.

Perrow (1984) *Normal accidents: living with high-risk technologies*. Basic Books, New York.

Reich, M. R. (1991) *Toxic politics: responding to chemical disasters*. Cornell University Press, Ithaca, NY.

Sagan, S. D. (1993) *The limits of safety: organizations, accidents, and nuclear weapons*. Princeton University Press, Princeton, NJ.

Turner, B.A. (1978) *Man-made disasters*. Wykeham Publications, London.

Zeigler, D. M., J. H. Johnson and S. D. Brunn (1983) *Technological hazards*.

CHAPTER 5

Social hazards: violence and the disasters of war

Life in a community is normally protection against physical destruction, but now [in war] it has come very close, just because one belongs to a community.

Elias Canetti (1973, 83)

In civil wars and other conflicts alone, the UN is providing humanitarian assistance to nearly 60 million people in about 30 countries.

Jan Eliasson (UN/DHA, 1994a, 9)

The problem of violence

Introductory remarks

Social hazards, the harm which one society or part of a society may do to another, are pervasive sources of danger and disaster. Uses of armed force against civil populations or that harm their settlements and habitats are not the only forms but will be our main concern here, for damage from and costs of armed strife outstrip the extreme natural and technological dangers considered above. This is so whether measured by loss of life, injury and disablement; social disruption or devastation; or the costs of armies and weapons. War has taken some 120 million lives directly in this century, the larger fraction being civilians. While the two world wars caused the greatest losses, there have been about 150 wars since 1945, each with more than 1000 violent deaths (Table 5.1).

A general survey of disasters in the 1980s, which excluded most wars, nevertheless found 'civil strife' responsible for almost half of all reported disaster deaths (US/OFDA, 1990). The second highest losses, almost 40 per cent, occurred in drought-related famines, most of them aggravated by civil strife and war.

Table 5.1 Global incidence of armed conflicts 1945–1992, showing the numbers and proportion of fatalities among civilians (after Sivard, (ed.), 1993, 21)

Region	Number of wars[a]	Deaths (millions)[b] Total	Civilians(%)	
Far East	36	10.9	6.6	(61)
Sub-Saharan Africa[c]	38	6.5	4.5	(69)
South Asia	13	3.4	2.4	(71)
Middle East	20	1.3	0.5	(38)
Latin America	24	0.7	0.47	(67)
Europe and (former) USSR	13	0.37	0.2[d]	(54)
North Africa	5	0.13	0.11	(84)
Totals	149	23.14	14.87	(63.5)

[a] wars with at least 1000 deaths from armed violence.
[b] rounded from Sivard (ed.) (1993, 21)
[c] includes South Africa.
[d] author's estimate.

More than 30 armed conflicts, which had each caused at least 1000 deaths, were in progress in any given year at the end of the 1980s and early 1990s, on the soil of at least as many countries (Table 5.2). Twenty-four of these wars had lasted more than a decade. They had killed more than 5.5 million civilians and forcibly uprooted more than 25 million people. There is no indication that with the end of the Cold War, the scale and geographical scope of these wars have declined, rather the opposite (Fig. 5.1).

As a rule, armed conflict is the province of military geography and history, of national security and defence establishments. Yet war hazards would seem to involve our field as severe and, in many ways, the most basic of dangers. On the one hand, civil life rather than the military has suffered the larger share of the casualties, human emiseration and destruction in wars of this century. On the other hand, the results of warfare resemble and, in some respects, typify the sort of risks and calamities that concern us. When used against settlements and

Table 5.2 Global distribution of armed conflicts in regions with at least one major war between 1989 and 1992 (after SIPRI, 1993, 86–7)

Region	1989	1990	1991	1992
Africa	9	10	11	8
Asia	11	10	8	11
Central and South America	5	5	4	3
Europe	2	1	2	4
Middle East	5	5	5	4
Annual total	32	31	30	30

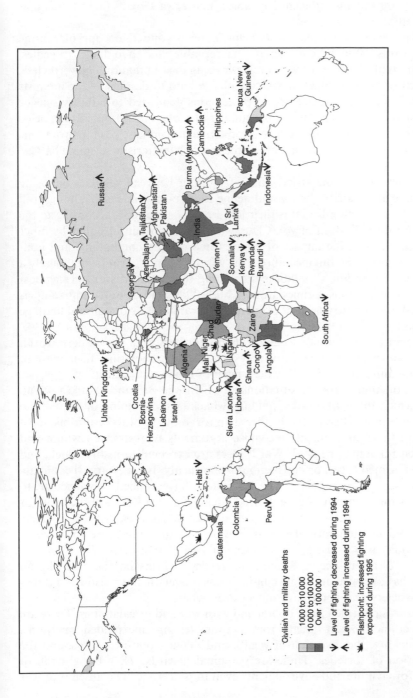

Fig. 5.1 The global distribution of major armed conflicts in 1994. Among the 36 conflicts, 14 started after the end of the Cold War in 1990. In the worst conflict of 1994, in Rwanda, estimates of people killed range between half a million and one million (after Project Ploughshares, 1995, 5)

non-combatants, life-sustaining land uses and the natural habitat, armed assaults bring very destructive events, long recognised as the 'disasters of war'. They are the almost exclusive subject matter of Goya's famous etchings with this title.

These are also events that overwhelm or lie beyond the scope of routine social controls. Widely considered unacceptable and avoidable tragedies, they often involve 'crimes of war' and 'crimes against humanity'. Nevertheless, such forms or consequences of armed violence tend to lie outside the interests of and to exceed the competence of professions dedicated to military science and national defence. That is, as one student of violent death puts it, the fields whose 'experts tend to think exclusively in terms of combat . . . chiefly in terms of major theatres of war and mainly of the most characteristic set-piece battles' (Elliot 1972, 133).

Indeed, warfare in the strict sense of fighting between regular, opposing forces is not a 'hazard' as we would define it, or even 'violence'. Though full of risk and pain, for soldiers it is their job. For the state, war, according to the oft-quoted rationalisation of von Clausewitz, is 'the pursuit of policy by other means.' Conversely, the impact of armed force on non-combatants and non-war items involves the full meaning of the word 'violence', that is, not merely a destructive process but a violation, something that overturns or goes against norms and expectations. Crimes are violent. Soldiers killing non-combatants is violence. Wanton destruction of property, natural habitats or cultural heritage is violence to civilised persons.

To a pacifist, and perhaps to any truly decent person in the modern world, all war is 'violence'. It violates, or necessarily ends in violating, their sense of reason, humanity and ethics. However, most of us are not pacifists. We believe, wisely or not, that there are situations in which one can and should take up arms, in which those who can should fight back. We champion or sympathise with one side in most conflicts. Rather than forgo the right to bear arms, most people and their governments recognise arguments to restrict by whom and how armed force may be used. War leaders are expected to justify which, and how much, armed force, or destruction of the enemy, is necessary. Illegitimate users and uses are criminalised. The restrictions have applied, in particular, to the hazards that concern us here. Indeed, ideas of a 'just war' and the 'laws' or conventions of war essentially define our concerns. They are not just disasters for, but also *of* humanity. There is, however, a growing sense that armed violence today is not just an 'instrument of policy' that gets 'out of control' when these disasters occur. Rather, the resort to arms has become like an epidemic that overwhelms whole continents as it overwhelms reason and threatens civil society and good government everywhere.

The hazards of armed force do not end with war and its calamities. There are other direct and indirect threats from weapons testing, unexploded munitions, arms trafficking, the diversion of wealth and talent from civil needs, and the militarisation of societies. These are perennial hazards. They bring a toll of disasters for civil life and environments even in peacetime. The absence of war from the soil of North America for over a century, and most of the wealthier industrial nations since 1945, has made these indirect or 'peacetime' hazards of armed violence the more immediate risks and causes of damage to their populations.

There is another dimension to the importance of armed violence as a form of hazard. Permanent military preparedness and investment are now on such a scale that they show war hazards to be the primary risk on the agenda of most governments. Apart from the frequency of wars, which has entailed some US$3 trillion in annual arms expenditure and global arms trade in recent years, related research and development employs more scientists and engineers than any other field. Research in many universities has been underwritten by defence-related interests. Arms manufacture and military procurement are important, sometimes decisive, factors in the commercial success or survival of many large corporations. Necessary or not, this activity takes away resources and talent from other dangers, and influences how every other hazard and disaster is treated. The armed forces are widely employed in actions outside declared wars and national defence, including response to all forms of disaster.

Geographers have only recently taken a concerted look at these matters as hazards. Yet their inclusion here would not have surprised earlier students of disaster. War has always been compared to and shared the language of great natural calamities, pestilence and famines. In a statistical study of famines more than a century ago, Cornelius Walford (1879, II, 108) found warfare a major factor, and added that 'the sword, pestilence and famine are now, as they have been in all time, the three associated enemies of the human race.' Pitirim Sorokin's (1942) pioneering work on disaster compared war, revolution, famine and pestilence as the main calamities of human history. Prince's (1920) classic study of the Halifax Harbour, Nova Scotia, munitions ship explosion of 1917 – perhaps the first from the perspective of contemporary social science – showed the huge indirect impacts that war may have even on civil populations remote from the fighting.

Following the Second World War, sociologists and psychologists treated natural, technological and war disasters as a common problem. Recently, the once secret but extensive wartime investigations of stress, psychological warfare and disaster among civilian populations, and the plans for targeting settlements and habitats, have become available. These include studies by key figures in postwar crisis and disaster research. In many ways, the social construction of ideas and practices in our field since 1945 reflects a mind-set that emerged in the victorious nations, especially in North America. 'Civil defence', developed mainly for wartime home security under air attack, became the basis for preparedness against peacetime natural, technological and social disasters. In most countries, these still share a common civil defence organisation.

Our main concern, then, is armed violence that destroys the bodies, settlements, means of existence, artefacts, institutions and habitats of unarmed resident populations. Dangers to the habitat also include harm to other living creatures with which we share the planet. Such concerns are in accord with the prevailing interests of human geography, which, in contrast to military geography, are about the material and cultural foundations of civil or 'peaceful' existence. In referring to 'civilians', the usual meaning of non-combatants and citizens is implied, but not to exclude disenfranchised persons, prisoners and 'captive' peoples, exiles, and refugees. As Heinrich Böll's short story of a wartime incident puts it, 'children are civilians too.'

Violent hazards

The hazards of armed violence are also most often defined by 'agents' that cause or initiate damage – weapons or weapons systems, types of armed unit, strategy or campaign (Table 5.3). Because more widely available and used, 'conventional arms' have caused the greatest harm to land and life. In military battles, machine guns and tanks can make a nasty mess of city streets, a farm or woodland. Gil Elliot (1972, 135–6) estimated that 10 million people had been massacred or executed by small arms fire in the first half of this century. This also caused about 14 million combat deaths. He estimated 18 million deaths from 'big guns' (i.e. heavy artillery) and a further 5.5 million from other forms of 'hardware'. This is about one-third of the violence-related deaths he identified, the remainder being due to deliberate and indirect privations brought about by violence. The situation has not improved since he completed his review (Fig. 5.2).

While all weapons are potentially a threat to civilians and their places, some are more likely than others to cause them harm or to be used against them. Weapons are dangerous not in terms of some abstract killing potential, and only incidentally because of their intrinsic properties. It is the purposes for which they are designed and most likely to be used that are more crucial.

'Dubious weapons'

In this century, the face of war has been transformed by the increasing availability and deployment of weapons known to cause indiscriminate destruction, to destroy settlements and crops, to terrorise and maim. Instruments intended for such purposes have been called 'dubious weapons' (Table 5.4). As a rule, they do the most concentrated and calamitous harm to civilians and their environment.

The greatest single disaster from the use of 'dubious forces' occurred in the Sino-Japanese War of 1937–45. In June 1938, the Chinese army sought to impede the advance of the Japanese invaders by destroying dikes of the Yellow River (Huanghe). Partly successful in its military objective and drowning several thousand Japanese troops, it was a vastly greater tragedy for the Chinese civilian population. Hundreds of thousands were drowned; several million were

Table 5.3 Hazards of violence: dangerous agents and methods of indiscriminate warfare

Weapons

'conventional'
(e.g. firearms, artillery, high-explosive bombs, aircraft, submarines)

'dubious'

anti-personnel	(e.g. fragmentation munitions)
incendiaries	(e.g. napalm)
chemical	(e.g. phosgene, biocides)
biological	(e.g. anthrax, crop blights)
toxins	(e.g. natural, artificial)
nuclear	(e.g. atomic, hydrogen, 'neutron')

Deliberate release of dangerous or 'dubious' forces

environmental warfare

fire-setting	(e.g. forest, grassland, firestorms)
hydrological	(e.g. dam blasting)
geomorphic	(e.g. avalanche triggering)
geological	(e.g. earthquake triggering)
weather modification	(e.g. rain making)

technological and industrial releases

fossil fuels	(e.g. oil spills and fires)
chemical	(e.g. pesticide plants)
nuclear power	(e.g. nuclear)
biotechnology	(e.g. infected materials, cultivated pests)

Armed units and weapons systems

strategic air power
ICBMs
space weapons
'special forces'
terrorist groups

left homeless. Many cities and thousands of villages were inundated; millions of hectares of farmland were flooded, destroying crops and ruining land through silting and soil erosion. The river changed to an entirely different course. The dikes and old course of the river were not restored until 1947.

This was an extreme example of disasters that may come from 'environmental warfare', or the deliberate 'release of dangerous forces' in, or into, the environment (Westing, 1990). This may be done either by triggering natural processes or by the release of industrial or artificially contained forces into it. Dams and dikes provide well-defined targets for the release of dangerous forces, causing mass destruction in areas below or protected by them. Other examples include triggering of avalanches or landslides on mountain slopes, setting fire to forest or grassland, poisoning water supplies, and deliberate spills or fires from oil facilities such as were seen on a vast scale in Kuwait in the 1991 Gulf War.

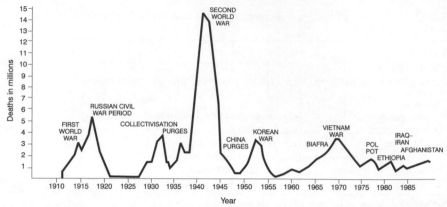

Fig. 5.2 Deaths from armed violence, 1912–1990. Estimates for the period to 1950 are those of Gil Elliot (1972, 195), and for 1950–1990 are based on war deaths in five-year periods from data in Sivard, 1993

Of all the releases of dangerous forces in war, the greatest overall destruction must be attributed to fire-setting. The use of flame in battle and laying waste is age-old. However, developments in incendiary weapons and new methods of delivery in this century have greatly increased its scope, and the scale of harm to habitat and human settlements. The peculiarly urban danger of uncontrollable mass fires has been a source of irreversible devastations and huge civilian casualties. In the Second World War, it was estimated that two-thirds of the destruction of British cities from air attacks was due to fire, although nearly all civilian casualties were due to high-explosive impacts. The single most destructive raid on the capital has been referred to as the 'Second Great Fire of London', recalling the one that transformed the face of the city in 1666 (Johnson, 1988).

Once the Royal Air Force adopted and became proficient in mass incendiary attacks, the fate of Germany's cities was decided by great mass fires. Of these, the 'firestorms' take the release of dangerous natural forces to the extreme (Table 5.5). A firestorm is a meteorological event brought on by extremely concentrated combustion and a fire reaching temperatures of more than 1000° C. It generates fierce winds, locally sufficient to blow people over and suck them into the flames. These fires, and the even more extensive 'conflagrations' with fire fronts that swept forward before the wind, overwhelmed firefighting, rescue and other civil defence services.

In the latter half of the twentieth century, the indiscriminate and uncontrolled devastation caused by incendiary weapons is associated with large-scale use of jellied gasoline forms, notably 'napalm'. It was developed for jungle warfare in the Pacific in 1943, as a means to destroy and clear otherwise hard-to-burn rain forest cover. However, napalm and related incendiaries proved very suitable for burning out villages and croplands, and setting uncontrollable mass fires in cities. In all, this and other oil-based incendiaries

Table 5.4 'Dubious' agents of war and other armed violence (see SIPRI, 1971; Geissler (ed.), 1986). The list gives a clear indication of the research and investment in these means to cause unusual suffering and indiscriminate death

Agents

Anti-personnel:
- firearms and projectiles (e.g. 'dum-dum' bullets)
- fragmentation: impact splintering metal ('shrapnel') or plastic, canisters or fragment-dispensing munitions (steel shot)
- blast weapons (concussion bombs)
- fuel–air weapons: to dispense explosive gas, aerosol or vapour droplet explosions
- 'high-tech': acoustic, electronic, laser, microwave devices

Incendiaries (inflammable constituents):
- metal (e.g. magnesium)
- oil-based ('Molotov cocktail', napalm)
- oil + metal ('goop' = magnesium + napalm)
- pyrotechnic (= combustible oxidising, e.g. 'thermite')
- pyrophoric (= spontaneous ignition in air, e.g. white phosphorus)

Chemical ('gas warfare'):
- choking (e.g. phosgene)
- blistering (mustard gas)
- nerve ('sarin')
- blood (hydrogen cyanide)
- 'tear' ('CN', 'CS')
- hallucinogens

Biological ('germ warfare'):
- bacterial (e.g. anthrax, cholera, bubonic plague, botulism)
- viral (dengue fever, yellow fever, smallpox)[1]
- rickettsial (epidemic typhus)
- fungal[2]

Toxin warfare:
by action:
- cardiotoxins = attack cardiovascular system (e.g. sea wasp toxin)
- dermatoxins = attack skin (e.g. Fusarenon-X)
- neurotoxins = attack nervous system (e.g. botulin, saxitoxin (= shellfish poison), tetanus toxins

by source:
- 'natural' – bacteria (anthrax toxin), protozoa ('shellfish poison'), fungal = mycotoxins (Fusarenon-X), plants (ricin from castor bean), amphibia (Columbian frog = batrachotoxin), reptilia (Chinese cobra = cobratoxin), fish (deadly death pufferfish = tetrodotoxin)

by impact:
- 'casualty' = to kill or permanently injure (e.g. phosgene, botulism)
- 'harassing' = temporary incapacity (e.g. 'tear gas', 'riot control')
- gene-altering = (e.g. dioxin)

Nuclear:
- fission ('atomic') = uranium or plutonium fissionable material
- fusion or thermonuclear = deuterium or 'hydrogen' (H-bomb)
- 'neutron' = enhanced radiation, reduced blast weapons

Table 5.4   contd.

Weapons

- all these agents may be delivered by projectiles, grenades, artillery shells, mines, bombs, rockets, 'cruise' or ballistic missiles; by land, sea, air, space weapons systems
- anti-personnel: include high-velocity rifles, scatterable mines, delayed action and booby-trap devices
- incendiary: improvised ('Molotov cocktail'), flame throwers, cluster bombs
- nuclear: warheads, ground ('dirty'), underwater, space, air detonation
- chemical, biological and toxin warfare may use 'natural' or 'commercial' delivery (i.e. poisoned water supply, air-conditioning systems, clothing, food, medicines)

[1]Today, the majority of BW agents, '19 out of 22...with highest suitability for operational use against humans' (Geissler (ed.), 1986, p. 22).
[2]Now generally considered anti-crop or anti-animal BW agents.

Table 5.5   Incendiary attacks upon German cities in which firestorms were generated, with estimates of civilians killed or bombed out, and built-up area razed by fire (after Hewitt, 1993)

Date	City	Deaths	Bombed out	Burnt out (km^2)
1943				
29/30:5	Wuppertal (Barmen)	3400	130 000	2.6
27/28:7	Hamburg	40 000	800 000	13.2
22/23:10	Kassell	7000	53 800	7.5
1944				
22/23:3	Frankfurt-am-Main	1001	120 000	1.5
11/12:9	Darmstadt	10 550	49 000	2.0
12/13:9	Stuttgart	1172	50 000	4.0
18/19:9	Bremerhaven	618	30 000	1.2
4/5:12	Heilbronn	7000	50 000	1.4
1945				
16/17:1	Magdeburg	16 000	190 000	3.1
13/14:2	Dresden	40 000	250 000	6.8
23/24:2	Pforzheim	17 600	50 000	4.5
16/17:3	Würzburg	4500	56 350	1.7
Totals	[12]	148 841	2 099 150	49.5

caused in excess of one million civilian deaths in the Second World War, mainly in urban attacks (see Chapter 11). The best-known and most intensive use, however, was in the Second Indo-China (or Vietnam) War. There, napalm was responsible for extensive destruction of habitat, fields and villages, for a large fraction of the more than one million civilian deaths, and for uncounted numbers with burn injuries. In recent years it has been widely used as a weapon of terror, civil and counter-insurgency warfare by air forces worldwide.

Given the nature of warfare, any weapon may be described as 'anti-person-nel'. However, the term is reserved especially for those known or designed to cause unusual pain and mutilating wounds, disfigurement and maiming, or indiscriminate death. If this includes virtually all dubious weapons, of particular concern are variants of 'conventional' weapons developed to cause these inhumane effects. Any bullet, shell or bomb might do so, but extraordinary efforts have been devoted to ensuring they will. Modern anti-personnel weapons emerged as responses to the bogged down battlefronts of the First World War, and colonial wars where small units sought to control and demoralise large populations. Developments included gas war, 'germ war' and the raiding of cities behind the lines. Manufacturers produced bullets that rip, mushroom or fragment on impact with human flesh; endless variants of 'fragmentation' grenades, mines and bombs; blast, concussive, fuel-air explosive and incendiary weapons. All of these increase indiscriminate killing power, terror and maiming. Efforts to outlaw them on the battlefield, even where successful in obtaining widespread backing of governments, have not prevented their stockpiling and frequent use against civilians.

Equally serious, from our perspective, is the way anti-personnel weapons are associated with both increased firepower and quantities of munitions used, and 'area' rather than precision targets. They are used to blanket areas with a hail of steel, to set mass fires, or infest whole landscapes with a myriad of mines and other explosive devices. Indiscriminate harm to the environment is inevitable, and civilians are at risk even when they are not the intended victims.

Anti-personnel, incendiary and environmental weapons have been the 'dubious' instruments responsible for the greatest harm to peoples and places, and are those most likely to be used. Nevertheless, the weapons commonly perceived as 'dubious' are nuclear and, especially, chemical, biological and toxin (CBT). Concern over these weapons is driven mainly by the potential of some to bring unprecedented annihilations, even threatening the extinction of human, if not all life. It is also a matter of the horrific effects upon the human body. Nuclear radiation and the action of CBT weapons on their intended victims involve peculiarly inhumane and extraordinary suffering. Increasingly, they have been developed to cause particular forms of pain and injury, or death through damage to particular organs. The weapons of choice are those most difficult to protect against, and for which medical treatment or even diagnosis is unlikely. Civilians will, in any case, have less protection and be more vulnerable than soldiers, and include persons whose age or condition makes them more susceptible to harm.

To the present time, independent investigations have confirmed 22 deliberate uses of these weapons in wars (Table 5.6). Civilians have been harmed in nearly all of them and, except in the 'gas warfare' of the First World War, have been part of or the main target.

These weapons also pose threats in peacetime. Their development, testing, stockpiling and disposal are all extremely dangerous ventures. They have already been the sources of disasters and public health crises in the industrialised nations of North America, Europe and the former Soviet Union.

Table 5.6 Uses of dubious weapons: chemical, biological and nuclear weapons in the twentieth century (after SIPRI, 1971; and Kidron and Smith, 1983)

Confirmed uses, worldwide	22*
Top confirmed users	
UK (chemical)	4
Iraq and Northern Russia [mustard gas],	
Oman [unspec.], Malaysia [herbicides]	
USA (chemical, nuclear)	4
Korea, Vietnam and Japan	
Portugal	3
colonial territories of Africa	
Alleged but unconfirmed	26
Testing and 'accidental' releases	
nuclear weapons testing, explosions:	
Total to date	approx. 1400
1944–1963	488
(USA, USSR, UK, France)	
1963–present	900+
(add China, India, [South Africa/Israel])	
Above-ground explosions	550+
known nuclear weapons systems accidents:	
involving the USA	150+
involving the USSR	16+
involving the UK	10+
involving France	5+
chemical and biological weapons accidents:	
major incidents	5+
minor**	hundreds

*includes four countries using poison gas in WW I; of Middle Eastern countries only Egypt, against North Yemen, 1963–1964; only confirmed biological, Japan in China.
** includes Canada.

Military units as hazards

There are severe risks even from the operations and preparations of those armed forces that scrupulously observe humane codes of conduct towards civilians and non-military areas. However, wars conducted without restraint, and in which armed units are deployed against non-military targets – admitted or not – have brought the greatest disasters for civil life.

A widespread source of atrocity and disasters are 'special forces', indoctrinated and trained to carry out armed actions against political opponents, ethnic groups, sometimes the whole civil population. They usually strike in pre-emptive 'actions' against unarmed people. The best known are perhaps the *Einsatzgruppen* or commandos – the long arm of the German security agency under Adolf Hitler. They carried out the racial and political agendas of the National Socialists and, in territories occupied during the Second World

War, abused, expelled, 'concentrated', executed, massacred or exterminated millions of civilians (see Chapter 12). Countless villages destroyed in recent Central and South American civil strife, the women and children maimed, the men who 'disappeared', have been victims of state forces and clandestine 'death squads' drawn from them.

However, the disasters of full-scale modern wars depend less and less upon how individual soldiers and units behave. They reflect ever greater use of targeting and warring 'at a distance'. This involves long-range, high-speed, automated and guided missiles; weapons of mass devastation; and strategies designed in distant centres and monitored remotely. They involve preplanned deployments of dubious weapons such as anti-personnel mines or systematic destruction of forest cover with biocides, in which soldiers in the field manage the machines more than fighting. These distanced methods, mediated by high technology, have not only proved an ever greater threat to land and life. They also promote and demand a geography of remoteness and abstract information that sees neither the pain of the victims nor the ravaged landscapes.

Hazards in the aftermath of war

Many, sometimes the worst, calamities of war for civil life continue or arise when hostilities end. Wartime and war measures may endanger civilians but often include provision for at least minimal rations, civil defence and employment in war work. Afterwards, especially in defeated nations, even these limited safety nets can disappear. For example, millions of people bombed out of their homes in cities of the Second World War had lived amid the rubble of devastated neighbourhoods for months or years of war, but the so-called 'rubble years' they remember most, and as most severe, tend to be postwar. Many of them spent further years in makeshift or partly damaged dwellings, often without adequate food, fuel, clothing or employment. They had to live in temporary, or old army, camps; by 'doubling up' in surviving housing stock; and in generally crowded and unsanitary quarters. More than a decade passed before a great many had a home of their own; two or three decades before they had accommodation comparable with their prewar homes. When we read of the distress and hardship faced by victims of hurricane or flood, after just a few days of evacuation, or in waiting a few months for replacement homes, the psychological and social impact on homeless victims of war speaks of a much greater disaster. Unfortunately, they and their plight are commonly the forgotten aspects of war, neither part of 'the fight' nor enjoying the return of peace.

Leftover weapons are hazards of warfare less easy to ignore: the unexploded bombs in battle zones and cities subject to massive bombardment, and the minefields and booby-traps or equipment thrown away by defeated armies. In heavily bombed cities like London and Birmingham, Berlin and Hamburg, after 1945, thousands of unexploded bombs remained buried in the ground. They continue to be uncovered to the present time. All too often, in the years following the war, they were found by children playing in waste ground or ruined buildings, with fatal results.

The 'mines plague' is one of the greatest scourges of the late twentieth century: vast numbers of unexploded mines, lying in regions of past and continuing conflicts (Table 5.7). Large tracts of land are infested with mines in more than 60 countries. They include fertile and critically needed areas for restoring the economies of war-torn countries. According to one report:

> In Angola, some of Africa's most fertile land lies fallow because farmers are afraid to tread on them. In Somalia, returning refugees prefer to sleep in tents rather than enter booby-trapped homes. In Cambodia, more than 200 people a month are dying or having limbs amputated by some of the 4–7 million land mines sown across the killing fields.
>
> UN/DHA (1994, 21)

A large fraction of some 19 million refugees and 25 million displaced persons at the start of the 1990s cannot return to their homes and former occupations because of this threat. Few other 'disasters' cause such widespread and sustained misfortune.

Hazards of the 'weapons cycle'

Many of the dangers of weapons and methods of making war exist at times or places where armed violence is not apparent. Modern weapons can have lengthy and intricate development requirements, involving places far from the battlefronts and in peacetime. Procurement of their ingredients, testing, manufacturing, stockpiling, shipping and decommissioning these weapons endanger civilians and the environment. Some of the most severely contaminated sites and habitats in the world are those where modern weapons have been fabricated. Unusual human health threats and their victims are concentrated around these places. Such hazards are especially identified with 'dubious' weapons. Nuclear, chemical and biological weapons materials have all involved lethal escapes, mishandling and careless disposal (see Table 5.6).

Table 5.7 The land mines plague: estimated numbers of land mines in 1993, for the most severely affected countries (after UN/DHA, Sept./Dec. 1993)

Afghanistan	9–10 million
Angola	9 million
Iraq	5–10 million
Kuwait	5 million
Cambodia	4–7 million
Western Sahara	1–2 million
Mozambique	1–2 million
Somalia	1 million
Bosnia-Herzegovina	1 million
Croatia	1 million

The sites and surroundings of the United States and former Soviet Union nuclear weapons complexes are severely contaminated, and major public health dangers (Fig. 5.3). This is apart from the test sites in Nevada, the Pacific islands and Central Asia, expected to be seriously contaminated for an indefinite period. Price tags in excess of US$100 billion and, according to some, over US$300 billion, are expected simply to clean up the US nuclear facilities. The essentially 'conventional' and commonplace wastes at Canada's military bases – leaked fuels and lubricants, unexploded bombs, PCBs, and garbage – are said to involve a host of toxic materials endangering surrounding communities and especially water supplies. In 1994 the Canadian government voted about US$200 million towards clean-up, but estimates go as high as US$700 million to complete the task.

Shipping of dubious weapons, their components and waste products also poses grave dangers to peacetime society. It has long been the practice to disperse key weapons-making or storage facilities and military bases to remote areas and among multiple sites. This is intended to reduce the dangers of attack, sabotage or espionage. As a result, however, the extent of shipments to produce and deploy given weapons is very great. Most military cargoes are transported along the same highways, waterways and railways, or through the same ports, stations and airports used by civilians. But this is only part of the interrelations between civilian and military technologies.

'Dubious' technologies?

Potential disasters and dubious weapons arise from the strategic importance and vulnerability of all dangerous facilities and industries in wartime. Thus, the next major international war will be 'nuclear' even if no nuclear weapon is used. In 1989 there were 429 nuclear power plants in operation in 26 countries, and over 300 'research reactors'. Another 105 power plants were under construction. A further 37 had been closed but still contain radioactive materials. There are dozens of associated uranium mines, and fuel assemblage, spent fuel and heavy water plants, not to speak of nuclear submarines and ice-breakers. None of the reactors is considered bomb-proof, and damage to other elements of their operation could lead to catastrophic failure. In a war there is no guarantee that, by mistake if not design, shells, bombs, missiles, out-of-control aircraft or ships, saboteurs, or terrorists will not hit 'peaceful' nuclear facilities.

This problem of the release of dangerous forces from industrial plants in war applies to virtually all forms of facility and production involving lethal and destructive agents. Major hazards attend chemical and fossil fuel energy industries, which concentrate large quantities of toxic, explosive or inflammable materials. These may be easily released or triggered by armed attack. They become weapons or objectives of 'economic warfare'. Moreover, industrial complexes cluster around oil-refining or coal-processing plants. A host of chemicals, pharmaceuticals, explosives, paint and dye products, toxic residues and other hazardous materials may be present. They are generally located in or

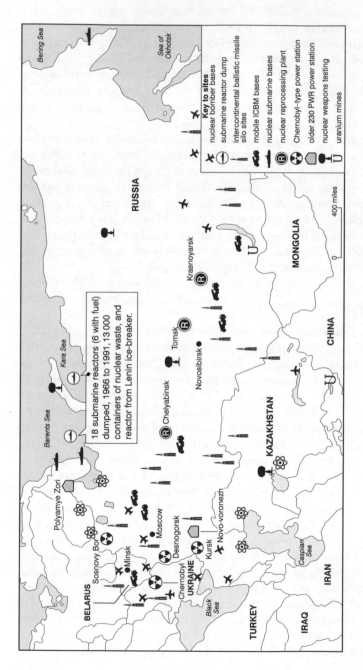

Fig. 5.3 Nuclear hazards from weapons, aging reactors and dumping in the former Soviet Union

near major urban centres, and along rivers and coastlines that would aid and direct the spread of dangerous substances. As Chernobyl and the 1991 Persian Gulf oil fires showed, airborne releases can affect whole regions and continents.

Peacetime disasters such as Bhopal or Three Mile Island are usually single incidents, widely separated in time and place. In war, all parts of a country and the countries at war may be under attack. Even in a single attack, there is a likelihood of multiple releases from many different kinds of dangerous facility. In a major war, the means and will to protect civil populations from these dangers can be greatly diminished.

Technological hazards that are peculiarly threatening to public health, the built environment and natural habitats are often involved with or inspire dubious weapons. Nuclear technology has been anything but one of separate military and peaceful endeavours. Both share the same sciences and laboratories, bureaucrats, technologies, mines, and industries. Less widely appreciated is the role of 'peaceful' industries in chemical and biological warfare.

The gas attacks against soldiers in the First World War are often taken as the start of modern chemical warfare. Highly toxic phosgene gas, or carbonyl chloride, was used first. However, before that time phosgene had, and still has, important industrial uses in making dyestuffs, polyurethane and polycarbonate resins, and a variety of organic chemicals. After the horrors and hatred surrounding its wartime use, how strange seems the accidental release of phosgene from a ruptured storage tank in a chemical plant in Hamburg, in 1928. Phosgene was an ingredient stored and used at Bhopal to make the even more lethal methyl isocyanate (MIC), which leaked to cause the catastrophe there in December 1984. No one has satisfactorily explained the enormous stockpile of these lethal compounds near a city – compounds whose storage and, in some cases manufacture, is banned in many industrial countries. If a substance is a banned *weapon*, should it be used and stored in the vicinity of any, let alone large, populations?

The infamous 'Zyklon B' used in the gas chambers at Auschwitz and other death camps was a powerful insecticide, developed in 1923 to fumigate seriously contaminated buildings. It is another cyanide compound like phosgene and MIC, a form of hydrogen cyanide, better known as 'prussic acid'. A defendant at the Nürnberg Tribunal claimed that he hit on the idea of using it for killing prisoners during a fumigation exercise. He had about 600 Russian prisoners of war and 250 infirmary patients locked in a basement and used some of the leftover 'disinfectant' to gas them (Reitlinger 1961, 154). The same man later had the job of picking up the 'Zyklon B' canisters, which were used to kill hundreds of thousands of Jews and others at Auschwitz, at a nearby factory (see Chapter 12). The Auschwitz-Birkenau death camp, place of the most concentrated, industrialised killing of civilians in war, has also been called 'a monument to science and technology' (Muller-Hill, quoted in Lifton and Markusen 1988, 103).

Such observations are pertinent to weapons not generally considered criminal. The history and development of nitrogen fertilisers is hardly separable from that of the various nitrogen-based high explosives like TNT (trinitro-toluene). Most vehicles for land, sea and air travel have closely related military models. Civil telecommunications, optical, electronic and other devices are variants of, were often first developed for or are readily converted to items for military use. The same sciences, laboratories, industries and engineering skills serve both purposes. It is very difficult to say where weapons of war potential end and peaceful, civilian production begins, if they are indeed separable. However, it is possible to identify where their development, storage, mishandling and deliberate use are major and unwarranted hazards for civil life.

Forms of damage and processes of harm

The hazards of violence, as of natural and technological forces, must be seen in terms of damage, not merely the agents that may cause them. It is an essential step to understanding the particular dangers for civil life and habitats. The forms and distribution of harm are indicators of their vulnerability to disaster.

Virtually every form of hurt and loss may occur in war (Table 5.8). Armed force is uniquely dangerous to human bodies, communities and habitats due to its focus, as well as to the scale of damage. A problem in assessing violent destruction is that it is generally horrible, and the settings are chaotic. This is apart from the tendency of wartime reportage to be censored so as to tell only 'good news', and secrecy to be used to hide ruthless measures. Nevertheless, warfare is always intended to kill and maim, to destroy and lay waste. It may be an excuse or cover for other forms of social violence such as arbitrary arrests, deportations and torture. There are increased risks from some or all other hazards. War measures take resources from or override public safety issues, diverting wealth and effort to support the fighting.

Deprivation and 'food as a weapon'

The most common and widespread consequence of conflict for civil life is to be deprived of basic supplies. Shortages are generally the chronic and 'routine' problems of wartime societies and the so-called home front. They may become life-threatening disasters for some, if not all, civilians. For them the important concerns are everyday life support, medical and social services, and resources and materials for basic consumer goods. Privations that derive from the destruction of homes, supporting infrastructures, the habitat and cultural arte-facts are considered separately.

Almost invariably the first deprivations for civil populations involve food. Food shortages may be simply an indirect consequence of the disruption and devastation of war. Everything in the embattled state tends to reduce total food availability and put pressure on what remains. Soldiers gone to war reduce the numbers in, perhaps the skills for, food producing, food distribution and pre-

Table 5.8 Forms of damage by violence: the processes of harm to which civilians, civil life and the habitat are subject and against which, without direct protection in law, by the state and its forces, they are largely defenceless

Processes of harm	*Forms*	*Examples*
Against persons and populations:		
1. Deprivation	scorched earth	Eastern Front, Europe, 1940
	blockage	Iraq, 1990s
	siege warfare	Leningrad, 1941–1944
2. Death and injury	bullet wounds, burns, amputation, 'shock'	(all wars)
3. Demographic violence	massacres of children	Bosnia, 1990s
	'rape as a weapon'	German SW Africa, 1904
	forced sterilisation	Nazi Europe
	'ethnic cleansing'	(former) Yugoslavia, 1990s
Geographic violence:		
4. Territorial violence	enforced displacement	Somalia, 1980s
	'concentration'	'Generalgouvernement', Poland, 1941–1945
	expulsion/exile	East Prussia, 1945–1947
5. War on cultures	'war on symbols'	'Baedeker' Raids, UK, 1942
	'ethnic conflict'	Chiapas, 1993–
	subjugation and absorption	Tibet, present day
	enforced religious/political conversion	Spanish conquest of Americas
6. Place annihilation	war on settlements	Lidice, 1943
	laying waste	Carthage, 146 BC
	counter city bombing	Tokyo, March 1945
Wars of extermination:		
7. Ecocide	biocidal weapons	Gruinard Island, Scotland, 1941
	environmental warfare	Vietnam, 1970s
8. Genocide	massacres, extermination	Armenia, 1917
		European Jews, 1942–1945
		Rwanda, 1994
9. Omnicide	weapons of mass devastation	Hiroshima, Nagasaki
	wars of annihilation	'nuclear winter' (potential)

paration industries, plus wholesale and retail businesses. International trade is disrupted, and so may be essential supplies of fuel, fertilisers, seeds and pesticides to farmers, along with fuel and other supplies for food industries. Food stocks for civil populations may be destroyed incidentally in warfare and are, in any case, diverted to armies.

After the Second World War, thousands of German and Japanese civilians from dozens of cities were interviewed in an Allied survey to determine the impact of bombing on their state of mind or 'morale'. Even in some of the most heavily bombed cities, the first thing most of them mentioned was food shortage. Many recalled poor or insufficient food as a wartime problem. Being hungry was often their main recollection. This was especially true of women, over half of those interviewed. For them it was not merely the quantity of food, but the poor quality, the inordinate amount of time they had to spend to get and prepare it, and the frustration of not being able to put a satisfying meal before their families. In most societies, it is women who remain responsible for the domestic economy and whose work and diet tends to suffer most, because they deprive or overwork themselves to feed dependants. They also tend to be hardest hit by the decline, high prices or disappearance from the market place of clothing, soap, disinfectant and other household supplies, and above all of medicines when children or elderly family members become ill. In cities, of course, problems of food are not independent of fuel supplies for cooking, or indeed of transport to shopping areas and food stores, all of which tend to be cut back, rationed or disappear in wartime. This is tellingly described in Masuji Ibuse's novel of Hiroshima *Black Rain* (1982, 63), in a passage attributed to the wife of the main character. She describes the long drawn-out trials and privations suffered by the residents of Hiroshima and especially a desperate food situation almost throughout the war. Her words not only contrast the banalities of everyday survival with the horrors of the A-bomb but also remind us of the continual miseries of ordinary people in a long war, usually hidden or ignored because they do not involve soldiers, the stereotypical heroes or famous people.

However, in this case the plight of Japanese families related to the policies of their own government and the war situation it had got them into. Denial of life support is also commonly used as a weapon against the enemy's people, or even one's own in civil wars. A study of the problem in recent African conflicts, and in relation to the huge refugee and famine crises there, concluded that:

> . . . deliberate intervention in food marketing, distribution and aid flows and attacks on [agricultural and pastoral] production have contributed to, and in some cases caused, successive famines in Ethiopia, Angola, southern Sudan, Somalia and Mozambique.
>
> Macrae and Zwi (1993, 308)

The decision to deprive an enemy of food, any use of food (denial) as a weapon, is bound to hit the civil population first and worst. Leaders will seek to prevent blockades affecting their ability to wage war, otherwise it is a victory for the enemy, but this means giving priority for available resources to the fighting forces. Civil populations, especially the least influential, will suffer deprivation before armies. Prisoners of war will suffer before fighting troops. Forces of occupation will satisfy their own needs before the people of the occupied territory. Powerful groups and army units may add to the problem by retaining or seizing food intended for their own citizens, including humanitarian food aid. Ruthless groups may withhold or seize food from dependent populations.

Starvation has always been a major threat for civilians in siege warfare. This is not confined to pre-modern times. It led to terrible privations for children and the elderly in the 872-day siege of the Russian city of Leningrad (St Petersburg) from 1941 to 1944. Almost one million people died, the majority of them civilians suffering from starvation, scurvy, and illnesses brought on by them. Fifteen years later, the city had still not recovered its prewar population.

'Deprivation' also affects other ingredients of material life. Industries, construction and businesses serving the civil population will be cut back or closed before war industries, often as the latter are expanding. There is some logic to this process in 'great struggles'. Civilians often willingly accept privations to assist the war effort, but it increases the chances that a disaster will occur for civil life – even in victorious states.

Violence to persons: death, injury, traumatic experience

Armed violence, especially acts of war, are intended, as Elaine Scarry points out, 'to alter (to burn, to blast, to shell, to cut) human tissue' (1985, 62). This is so often missing from, if not deliberately avoided in, 'serious' as well as popular accounts. Thus, Gil Elliot (1972, 2), writing of the world wars, found a 'lack of historical focus on those who get themselves killed.' He suggested it involves a larger pattern of removing human beings from the study of conflict, or marginalising the majority of people in the record of human history. Surely, however, this is also to ignore the greatest hazard in the disasters of war. If we ignore the pain and maiming behind the statistics, the children who lose their parents, a limb or their lives – at least, their 'childhood' – then we have missed the real content of these disasters, and reasons to seek an end to war.

Of course, all the forms of violence include or may lead to death, injury and terror, but we also need to consider directly what can happen to the bodies and minds of non-combatants who are its victims. Casualties among civilians involve distinctive profiles of risk. If the wounds and forms of death are like those of soldiers' battles, there is a profound difference in whom they affect and where they happen. A great gulf separates bomb casualties in a populous city, or villagers massacred by 'death squads', from a Battle of Waterloo, Verdun or Iwo Jima. In the former, women and children, the elderly and infirm, are involved and generally the main casualties. The social geography and distinctive vulnerabilities of civil life, its different members and contexts, come into play (see Chapter 6).

In a study of the impact of war on Lebanese children, Mona Macksoud (1992) identified nine main classes of trauma and 'traumatic events' (Table 5.9). She concluded that 96 per cent of children in Greater Beirut had suffered at least one traumatic experience, and an 'average Lebanese child' had suffered five or six. From a sample of about 2000 children, she found over 90 per cent had undergone shelling and/or been present where fighting took place. Almost 70 per cent had been driven out of their places of residence. Over half had suffered extreme privation and had witnessed acts of violence. A more detailed breakdown showed more than half suffering bombardment of their homes and lack of basic foodstuffs. She also concluded that 'children from lower socio-economic levels experienced a greater number of traumatic events' (p.10). Boys, more often than girls, were victims of direct violence and war injuries. Older children had generally experienced a greater number of events.

Table 5.9 Traumatic experiences of children in a war zone: from a survey of children 3–16 years old living in Greater Beirut, late 1980s (after Macksoud, 1992)

Percentage distribution of sample by types of traumatic events experienced		
Type of traumatic event	%	No. surveyed*
Exposure to shelling	82.9	2164
Exposure to combat	62.6	2145
Forced residence change	60.5	2185
Deprivation of basic foods	54.5	2189
Home bombarded	53.3	2166
Exposure to bomb explosion	46.0	2158
Witnessed panic reaction in immediate family	39.5	2164
School bombarded	29.7	2048
Forced school change	27.7	2181
Death in extended family and others	21.9	2166
Witnessed injury in extended family and others	14.9	2155
Witnessed killing in extended family and others	10.6	2169
Kidnapping in extended family and others	9.9	2161
Forced immigration	9.7	2177
Separation from father (at least six months)	9.5	2074
Witnessed panic reaction in extended family	8.9	2158
Witnessed injury in immediate family	6.7	2156
Witnessed intimidation of extended family and others	5.7	2166
Suffered serious physical injuries	5.6	2179
Witnessed intimidation of immediate family	5.4	2166
Death in immediate family	4.8	2167
Kidnapping in immediate family	4.2	2161
Victim of intimidation/torture	3.1	2181
Separation from mother (at least six months)	3.0	2078
Held in detention	1.4	2166
Witnessed killing in immediate family	1.0	2171
Developed physical handicap	0.8	2177
Carried arms	0.2	2181

* Sample size varied due to some ambiguous or missing responses for each type of traumatic event.

'Demographic violence'

Populations or particular peoples may be deliberately targeted on racial, religious, political or ethnic grounds. The attack may be driven by an overt 'social Darwinist' policy, in which the survival and flourishing of one people or group is supposed to depend upon suppressing, weakening or eliminating others. Boys may be singled out for destruction or removal in order to prevent them becoming future soldiers. Rape of women and child massacres may be encouraged, as in 'ethnic cleansing' in former Yugoslavia in the 1990s. Another form of such criminal violence has been systematic euthanasia or 'medical killing' of the elderly and infirm. People have been defined as 'undesirable' or 'unfit to live' on the basis of habits, sexual preferences, appearance, history, or religious and political beliefs – something more readily practised and hidden in times of crisis and war. In National Socialist Germany and the territories it occupied in the Second World War, all these were used as grounds for systematic violence. Unfortunately, the welcome defeat of Nazi power did not put an end to such practice in other parts of the world. Often, as there, they have led to genocide (see below).

Perhaps the greatest single 'man-made' disaster in our century was the Ukraine famine of 1932–33. In it, more than seven million people died of starvation and related diseases, three million of them children. It is generally regarded as a 'terror famine', due not to insufficient food but the ruthless demands of the Soviet state, and some intention to weaken and subjugate the Ukraine by Moscow. According to Robert Conquest (1986), we must add to this 'conservative estimate' at least a further seven million deaths of rural people between 1929 and 1933. These were caused by state violence in the process of collectivising Soviet agriculture.

Geographic violence

Forms of uniquely 'geographic' violence involve the targeting of places and living space, and of people on the basis of where they reside. Deliberately or by default, this tends to happen in all war zones but especially where there is civil war, ethnic conflict and misrule. These have served to generate huge refugee crises, especially in Africa, and South and Southeast Asia, in recent years (Figs 5.4, 5.5). Violent geographical change is brought about especially by policies of enforced uprooting, expulsion or relocating of resident populations, and the elimination of habitats, cultural landscapes and settlements. These change the map of human geography, and in ways that would not occur without violence (Table 5.10). The use of weapons of mass destruction against human settlements works towards and may result in 'place annihilation' – the irrevocable destruction of an historic and living association of people and place. The same may occur if all the resident community is removed or exterminated, because the survival of the physical environment alone does not make it a place. Examples of both forms of place annihilation are described in Chapters 11 and 12.

133

Fig. 5.4 Regions of conflict and displacement. Main areas of armed conflict, refugee movements and places of refuge in Africa during the late 1980s and early 1990s (after Sivard (ed.) 1986, 1989; Rogers, 1990; Francois (ed.) 1992, 1993)

'War on culture' is a strategy directed at civil life, in which buildings and institutions, and sacred or highly valued places and landscapes, are attacked because of their historic and symbolic significance. It is sometimes referred to as 'war on symbols', reflecting the central role of symbolic communication and surroundings in the ordering and sense of belonging to a community. The attacks may be simply in hope of depressing the morale of enemy people and leaders, or 'teaching them a lesson' by harming things that they cherish. For example, the German 'Baedeker' raids on British towns like York and Bath in 1942 – named after the Baedeker tour guides – were intended to 'punish' Britons by harming places of great historic value although of negligible military importance.

War on culture may be part of a systematic programme to eradicate a people's heritage. For subjugated or captive peoples it can involve outlawing

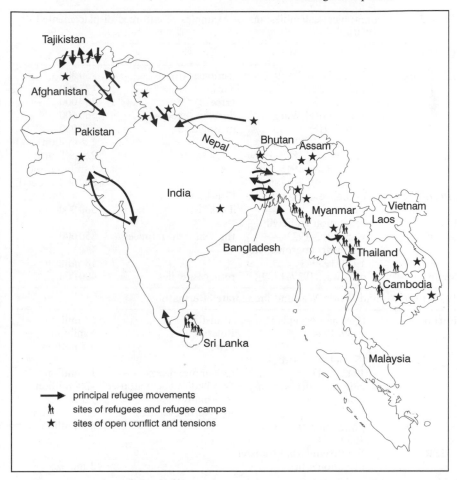

Fig. 5.5 Regions of conflict and displacement. Main areas of armed conflict, refugee movements and places of refuge in South Asia during the late 1980s and early 1990s (after Sivard (ed.) 1986, 1989; Rogers, 1990; Francois (ed.) 1992, 1993)

their language, religion and customs and an enforced assimilation or confinement. The native peoples of the Americas, from northern Canada to Tierra del Fuego, have been massively harmed by such policies, from the Spanish Conquest to the present. Again, this is often a stepping stone to genocide. A war on places and buildings of historic significance has been an integral part of the 'ethnic cleansing' in former Yugoslavia.

'Laying waste' or the 'ecology of devastation' is as much a geographic as an ecological assault, endangering and destroying living space and life worlds. Dubious weapons are an integral part of ecological devastation, a distinctive development of industrialised warfare, especially in Vietnam. There, modern technological instruments such as biocides and defoliants, heavy earth-moving

Social hazards

Table 5.10 Geographical calamities: major examples of enforced displacement, 1900 to 1990

To 1939

SE Europe	Bulgaria, 1878–1912	(refugees in)	250 000
	Balkan wars,	Turks	177 000
	1912–1913	Greeks	200 000
	First World War,	Serbian 'exodus'	500 000
	1914–1918	in Bulgaria	300 000
	postwar, 1922–1939		2.2 million
Turkey	Armenians, 1890–1919		2.5 million
	Armenians, 1915		2 million (incl. 800 000+ dead)
Greco-Turkish	exchange of minorities,	Greeks	1.3 million
	1919	Turks	500 000
USSR	revolution and civil war, 1917–1922	the 'Nansen' refugees	850 000
	civil war, pogroms, etc., 1928–1939		18 million
Spain	civil war, 1937–1939	refugees, exiles	400 000

1939–1947, Second World War and immediate aftermath

Europe	(German-occupied),	'transfers,'	18 million
	1939–1944	Poles	6 million
		Jews	6.5 million
	(Soviet-occupied),		
	1945–1947	German expelees	13 million
	cities and the bombing	Bombed out – 'dehoused'	12.5 million
	war:	(Germany)	7.5 million
		long-term evacuees	20+ million
SE Europe	population transfers, 1939–1947		8.9 million
USSR	Punitive internal transfer of nationalities		2 million
	(*Total European transfers, refugees 1922–1947, approx.*)		78.5 million
Pacific Theatre	China, mainland peasants		30 million
	Bombing of cities:	bombed out	15 million
		(Japan)	8.3 million
		evacuees	30+ million
Partition of India	1947–1950		14 million

Table 5.10   contd.

1950–present		
1950s datum	European refugees registered, 1951	404 567
Afghanistan	external refugees, 1980	1.519 million*
	war refugees in Pakistan and Iran, 1985	5+ million
	1986, external refugees	5.751 million*
	internal refugees	2 million
	1990, external refugees	6.027 million*
	internal refugees	2 million
	external refugees, 1993	4.286 million*
Angola	internally displaced refugees, 1986	3.5 million
Bosnia-Herzegovina	external refugees, 1993	940 000
Chad	external refugees, 1980	396 000
Croatia	external refugees, 1993	350 000
Ethiopia	1980, external refugees	1.967 million
	internal refugees	1.8 million
	1986, external refugees	1.122 million*
	internal refugees	0.7–1.5 million*
	1990, external refugees	1.066 million*
	internal refugees	1 million
	external refugees, 1993	746 700
Iran	internally displaced refugees, 1980	1 million
	internally displaced refugees, 1986	1 million
Iraq	external refugees, 1986	400 000*
	external refugees, 1990	529 700
Kampuchea	war, bombing, 1970–1972	2 million
	forcible relocations, 1972–1974	5 million
	refugees, 1975–1981	850 000
	permanent exile, 1975–1981	116 000
Lebanon	internally displaced refugees, 1986	400 000–800 000
	internally displaced refugees, 1990	800 000
Liberia	external refugees, 1990	729 000
	external refugees, 1993	599 200*
Mozambique	1986, external refugees	917 000*
	internal refugees	2 million
	1990, external refugees	1.428 million
	internal refugees	2 million
	external refugees, 1993	1.725 million*
Palestine	external refugees, 1980	1.844 million
	external refugees, 1986	2.218 million
	external refugees, 1990	2.428 million
	external refugees, 1993	2.658 million
The Philippines	internally displaced refugees, 1990	900 000
Somalia	external refugees, 1990	454 600*
	external refugees, 1993	864 800*
South Africa	internally displaced refugees, 1986	3.57 million*
	internally displaced refugees, 1990	4.1 million
Sudan	internally displaced refugees, 1986	1.5 million
	internally displaced refugees, 1990	4.5 million

137

Table 5.10   contd.

Vietnam	war, 1965–1975: S. Vietnam pacification, war uprooted	17 million
	postwar, internal relocations, 1975–1985	2.1 million
	postwar, refugees, 1975–1985	709 570
Zimbabwe	internally displaced refugees, 1980	660 000
	internally displaced refugees, 1990	750 000

*Significant variance among sources.

equipment, and attempts at artificial weather modification were used. The purpose, apparently, was to destroy the environment in which the enemy was thought to move, agricultural areas they were thought to control, or communities they might obtain supplies from. There was extensive burning of forest with incendiaries, injury to vegetation by shrapnel and artillery fire, and 'craterisation' of the countryside by high-explosive bombs. If the commitment to these ecocidal methods and their concentration in one region were unprecedented, the principle is much more widespread and continues to be applied.

'War of extermination' and ultimate risks

Taken to their extremes, these forms of violence may become 'total' and lead to great catastrophes. They may be used to exterminate all of a given people (genocide) or all of a given habitat ('ecocide') and with that, the culture and ways of life rooted in them. Such things have actually happened, if comparatively rarely over human history, and less often than intended. Modern notions of enlightened and civilised behaviour have sought to class such actions as 'barbaric', ancient warrior practices that should disappear but have not. The notion that everything together might be irrevocably destroyed, or 'omnicide', seems a peculiarly twentieth century notion of hazard. It is seen as the ultimate threat from weapons of mass devastation.

Geographers have been involved in efforts to predict the likely consequences of a major nuclear exchange. There would be enormous threats from the explosions, radiation, fires and other direct devastation, but many believe the greatest threat of 'omnicide' – annihilation for civilisation, possibly all human and most other life forms – would come from the immense quantities of smoke and other combustion products lofted into the atmosphere. These could cause a 'nuclear winter', shutting out sunshine from much of the Earth over one or several years. Destruction of the ozone layer might follow, bringing irradiation of the Earth's surface by damaging ultraviolet rays. The only parallels for such devastation appear to be mass extinctions in distant geological eras, perhaps caused by large meteors or comets striking the Earth. Some biological and toxin weapons have the potential to cause pandemics equal to or worse than the Black Death, which was the greatest disaster, affecting humanity world-wide, for which we have historical evidence.

Hopefully, these are speculations about future omnicidal catastrophes that we may avoid. However, they will hardly be more painful for those involved if more widespread than genocidal, ecocidal and geographic catastrophes that have already occurred in this century. We will examine in Chapter 12 what actual events of this sort have meant.

Concluding remarks: yardsticks of calamity?

The social hazards that concern us are not only physical dangers. They are matters of ethical concern, affecting the possibility of good order and civil life, if not of species survival. In fact, human rights and 'laws of civilised warfare', the effort to outlaw dubious weapons and follow 'just war' principles actually serve to define and circumscribe the worst hazards of armed violence. If that is the special focus of humanitarian, peace and disarmament efforts, it also has a material and conceptual significance for our field, serving to identify as well as to condemn the most severe disasters that we encounter. Meanwhile, this is to situate ethical and humanitarian values at the heart of the definition and treatment of these risks.

Accident and disarray, hatred and despair, are indeed bound to occur when armed actions are undertaken. They are a major reason for demanding the highest levels of discipline in armed forces, and seeking other methods to settle disputes or to compete for wealth and power. Yet such unplanned aspects of violence are a small part of the problems we confront. More clearly than any others, the concerns here are with planned and deliberately undertaken cala-mities, tragedies that 'do not simply happen, nor are they sent; they proceed mainly from actions, and those the actions of men' (Bradley, 1906, 11).

Sources

Organisations that monitor and publish updates on violence include Amnesty International; the Stockholm International Peace Research Insti-tute; Project Ploughshares *Monitor* (Waterloo, Ontario); World Priorities, Washington DC (Sivard, R. L. (ed.)); the Minority Rights Group Reports; *Cultural Survival Quarterly* (Cambridge, Mass.); Survival International reports and newsletter (London); and *Bulletin of the Atomic Scientists*. The United Nations Department of Humanitarian Affairs publishes *DHA News*, a bi-monthly publication that considers violence and its victims as well as natural disasters. The International Federation of Red Cross and Red Crescent Societies (IFRCRCS) now produces an annual 'World

Disasters Report' dealing with all types of disasters that require humanitarian relief. It is published by Martinus Niihoff, Dordrecht.

Disaster studies that include armed violence

Westing, A. H. (ed.) (1990) *Environmental hazards of war: releasing dangerous forces in an industrialized world*. Sage Publications, London.

Ahlström, C. (1991) *Casualties of conflict: report for the world campaign for the protection of victims of war*. Department of Peace and Conflict Research, Uppsala University, Sweden.

Amnesty International (1975) *Report on torture*. New York.

Elliot, G. (1972) *The twentieth century book of the dead*. Penguin Books, Harmondsworth.

Harrisson, T. (1976) *Living through the Blitz*. Penguin Books, Harmondsworth.

Kirby, A. (ed.) (1992) *The Pentagon and the cities*. Sage, Newbury Park, Calif.

Nietschmann, B. (1986) Militarization and indigenous peoples: introduction to the Third World War. *Cultural Survival Quarterly*, **11**, 1–16.

Zwi, A. and A. Ugalde (1991) Political violence in the Third World: a public health issue. *Health Policy and Planning*, **6** (3), 203–17.

CHAPTER 6

Vulnerability perspectives: the human ecology of endangerment

Disaster is the actualisation of social vulnerability . . . we interpret the notion of social vulnerability as an independent factor (predictor) of risk.

Pelanda (1981, 69)

Vulnerability is the state of powerlessness in the face of a known or unknown hazard.

O'Riordan (1990, 295)

The challenge is to create ways of analysing the vulnerability implicit in everyday life.

Wisner (1983, 16)

The idea and scope of vulnerability

Over the past decade or so, the idea of vulnerability has come to identify a distinctive view of risk and disaster. While I see vulnerability studies as a necessary complement, rather than an alternative, to a hazards perspective, they involve some profound shifts in the balance and range of our concerns. Their emphasis is upon how communities are exposed to dangers or become unsafe, rather than the character of natural or technological agents. When people are in danger, the conditions that influence their protection and coping capacities are the main concern, rather than the severity of a damaging agent. When disasters do occur, the focus is especially upon who is affected and their ability to withstand, mitigate and recover from damage.

In this perspective, risk is seen to depend primarily upon on-going societal conditions. Society, rather than nature, decides who is more likely to be exposed to dangerous geophysical agents, and to have weakened or no defences against them. Failures to exclude dangerous technologies and practices, or to

141

provide protection from them, are interpreted in terms of conditions and decisions embedded in the social order. Meanwhile, natural and technological hazards appear primarily as agents that reveal pre-existing weaknesses, or lack of protection, through a disaster.

Vulnerability can be thought of as the other face of safety or security. Increased vulnerability means decreased safety. In that sense, the idea seems straightforward enough. It is surely commonplace to find that some persons are less robust than others, and more at risk from a variety of dangers as a result. Insurable risks, and the bases of variable insurance premiums, are a precedent for the vulnerability approach. Some types of individual or group are found to have proportionately more motor vehicle accidents, or to be more likely to suffer premature death or debilitating injury, than others. Some buildings or activities in them are more often subject to fire damage than others. Some cargoes are more at risk on the high seas, or give rise to exceptional losses in maritime accidents. Premiums reflect, among other things, this differential vulnerability of people, structures and activities.

However, work within the hazards perspective has tended to ignore or to treat such notions as irrelevant or incidental in disasters. To the extent that damage does not reflect the nature and impact of the damaging agent, losses have been viewed as 'indiscriminate' and survival as a matter of chance. Even efforts to define for insurance purposes the more extreme natural and technological hazards have favoured geophysical and technical measures. The likelihood and magnitude of earthquakes, or of a meltdown at a nuclear power plant, are the sorts of measures stressed, rather than who and what will be endangered by them. Proponents of a vulnerability perspective challenge this view. Hence, vulnerability has also become a rallying point for alternative visions to the dominant hazards paradigm. Large losses for the insurance industry in recent earthquake, hurricane and oil spill disasters are also behind a growing interest in questions of exposure and vulnerability.

Particular forms of vulnerability can be related to, or are brought out by, particular hazards, yet neither these nor any forms of vulnerability stem directly from the natural or technological hazards discussed earlier. This too involves a departure from the hazards perspective. On the one hand, the forms of vulnerability that concern us are created largely by and always modified by human actions, whether deliberately or by default. Few are incapable of being limited and offset, if not prevented. On the other hand, many forms of vulnerability arise more or less independently of where and whether flood, storm, epidemic disease or explosion may occur. In many communities, for instance, it is not only, often not at all, fire risk or snow loads that decide the materials used for construction; rather, cost and availability, the tools and practices used in construction, traditional or innovative uses of, and preferences in, buildings tend to be more important.

Vulnerability itself is a property or a circumstance of persons, activities and sites. The forms and degrees of vulnerability to storm or famine are often related to gender and wealth, to influence or lack of it. In wartime, or in industrial or transportation 'accidents', occupation and life-style may be critical to vulnerability. Age group may be decisive in how people are affected by economic trends or urban congestion. These in turn arise from human conditions having little or nothing to do with the causes and patterns of flood and drought, or types of industrial process, yet they may be critical to risks from these dangers. In a modern or modernising economy, who lives, builds and works where reflects the larger spaces of economic and political organisation. Local societal conditions and protections are more or less adapted to and dependent upon them.

These observations begin to show how vulnerability has required a fundamental departure from the established hazards view. Here is an interpretation of risk that recognises the important extent to which disaster depends upon the social order, its everyday relations to the environment and the larger historical circumstances that shape or frustrate these matters. This approach underscores the importance of human defences and resourcefulness in the face of dangers. In disasters, the emphasis is upon conditions and means that influence how well people cope and how rapid and complete, or otherwise, is their recovery.

There is no form of vulnerability that some or all societies do not deliberately seek to reduce and offset, or against which they do not protect some, at least, of their members. There may be some 'absolute' vulnerabilities. Civil societies, perhaps all of the biosphere, seem totally vulnerable in the face of a nuclear war, a comet impact or a supernova in our part of the galaxy, but in almost every other risk, there is something humans can do, and do do, to mitigate or reallocate vulnerability.

That said, as the term itself emphasises, we mainly examine the ways in which people are actually at risk: their frailties, lack of protection and limited survival capacities. Since these are largely a reflection of social conditions, however, they have to do with *the making of vulnerability by human activity*. If this in turn is largely a product of on-going and everyday life, we must consider the processes and conditions of *endangerment* acting in and upon human groups. This has profound implications for, but is distinct in its focus from, the ways in which societies reduce and adjust to dangers – the subject of the next chapter.

These arguments are of special interest here. Vulnerability assessment is, essentially, about *the human ecology of endangerment*. Moreover, we can show how vulnerability is embedded in the social geography of settlements and land uses, and the space or distribution of influence in communities and political organisation. Finally, vulnerability is surely decisive for the growing concern over 'sustainable' human communities and environment relations.

The anatomy of insecurity

The ways in which people may be vulnerable, and become more or less so, involve several broad concerns (Table 6.1).

143

Table 6.1 Forms and conditions of vulnerability

Forms

exposure to hazards: through occupation, life-style, location.
weaknesses and susceptibility: genetic predisposition, disability, poor design of buildings, insecure practices.
disadvantage, or 'structural weakness': poverty, dependency, lack of capabilities and rights.
defencelessness: lack of protection or aids to counter or avoid danger.
lack of response capabilities: limited and impaired resilience, actual or recognised options, creative responses.
powerlessness: inability to influence sources of danger or protection.
enforced vulnerability (see Table 6.2).

Locus of vulnerability

individual: risk-takers, personality and choice, luck, (lack of) experience, training.
domestic: family situation, inheritance.
gender: patriarchy.
social space: communal, sectoral, urban, rural, class.
economic: (lack of) skills, employment security, resources owned or entitled, subsistence, commercial.
ethnicity: minorities, indigenous peoples.
cultural: linguistic, religious, ethical (e.g. pacifist).

and all of the above:
geographic: local, regional, national, rural versus urban, natural regions, centre–periphery, 'North–South' (see also Table 6.3.)

Vulnerability 'syndromes' (i.e. assumed determinants)

political economy: stages of development, underdevelopment and marginalisation; political form (socialist, fascist, etc.); class and power.
'worlds': First, Second, Third, Fourth 'world system'.
impersonal forces: 'population', environmental, economic, racial (see concluding remarks).

Exposure to hazardous agents

Exposure is an important criterion for assessing risk and its geography. For example, otherwise safe structures, if located close to the shoreline, may be exposed to destruction by tsunamis or storm surges. Expansion of a certain type of rain-fed agriculture into regions of smaller or less reliable rainfall makes it more vulnerable to drought-related crop failure. In fact, this is how most physical agents, such as a flood or a drought condition, become actual hazards. With hazardous materials or high-energy processes, exposure is the critical factor. Every human is vulnerable to death or severe injury if exposed to such highly toxic chemicals as methyl isocyanate. However, when this material leaked to form a deadly cloud at Bhopal, India, in 1984, the huge numbers killed and permanently injured came largely from poor families allowed to settle around the chemical plant.

Exposure is described by the geography of social and material life, and the contexts in which people concentrate in particular places. They may be factories or homes, fields or downtown offices, along shorelines or around water holes. However, more fundamental issues are involved than mere location or what can be mapped. Exposure in any given locality is also a matter of activities and responsibilities. Social values and the roles that people are expected to fulfil are equally relevant. In many societies, for example, women have a primary responsibility for domestic space and spend a large part of their lives in or near the home. They often have much, if not sole, responsibility for dependent family members, whether children, the elderly or disabled. They are at greater risk from dangers affecting homes and domestic life. The duties and attachments or habits of caring, no less than mere functions, can expose them to a double jeopardy. If there is danger from water-borne diseases or pollution in societies where women do most of the water fetching or washing of clothes, they are more at risk than others. Again, such vulnerabilities as they may have personally and as women are compounded in a crisis by the demands of caring for and saving dependants. Although we have no exact numbers, a great many women have died in bombing raids and other acts of war while trying to rescue, or to escape with the added burden of, small children and elderly family members. These are extreme examples of a more general principle that exposure, and the reasons for being exposed, reflect people's participation in the activities and concerns of their society.

'Inherent' weakness or susceptibility

Some persons or activities are unusually vulnerable due to biological, physical or design characteristics. There is, of course, an actual or implied comparison with others less vulnerable, even some implied 'norm' or desirable degree of strength. Small children and the elderly, and people who are ill, disabled, undernourished or recovering from some other misfortune may be inherently less able to withstand certain assaults. Children tend to be more sensitive to nuclear radiation, as was tragically shown in the victims of Chernobyl. Since they have smaller organs, in particular thyroid glands, a given exposure to radionucleides is proportionately more harmful to them than to an adult.

Certain forests and grasslands can be unusually susceptible to fire, or can become so in seasonally hot, dry weather. Certain crops are susceptible to damage by particular blights, pests or weeds. Settlements where buildings of wood and other plant fibres are common are specially vulnerable to the development of mass fires. In the Second World War, air forces attacked the crowded central districts of Japanese and Chinese cities with incendiary bombs because the buildings were exceptionally inflammable. Raids were carried out in dry and windy months because historical evidence showed that mass fires were more likely to develop then, and would be harder to contain (see Chapter 11).

Before the first Europeans arrived in the Americas and the Pacific islands, the peoples living there had little or no previous exposure to diseases widespread in Eurasia. These included measles, scarlet fever and smallpox. Some are severe or fatal for adults but rarely so in childhood, when the disease is usually caught where it is endemic. Among the indigenous New World populations, lack of immunity and widespread exposure of adult populations led to the deaths of many millions. Whole cultures never recovered from the scale and scope of loss in these sudden epidemics. Their ecological and spiritual balance was radically upset. This high level of susceptibility to European diseases greatly magnified their emerging vulnerability to the weapons and colonial policies of the invaders.

Proneness to particular diseases, especially endemic ones, does raise more complex issues of vulnerability and its interdependence with such factors as past exposure, acquired immunity and modified behaviour. Nevertheless, the point remains that it is not just a function of the presence and virulence of the disease organism. We must also consider behaviour that can bring a person into contact with the latter, a health profile that can apply whether or not it is present, and social conditions that affect such matters.

It may be added here that the forms of vulnerability that seem most to suit a hazards perspective are the two already discussed, exposure to dangerous agents and 'inherent' or structural weakness. In fact, 'vulnerability' is still most often taken to mean simply exposure to natural and technological hazards – 'being in the wrong place at the wrong time'. This is how it appears in what are called 'high-hazard locations' and 'harsh lands'. These define a geography of vulnerability in terms of human occupancy of flood plains, drought-prone areas and seismic zones, for example, or of proximity to dangerous industrial plant or to the targets of war. Clearly they are important concerns, yet this is to miss what is distinctive about a vulnerability perspective, and the more important or decisive sources of vulnerability, to which we now turn.

Disadvantage or 'structural' weaknesses

It is important to distinguish inherent weaknesses of individuals from social disadvantage. This is the kind of vulnerability associated with, say, landlessness or urban overcrowding, or a lack of access to training and education, better-paid employment, and a variety of resources and services. It relates to disadvantages that arise from more or less permanent social conditions. They can be called 'structural' vulnerabilities, by analogy with 'structural adjustment' or 'structural violence'. The vulnerabilities arise from the fabric of social life rather than hazardous conditions or accidental changes.

In many, if not all societies, being born into a certain type of family or group, being born female or during a period of family if not societal crisis, can make one more vulnerable to a host of dangers. Being part of a lower class or caste, or a member of certain religious, ethnic or 'visible' minorities, may place one at a severe disadvantage and close off means and opportunities for greater security. This is to discover vulnerability as *constructed* in everyday life and arising from on-going social conditions. It is, however, less a matter of whole societies or populations being vulnerable than the greater vulnerability of particular persons, occupations and groups within communities or institutions.

Vulnerability studies have exposed the dangers of stereotyping groups at risk, of lumping whole societies, if not countries and continents, together. On one side, there is Livermore's (1990, 32) point that

> the most vulnerable people may not be in the most vulnerable places – poor people can live in productive biophysical environments and be vulnerable, and wealthy people can live in fragile physical environments and live relatively well.

On the other side, disadvantage is relative not only to others in one's own society, but to the kind of society. If large differences in wealth are usually associated with large differences in vulnerability, more modest ones, and radically different ways of life, may not be. This is a particular problem when comparing different cultures, or rural and urban folk. Poverty or ethnic status are often indicators of heightened vulnerability, yet the 'poor', sometimes the poorest of them according to standard measures, are not always the most vulnerable. They may have special skills and a flexibility that allows them to avoid or pull through a disaster. They may know how to take advantage of social safety nets more readily than others, less materially lacking at the outset. Sometimes ethnic minorities are powerful or controlling groups.

Humans everywhere are capable of seeking out and seizing opportunities. Communities are complexly arranged, with many cycles of opportunity. This creates options to appeal to better-placed persons or flexibly seek out other resources. Thus people can, and usually do, offset particular disadvantages. As a result, however, *multiple, reinforcing disadvantages and absent defences* single out the truly imperilled members of society.

We can, perhaps, define the argument best by looking at the other end of the spectrum, at wealthy groups and safety in 'high-tech' contexts. Wildavsky (1988, 58), for example, concludes that 'richer is safer' because wealth gives access to the 'general resources upon which our safety mostly depends.' Studies of complex, potentially very dangerous technologies and activities that, nevertheless, are 'nearly error free' reinforce the point (LaPorte and Consolini, 1991). They include such matters as air traffic control or peacetime safety of nuclear weapons. While design, performance and high-quality personnel may be necessary, even more critical is that expense be no barrier to safety measures or the development of a 'high-reliability culture' in the responsible organisation. And particularly relevant to the present argument, long-term safety is seen to depend upon multiple 'safety loops'. Their significance is indicated by referring to 'high levels of redundancy in personnel and technical safety measures' (Sagan, 1993, 17). In other words, even in the most sophisticated institutions and technologies, safety, and especially that bearing upon rare, extreme failures, requires multiple levels of back-up to the first line of security.

For similar, but reverse, reasons, the disadvantages most relevant to endangerment are multiple and compounded – a wide spectrum of personal and social impairment. In a disaster, those most likely to be harmed are distinctly more disadvantaged, and in multiple respects. Some people seem driven into, or trapped by, a vulnerability 'syndrome'. Rather than some single weakness or category, they are victimised by a whole social context. Wisner (1993, 22) provides the following illustration:

> . . . women may not be particularly vulnerable qua women . . . but more commonly poor women (e.g. class + gender), old, poor women (age + class + gender), or old, poor, minority women (age + class + ethnicity + gender) are most vulnerable. It is highly probable that, everything else being equal, the addition of disability (blindness for instance) would create a concatenation of vulnerability factors that more or less assure that this person will be most severely affected by most hazard events and, if she survives, will find it hardest to recover.

The sense in which risk is also a *gendered space* has been taken up especially in vulnerability studies. They recognise the social construction of vulnerabilities through gender-based values and actions as a major feature of the shape and meaning of danger and disaster. Women's vulnerability is often aggravated by their 'invisibility' in male-dominated societies, and where women are excluded from public life. To be sure, where they do not or cannot visit poorly protected public and work spaces, women may be less vulnerable to certain dangers, yet to lack a voice in family or public affairs is to be permanently vulnerable to conditions and decisions that ignore one's predicament. Various studies have found that where women are normally disadvantaged and suppressed, they are likely to suffer disproportionately in crises. If girl children are treated badly most of the time compared, say, with their brothers, they are more likely to be 'sacrificed', exploited or abandoned in a disaster.

However, thinking in terms of uniform or universal stereotypes of vulnerability is again a problem. Studies of women and famines in India, for example, find their plight and their roles vary enormously from place to place, and even from family to family in the same place.

Similar observations are relevant to the vulnerability of the elderly or children. Last (1994, 193) points out how a tendency not to notice or to subordinate the needs of children in public issues can be a special problem when they are at risk:

> . . . children, even in good times, are often unheard but in distress they are apt to be simply not seen . . . [especially] children in families have an invisibility that requires unusual perceptiveness (and concern) to penetrate.

In all such cases, however, the issue is not only the disadvantages of those who may be exposed to disaster, but whether and why they lack otherwise available protection from dangers.

Defencelessness

Defencelessness introduces a set of terms having to do with protection. The victims of disaster may well be those exposed to dangers to which they are inherently and 'structurally' weak or susceptible. But why are some people exposed, although they lack defences, while others, in the same society or elsewhere, are well-protected? Most societies try to counter, if they are not primarily organised so as to reduce, inherent and even structurally derived vulnerabilities. There are wide and complex arrangements to protect frail and weak members, and the more valued 'risk-takers'. This is commonly the case for children and those who are temporarily ill or impaired. Are not such patterns of caring for valuable members or vulnerable stages in life the hallmark of 'social animals'? Modern societies have introduced farm subsidies, baby bonuses and corporate 'bail-outs' to protect valued members or favoured sectors. Prominent personalities, works of art or holy places receive special protection and privileged treatment. In these ways, potentially very vulnerable persons, institutions and activities are made much less vulnerable to storm, disease, social violence or industrial hazards than others, inherently more robust.

In a famine or war zone, dependants, whether children, the elderly or persons who are ill, may be the first to receive humanitarian assistance. Those taking care of them may be given preference in relief activities, but they may also be treated like everyone else, or even ignored. Then they are shown to be unusually vulnerable. One woman, among millions thrust about Europe in the huge evacuations and expulsions of civilians during the Second World War, made the point:

> The roads inside the camp were indescribably dirty, my children lay ill in straw in the hut. The Welfare organisation was a complete failure. *Only people who were alone and healthy*, could, with any chance of success, queue up the whole day for bread and watery soup. *I could not leave the children so long alone* and no longer had any utensils to fetch food in.
>
> (quoted in Hewitt, 1987, my italics)

Vulnerability and the disasters of war

The idea of 'defences' is identified above all with war and war risks. Unfortunately, the focus of wartime and war histories tends to be on military defence, yet the risks and harm to civil populations, as described earlier, most often relate to problems of defencelessness. While the treatment of non-combatants, non-military areas, the environment and prisoners in war is usually viewed as a moral and legal question, it is equally about vulnerability. The disasters of war, described in the last chapter, involve persons and places especially vulnerable in themselves, and assailed though they lacked any or sufficient protection.

Civilians and settlements are in extreme peril whenever exposed to armed assaults and not adequately defended by their own forces. Unarmed, they are

149

always 'weak' with respect to armed violence, the more so if they lack protection. Much the same applies to most of the built environment and livable habitats. These cannot survive the impact of high-explosive and incendiary bombs or the agents of chemical, biological and nuclear warfare. If unprotected, they will be destroyed. Against modern weapons of mass destruction, the only real protection is prevention. Civil life and natural habitats simply must not be exposed to such assaults.

However, some people and places are much more vulnerable to particular kinds of war situations than others (see Chapter 11). In relation to air power, Air Marshal Lord Trenchard argued that cities were:

> . . . the points at which the enemy is weakest . . . The great centres of manufacture, transport and communications cannot be wholly protected. The personnel . . . who man them are not armed and cannot shoot back. They are not disciplined.
>
> Webster and Frankland (1961, I: 73)

He used this as an argument for having Royal Air Force bombers attack them. To be sure, this comes from a country whose single, most famous air triumph was the Battle of Britain, a defensive fight against the enemy's war planes. At least it indicates a prevailing and justified feeling that, if actually attacked, civilians are especially vulnerable to mass bombings.

During wartime blackout, when streets and homes are kept in darkness as a precaution against night-time air raids, the elderly are found to have more accidents from tripping or falling down stairs. There may be a rise in accidental choking and suffocation deaths among babies or small children sleeping in darkened rooms. In air raids, a special 'shelter-death' of the elderly has been recognised. Heart attacks and strokes may take them, even in otherwise safe shelters. Often they have died unnoticed in the atmosphere of fear and terrifying noise of bombs, their death perhaps brought on by that.

In sum, whether during wars or in peacetime, greater vulnerability is identified with combinations of exposure, weakness, disadvantage and lack of protection. But are these, and especially the seeming inability to offset them, or gain protection, merely passive features of victims or populations?

Lack of response capacities

Every person and society is in some sense vulnerable, at times, perhaps, in all of the ways that are relevant to our concerns. Yet most have the means, or find ways, to offset or be rescued from their predicament. They cope, survive and bounce back. What about those who do not? Vulnerability studies have given as much emphasis to people's active capabilities or resilience in relation to dangers as to weakness. Direct and longer-term involvement with communities at risk has shown how survival turns upon community values and arrangements. Often there are remarkable adaptive and coping capacities in the face of

150

stress or damage. Nevertheless, the notion of vulnerability itself directs us to the limitations or loss of these capacities. Anderson and Woodrow (1989) expressly contrast 'vulnerability' with 'capability', the latter meaning people's abilities to protect themselves and those they care for, to cope with emergencies and to recover from disaster.

The problem of vulnerability, in this sense, is rooted in impaired adaptive capabilities. It directs us to people who are not just suffering from weaknesses, defencelessness and structural disadvantage. They are unable to cope and lack well-known forms of protection. In this way, vulnerability also identifies impaired skills and practices.

An analysis of hazards in Lima, Peru, emphasising vulnerability, concluded:

> The people most vulnerable to the effects of an earthquake in the city are those with limited options in terms of access to housing and employment. The inhabitants of critical areas would not choose to live there if they had any alternative, nor do they deliberately neglect the maintenance of their overcrowded and deteriorated tenements. Low income families in Lima only have freedom to choose between different kinds of disaster. Within the options available, people seek to minimise vulnerability to one kind of hazard even at the cost of increasing their vulnerability to another [such as earthquake].
>
> Maskrey (1989, 12)

In this view, the capacity to respond to risks and cope with disaster is interpreted as strategically embedded in overall relations to natural, technological and social environments. Patterns of everyday living prefigure and constrain the readiness, capacities, resources and values that will apply in a crisis. Vulnerability is indicative of severe societal impairment, most often due to externally imposed constraints. The actual, or worst-hit, victims of disaster may be severely weakened persons whose vulnerability goes unprotected or is ignored in social decisions.

This in turn raises the most fundamental departure from the prevailing hazards viewpoint. It is to see risk and disaster as originating, via vulnerability, in a lack of ability to influence the decisions and direction of a society in those matters that determine one's security. Here, the key to vulnerability is found in *powerlessness*, and relative security in its opposite. Thus powerlessness is not only another type of vulnerability. It is also seen to underlie all the forms of endangerment: an *interpretation* as much as a *form* of vulnerability.

Interpreting vulnerability

Powerlessness and vulnerability

In various theories of vulnerability, *powerlessness* is seen as crucial to, and the most likely reason for, *defencelessness*. The mild version is expressed by O'Riordan (1986, 272–3):

> Vulnerability to hazard is not always the result of foolhardiness, ignorance, or misunderstanding; it may be the result of involuntary pressures forced upon people with or without their knowledge. The manner in which individuals and various social institutions cope with both predisaster preparations and postdisaster relief should therefore suggest something about the relationship between political power and vulnerability to environmental risks.

This is discussed in terms of the relations of wealth and social status or whatever else contributes to influence, subject only to systems of rights and decencies that may restrain its use. A much stronger statement, in relation to food insecurity and famine, is Watts and Bohle's (1993) critique and extension of the entitlements argument. They find that

> . . . hunger is a sort of silent . . . violence imposed on the powerless . . . [and] delimits those groups of society collectively denied critical rights within and between . . . three domains: the domestic (patriarchal and generational politics); work (production politics) and the public–civil sphere (state politics). (p.49)

They define 'the social space of vulnerability' as arising in three intersecting areas (p.54):

- '*vulnerability as an entitlement problem*' (see below) or 'lack of command over food (endowments)';
- '*vulnerability as powerlessness*' or 'disenfranchisement';
- '*vulnerability through appropriation and exploitation*' or a 'political economy' that deprives people of 'capacities' and makes them 'crisis-prone'.

Various examples cited by them show modernisation adversely affecting the vulnerability of those who suffer in famines. They include settings in which there have been large overall increases in food production per capita, and in agricultural exports.

More generally, people at risk who are less well-provided with rights and protections, *and* lacking in adaptive capacities, are found to be *politically* vulnerable. They have limited or no power to influence national and local policies with respect to land, development and social security.

It should be said that those who are comfortable with the dominant hazards view tend to feel uncomfortable with, if not positively opposed to, such arguments. A technocratic approach commonly goes along with a vision of social reality in which 'politics' and social struggles somehow lie outside of, or have no place in, the understanding of practical applications of science. Secular, let alone religious ethics, and words like power, exploitation and political economy, are rejected as signalling unscientific or 'subjective' arguments. That being the case, the researcher must reduce vulnerability to impersonal variables. This is to ignore the fact that humans, unless severely impaired, are first and foremost active, socialised and intelligent participants. They have minds and shared interests of their own. This is most likely to shape behaviour and concerns in matters affecting the well-being, perhaps the survival, of self or those we care about. Some are happy to think of this as 'instinct', but that idea pales before the diverse social and shared, or competitive and exploitative, manifestations of organised and collective life. Modern states and cities, most obviously, submerge mere biological drives in elaborate and socially constructed arrangements and powers. These have a huge influence upon all aspects of well-being and endangerment.

Vulnerability, as a question of capacities and freedom of decision, also involves the danger of too much control or dependency, as well as deprivation or lack of protection. The tendency of hazards work has been to dwell upon specific solutions in keeping with technical and centrally administered practices – and, commonly, the creation of dependency on them. However, vulnerability, or critical aspects of it, arise from social conditions and forces beyond the control of those at risk and their local environment. Then, they may suffer most from impaired self-defences, or need protection from social forces more than against flood or fire. Lack of active capacities is as likely to be due to too much, rather than too little, imposed arrangements and constraints. Again, *empowerment* may be much more critical to reducing the vulnerability of such people than any particular tools, information or regulations to combat a hazard.

The reproduction of vulnerability and 'positive disadvantaging'

Vulnerability due to powerlessness arises from the social order. If it is to remain a fact of life it has to be actively *reproduced* through social action (Table 6.2). Maskrey (1994, 47) has suggested that vulnerabilities derive from *social, economic, cultural* and *institutional* variables. Lavell (1994a, 86–93) added 'physical' vulnerability (essentially exposure as defined above), and *political, technological, ideological, educational* and *ecological* sources. Risk is seen to involve, primarily, the human allocation of endangerment through these aspects of communal existence. Vulnerability is maintained by economic and other conditions. It is reproduced by the activities that sustain unsafe living conditions for some, or disempower them, and changes only if these conditions are transformed.

In an industrialised, monetary economy, economic vulnerability can relate mainly to having or not having a job; its pay scale, benefits and status; and the social security system for protecting working people. These drastically affect the means and choices available to householders that will render self and family – even neighbourhoods, city districts or whole regions – less or more vulnerable to a host of dangers. Better-off and more secure workers will be able to take advantage of, or choose among, various places of residence and ways to protect family members.

In more traditional agrarian economies, economic vulnerability is likely to revolve around land ownership. To be landless is generally to be most vulnerable. If one has land, then vulnerability for self and family depends upon how much, of what quality and the constraints upon the disposal of its produce. People obliged to occupy poorer land, or to farm without government or other

Table 6.2 The socially reproduced and active forms of vulnerability, with particular stress upon those that are socially imposed and enforced

Voluntary
- *risk-taking*: undertaken or sought-after dangers.
- *risk-accepting*: agreed, contractual, participatory risks.

Involuntary
- *systemic or everyday vulnerability* ('structural violence'), which can include:
  - *material poverty*: absolute and relative lack of means or 'endowments'.
  - *lack of rights*: formal constraints on the access to available resources, protection, etc. (e.g. 'entitlements', Sen, 1981).
  - *disenfranchised*: inability to influence political direction.
  - *marginal*: situated outside or unrepresented in the groups or institutions that decide economic and political direction.
- *enforced 'marginalisation' and defencelessness:*
  - *denial*: of resources, land, rights, protection (i.e. of 'entitlements')
  - *exploitation*: taking away an unjust portion of people's 'earned' wealth and means of existence (e.g. 'surplus appropriation', Watts and Bohle, 1993), so as to expose them to higher risk and deprive them of defences against hazards.
  - *subject and captive peoples*: serfdom, slavery, forced and indentured labour, occupied territories, stigmatised and marginalised minorities.
  - *Marginalised areas and peoples*: dependent territories, regions of economic decline, disenfranchised or unfairly discriminated against indigenous peoples, 'tribal' or minority 'nations'.
  - *Displaced peoples*: populations (recently) relocated in an unfamiliar environment, stripped of adaptive knowledge, well-tried practices and the advantages of belonging.

assistance in normal years, are more at risk from crop failure and its worst consequences in bad years.

A majority of those who have died in disastrous cyclones and riverine floods in Bangladesh in the past three decades have been landless or land-poor peasant families and migrant workers. Summarising features of one of these disasters, Burton *et al.* (1978, 11) concluded ' . . . inequity was not randomly distributed. It was not mere chance that the burden fell most heavily on the landless labourer of Bangladesh.'

Around the world today, the most vulnerable members of a society often confront small landed or urban elites, backed by a pliable military or repeated military rule. The latter determine the political agenda, often with external economic and military assistance. However, even where the political system is not grossly inequitable, vulnerability may still depend upon who possesses and can more effectively exercise influence and political power. According to this view, unusual vulnerability is imposed as *positive disadvantaging* by social actions that take away the options or self-determination of others.

Attitudes and a history of prejudice and exclusion can permanently relegate certain persons and groups to a more vulnerable position. The Romanies ('gypsies') of Europe have been frequent targets of persecution. Their way of life makes them specially at risk in, and vulnerable to, conditions in the modern state. Sedentary nationalistic populations, strongly enforced land ownership laws and lost commons have situated such nomadic or semi-nomadic peoples legally, psychologically and physically beyond the pale of civil protection. In many countries the Romany have responded by taking on 'pariah' jobs, looked down upon or feared by most others. They have become entertainers and fortune-tellers, titillating the fears and fantasies of visitors to fair-grounds and circuses. These also helped to develop caricatured images and expose them to self-righteous harassment and false accusation in times of crisis. While showing great resilience and enterprise themselves, they have been left defenceless where dangers and hardships exceed their communal resources. Capitalising upon the fears and hatreds surrounding them, the Nazis felt no compunction at exterminating some 240 000 Romany and Sinti people of Europe, many of them in the Death Camps. Even more died of privations, persecution and random killings between 1939 and 1945. Unprotected by state laws and ignored by the international community, few persons lifted a hand to help them, even in countries not occupied by German forces (see Chapter 12).

Vulnerability and entitlements

All societies have a range of formal or legislated, and informal or customary, arrangements that members, or some of them, can count on for greater safety and protection. Some may be deployed on a needs basis, in times of hardship, or to help offset and recover from misfortune. The existence of, and access to, such arrangements is generally critical for people's ability to cope with and survive family if not communal crises. Systems of social security and benefits to the needy can offset inherent and structural disadvantages, and reduce vulnerability to a range of dangers. Greater vulnerability within or between societies most often relates to the lack, or the undermining, of such safety nets. A growing number of studies reveal how damage in disasters is concentrated among those members of families, communities, nations or the global scene who are deprived of, or not given access to, known forms of social assistance and opportunities that protect others.

The work of Amartya Sen (1981) bears most directly upon these arguments. He has identified the bases of food insecurity, and specifically defencelessness against famine in rural Africa and Asia, in terms of 'entitlement relations'. Sen (1981, 36) defined entitlements as 'the set of different alternative community bundles that a person can acquire through the use of various legal channels of acquirement open to someone in his position.' If the word 'bundles' seems vague, it indicates that we are not considering just one kind of benefit or right. Sometimes the critical issue is a person's direct claim over the material products of his or her labour, or owned resources. In emergencies, the benefits and rights that one may be entitled to as a member of a family, or as a citizen, can be decisive. As a view of famine, this shifts the focus of risk from drought

155

and other agents that reduce food availability, to the societal arrangements for food allocation. The issue is access of members of a community to the food available there, or within the food systems of which it is a part. It may be a question of the ability to pay the market price for foodstuffs, and the nutritional value of those that can be afforded. But for people who *cannot* afford to buy food or who reside in areas of absolute, local shortage, a range of other entitlements becomes critical. To be sure, extraordinary measures may be needed to distribute available food. Seeing that it reaches the hungry, or keeping prices within their means, may require unusual vigilance and organisation, but these are rarely, if ever, beyond the capabilities of modern states or the international system of humanitarian assistance.

The entitlements argument has not received much attention in relation to dangers other than hunger and famine, yet it is surely of general relevance. In cities, or mass societies generally, the system of laws and rights is crucial to the well-being of those who may be at risk. The entitlements notion identifies basic issues of social organisation directly affecting vulnerability but not necessarily apparent in formal measures of wealth, costs, demographics, hazards or damage.

However, too much emphasis upon formal and legalistic aspects of vulnerability or adaptive capabilities may itself hide an even more basic issue. This is the active and creative features of human life that, as noted earlier, some people seem to lack or be deprived of. It involves the difference between the spirit or letter of 'rights' and the actual use or abuse of powers needed to guarantee them.

Vulnerability in everyday life

The discussion thus far indicates that the forms and sources of vulnerability arise mainly as aspects of everyday activity and social problems. They are not special to what happens during an earthquake or fire. Rather, the reproduction of vulnerabilities directs us to a domestic and local space of risk, to work and leisure. It may be related to life stages and responsibilities, gender and experience, or occupation and available cash. Even the vulnerability that appears in major disasters arises largely within the local, interpersonal and often domestic spaces of everyday life. Sometimes, there can be quite striking differences in the ability to cope with and survive disaster within families. In this way vulnerability raises questions of 'micro-organisation' and 'micro-politics', meaning human action and authority at the level of families, villages and neighbourhoods.

In such terms, Wisner, (1983, 16) argued that we will fail to get to the heart of the matter of risk unless we 'create ways of analysing the vulnerability implicit in everyday life.' This is a basic challenge to the hazards paradigm and to abstract, systems thinking. Of course, 'everyday life' is subject to unprecedented difficulties and changes arising from economic globalisation and deliberate intervention by the state. Wisner (1988) and others with similar concerns are well aware that these have become critical aspects of the vulnerability of 'ordinary' people, and they direct us to 'underlying processes', often extraordinary and novel, of vulnerability. However, if it depends upon the social order, danger is found and felt mainly in domestic and everyday living. The social geography and uses of living space thus provide vehicles to describe and assess vulnerability. They emphasise its variable distribution within and between societies.

Vulnerability and disease

All the categories of class, race, gender and ethnicity which differentiate privileges and oppressions also affect vulnerability . . . Exploited and stressed individuals and populations are at greater risk of emerging and resurgent diseases. In addition, because they cannot avoid exposure to unsafe conditions in their social or physical environments at home or at work or play, they are often constrained by circumstances into unwanted and unsafe choices and offered limited access to therapeutic interventions.

Wilson *et al.* (eds) (1994, 334–5)

All the notions of vulnerability described above appear in the language of health hazards and exposure to disease. Predisposition or susceptibility to infection; (undue) exposure to a communicable disease; impaired resistance or weakened defences; lack of preventive measures or access to medical care – these are all expressions of vulnerability. However, they become of particular concern for public risks and disaster when they apply to communities or whole populations. In the modern world, increased vulnerability to disease accompanies many forms of social and environmental change or development pushed forward with little or no regard to health risks.

Rapid urbanisation commonly leads to greater public health risks. They occur, especially, where new arrivals or growing communities are forced into crowded living areas with few or no public services (see Chapter 10). Irrigation projects in the tropics have been found to spread the debilitating, and if untreated fatal, schistosomiasis (or bilharzia). The schistosome parasite is generally acquired from snails, which flourish in impounded waters and irrigation channels. It is perhaps the second most prevalent and serious parasitic infection in humans after malaria, yet concern over such health risks has been absent from, or low on, the priorities of those promoting water resource projects, sometimes with horrendous consequences. There was, for example, a huge increase in cases in Egypt following construction of the Aswan Dam.

Irrigation schemes and cultivation based upon them is widely associated with increased incidence of malaria, arboviruses such as those that cause 'Japanese encephalitis', and various forms of filariasis. The latter are a group of tissue-destroying infections of nematodes (tiny worms) that include 'river blindness' (onchocerciasis) and 'elephantiasis', in which the legs swell to huge size. Again, they enter our concerns because they tend to affect substantial numbers, sometimes the majority of communities. Where they are medically out of control, they result in debilitation, high mortality in some age groups, lowered ability to work and, usually, more serious vulnerability to other hazards.

However, the hazards paradigm is firmly in the driver's seat in medical science and public health measures. The virus, bacterium, parasite, or 'condition' itself; the disease process, vector or state; each tends to be the focus of the greatest amount of research, and measures for prevention or cure. No doubt there are important benefits, yet there is no lack of evidence that environment and way of life play major roles in exposure to disease and whether, whom or when most illnesses strike. In families and communities, and their surroundings and social conditions, life-style and wealth can be decisive for the general level of health and longevity. There is a growing recognition that better public health and greater life expectancy in the West over the past century have depended upon broad improvements in living conditions. These include improved means and habits of personal and communal hygiene, better information about the causes of disease, and access to health care. Equally important, notably in cities, were improvements in sanitation systems, potable water, and the collection and disposal of human waste.

In all but a few epidemics, who dies ('mortality') and, often, who contracts the disease ('morbidity') is a function of age, gender, previous health status, occupation, environmental setting and material well-being. Communities exposed to particular pollutants, or groups with a particular diet or habits, may have high levels of susceptibility to certain diseases.

In addition to local conditions and social interaction, broad, including global, interchanges and contacts may lead to major health crises or spread of infectious disease – as they may also promote the spread of improved medical understanding and treatment. Some of the greatest catastrophes of epidemic disease arose through social contact between previously separated peoples. European contact with and conquest of indigenous peoples in the Americas provides some of the most calamitous examples in history. In that case, mere exposure to exotic diseases played a large, perhaps the largest, role. Yet it is increasingly recognised that the particular demographics, social and cultural features of native American peoples, and the conditions of contact – often bloody, treacherous and disruptive – increased their vulnerability to infection. Incompatibilities in the values and goals of conqueror and conquered people were involved, not least the military, economic and environmental ruthlessness of the former. However, it is not only past history that shows indigenous Americans made unduly vulnerable to disease. Many of their communities today are severely undermined by higher incidence of a variety of diseases that are identified with impaired living and high rates of mortality.

The HIV/AIDS pandemic

The late twentieth century plague known as AIDS is an out-of-control disaster, rooted primarily in human behaviour and volatile social and economic conditions. It should be stressed that this is not a disease in the air, the water or the food we consume. Nor can it be passed on by most forms of social behaviour. It seems people can be infected only by particular types of human contact or actions, usually in particular social contexts. The epidemic can continue if, and only if, those already infected engage in quite specific types of behaviour with a widening circle of others who are uninfected. In the spread of the disease, social exposure relating to everyday habits and weaknesses; lack of protection, rights and information; and public indifference or stigmatising, have played large roles.

Tragically, once a person is infected by HIV, the human immunodeficiency virus identified with the disease, there is no known way to prevent AIDS, acquired immunodeficiency syndrome. Its progress may be slowed or halted for a year or two by proper care. Persons already weakened by poor diet, ill-health or drug abuse seem likely to succumb sooner, but few victims live more than two or three years once they enter the AIDS stage. They are at grave risk from other communicable diseases around them and also from disease-causing organisms that we carry in our bodies, which their immune systems cannot hold in check. AIDS victims may be severely or completely vulnerable to otherwise treatable diseases. They are at unusual risk from new strains of disease, and from the growing range of drug-resistant ones, notably tuberculosis. Increasing numbers of persons with both tuberculosis and HIV are being identified. Many of the same persons are also at risk from, and are likely to have, other sexually transmitted diseases.

Globally, in the 1980s, more than 14 million adults became infected with HIV. About 40 per cent were women and about 1 million were children (Fig. 6.1). It is thought that perhaps 6000 persons are newly infected every day. By 1992, almost 1.9 million adults and 560 000 children had died of AIDS. Projections for the year 2000 range from a lowest estimate of 5.9 million to as many as 20.4 million (Mann (ed.) 1992, 132). While the great majority will continue to be in sub-Saharan Africa, the numbers and share in the rest of the world are growing. Unlike the world situation, the ratio of HIV-positive men to women is 8 to 1 in North America, and 5 to 1 in Western Europe. However, the global ratios, dominated by sub-Saharan African figures, where men and women are about equally infected, may be a better guide to the future of the disease in industrial countries.

The largest fraction of victims, worldwide, have been infected by unprotected sexual intercourse. Persons specially at risk are those whose work or situation leads to casual sexual contacts with strangers. Mobile individuals and those drawn into casual sexual liaisons in places remote from home are more likely to become infected. In this part, the main groups involved have been soldiers, overseas workers, travellers, tourists, long-distance truck drivers and

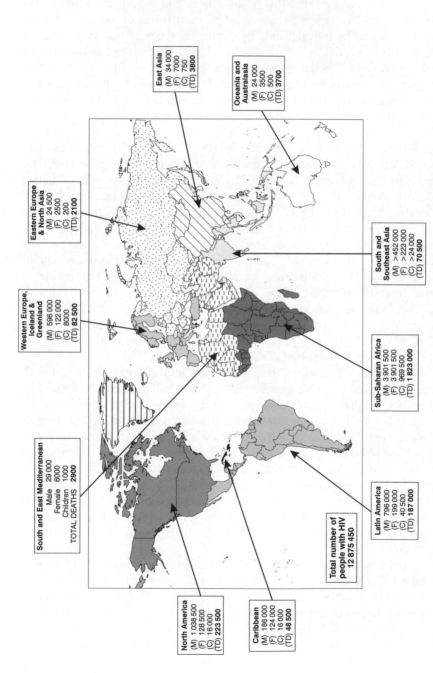

East Asia
(M) 34 000
(F) 7000
(C) 750
(TD) **3800**

Oceania and Australasia
(M) 24 000
(F) 3500
(C) 500
(TD) **3700**

Eastern Europe & North Asia
(M) 24 500
(F) 2500
(C) 200
(TD) **2100**

South and Southeast Asia
(M) >452 000
(F) >223 000
(C) >24 000
(TD) **70 500**

Western Europe, Iceland & Greenland
(M) 596 000
(F) 122 000
(C) 8000
(TD) **82 500**

Sub-Saharan Africa
(M) 3 901 500
(F) 3 901 500
(C) 969 500
(TD) **1 823 000**

South and East Mediterranean
Male   29 000
Female   6000
Children   1000
TOTAL DEATHS **2900**

Latin America
(M) 796 000
(F) 199 000
(C) 40 500
(TD) **187 000**

Total number of people with HIV
12 875 450

North America
(M) 1 038 500
(F) 128 500
(C) 16 000
(TD) **223 500**

Caribbean
(M) 186 000
(F) 124 000
(C) 16 000
(TD) **48 500**

Fig. 6.1 The HIV/AIDS pandemic. Estimates of victims to 1992 by major world regions. The figures are for those infected with HIV by gender and for children, and deaths from AIDS (after Mann *et al.* (eds) 1992, Fig. 2.52; Tables 2.4 and 3.13)

others 'on the road', and, of course, 'sex workers'. The vast majority of sex workers are women. This, and the fact that they are rarely able to determine the behaviour of their clients, puts them at unusual risk. Later, it is the mobile individuals who, when returning home, spread the disease into otherwise unaffected groups and areas.

Early, rapid spread of the disease in North America was somewhat different, but still essentially 'social'. It was mainly among homosexuals and hard drug users. Both proved unusually vulnerable groups. Anal sex seems more likely to cause transmission of the disease, magnifying the risk for homosexual and bisexual males. They are the principal victims to date in North America (60 per cent) and some other industrialised nations. Drug users by injection are at unusual risk from shared and inadequately sterilised needles and syringes, a danger increased by dependency and loss of control over their habit. The rapid spread of hard drug use in recent years has helped spread the disease into every continent and major city (see Chapter 10).

Although a small factor in the spread of HIV/AIDS overall, the infecting of persons through medical practices, specifically through contaminated blood and blood products, or unsterilised syringes and surgical equipment, poses a special threat. This is one sure way in which unaffected groups can be infected. Then, spread becomes socially indiscriminate and independent of people's own efforts to avoid dangerous activities.

A growing side-effect of globalisation, and especially the global oil system, is AIDS transmission through the huge expatriate or 'guest worker' labour market. This involves men, sometimes women, residing in foreign places for months and years without family or close friends. Many eventually seek sexual relations in casual encounters and with paid sex workers. Expatriate women workers may be forced into a relationship to keep their jobs. However, these developments become a huge danger when guest workers return home, temporarily or for good. AIDS has been taken back to families and whole countries where it had not existed before. Families in Pakistan, for example, are currently paying this deadly price for years of large, hard currency payments sent home by guest workers in the Gulf states. As in so many places, the disease is generally spread first to the spouse, perhaps the unborn children, of the returned worker.

While it is terrible for anyone, there is a special tragedy for the millions of victims infected not from some unusual or even unsanctioned social action but in the home and as a result of their domestic obligations. Here, ordinary women have proved especially defenceless. Even where fewer of them are infected, at least for the moment, than men, women are more vulnerable to HIV infection:

In part, this is because the direction of sexual spread favours male-to-female transmission. Yet, a much more important cause of vulnerability is the socio-economic and cultural contexts in which women live. Without equal status, they have no decision-making power on issues that determine their health and

161

welfare, the welfare of their children, and the relationships in which they are sexual partners.

M. M. Mhloyi (1992, 373)

This is a clear example of how vulnerability is associated with disempowerment. Most often it is women who stay behind while the man travels, works or finds pleasure elsewhere. Few have the strength or rights to control their sexual relations, even within marriage. Worse, in patriarchal societies they are often blamed and rejected when the man infects them. Then again, left alone for years, they too may have found other partners, secretly, and become part of the chain of infection. However, the wife usually ends up caring for the AIDS-stricken husband and child, often hiding the fact from relatives and community. But when the husband dies, the extended family may throw her out, especially if she has the disease. She may be left without resources or anyone to care for her when AIDS takes hold. Thousands of new victims of this double disaster appear in South Asia every day now. Already, hundreds of thousands of women have lived and *died* in this terrible predicament in sub-Saharan Africa. They prove to be the most vulnerable of the vulnerable.

Economic changes, marginalisation and displacement of populations may also be precursors of exposure and vulnerability to the disease. In Southeast Asia, they are associated with flight to the cities. Once there many drift into crime, prostitution and drug abuse, especially young women and unattached men, sometimes children. Such was the background to a sudden, rapid rise in HIV infection among sex workers in Thailand from the late 1980s. 'Sex tourism' has been an important factor in the spread of HIV here and elsewhere. Sudden economic pressures and opportunities also seem to explain a dramatic rise in HIV-infected drug users in Bangkok, seemingly in a single year, 1988, from 2–3 per cent to 35 per cent; and it has remained high since.

In sub-Saharan Africa, where nine out of ten HIV/AIDS victims live, the disease has blazed a grim trail alongside social upheaval and migration due to political and economic change. It has followed the routes of military movement, the many battle zones and places of military occupation. Long-lasting or recurrent conflicts in Angola, Mozambique, Zimbabwe, Rwanda and Uganda are clearly identified with the transmission of HIV. Very high rates of HIV infection are reported in 'commercial sexual workers' in Kenya (88 per cent in 1990), Rwanda (80 per cent), Malawi (55 per cent in 1984) and Tanzania (38 per cent in 1988).

War and other violent upheavals lead to the unravelling of communities, the displacement and impoverishment of families. Meanwhile, more wealth and a greater share of it goes to the military and *its* infrastructure, not just the war-making. Army barracks become places where the poor, unemployed and those with goods to trade congregate. One of the 'commodities' most in demand is sex; sexually transmitted diseases have long had an 'affinity' for military bases. Women come there as sex workers to support family members, hoping to return to a normal life. Then, their saddest role may be to bring infection back to those for whom they had given up so much to assist.

Wars and other armed violence are also closely associated with the growth and redeployment of drug trafficking and drug-taking. Drug abuse among soldiers and veterans of recent wars has been a factor in HIV spread. The drug trade often follows them home.

The hatreds or brutalisation of protracted wars, or those with ethnic and urban–rural conflicts, have made rape another epidemic. In 'militarised rape', women or young boys may be assaulted many times by different men. This increases the chances of contact with an infected person. Violent, resisted or multiple sex acts cause worse lesions, by which infection is passed.

The actualities of the spread and dangers of AIDS involve complicated and seemingly intractable problems of social change, human behaviour, rights and organisations, but they are rooted in social conditions and vulnerability rather than in biology, the virus itself, or medical matters. 'In principle' the epidemic appears eminently stoppable. Means to do so are well-known. What to avoid, to prevent HIV infection – with the singular exception of 'medicalised' transmission through contaminated blood and instruments – involves knowing some obvious and readily communicated facts:

> AIDS is the only epidemic for which the means of prevention are available in the local supermarket, are cheap, and are easy to use. The resistance to using condoms and bleach or alcohol [steriliser] is, in one sense, extraordinary.
>
> Perrow and Guillen (1990, 38)

The means to 'safe sex' and HIV-free needles and syringes, the overwhelming modes of transmission, are available elsewhere, or could readily be made so. What singles out all the persons or groups most likely to get infected and spread the infection is that they do so in places where they are engaged in commercial transactions, whether their employment, for sex or drugs. Almost all will be working for cash and, in many cases, are potentially easy to reach and inform.

These statements, and the quotation above, are too glib, of course, and entirely miss the predicament of those who are most vulnerable. However, they highlight where the actual complexities lie. It is a striking case of how unhelpful the medical hazards paradigm, based on the character of the disease or research and actions directed at it, has been. Vulnerability, and the social conditions and behaviour associated with it, defines, if it does not cause, the disaster. In North America, the debate has become strongly polarised between those concentrating on care for the victims, or persons likely to become so, and those interested only in a medical cure, or who believe the people at risk deserve to be so. It is typical of a vulnerability perspective that increasing numbers of AIDS workers advocate measures not merely to help persons and groups most at risk, but also to empower them. They see that their only hope, in the end, is to have the means and the rights to help themselves or, if infected, to gain the care they need without loss of dignity and humanity.

Geographies of vulnerability

> vulnerability must be seen as a system property whose locus is the social structure.
>
> Wilson *et al.* (eds) (1994, 335)

Dangerous places and spaces

No place or group of people is entirely safe, but the forms and severity of risk vary markedly among different areas and groups of people, between different parts of the world, and even within any local community. As a part of this, each of the forms and sources of vulnerability has a geography, a presence, a mix and a severity that vary from place to place. The map of vulnerability may be described in terms of the various classes and processes discussed above (Table 6.3).

Clearly, people are at risk from particular hazardous agents to the extent that part or all of their activities are in locations likely to be affected by those agents. A first assessment of the geography of risk from given hazards is to show who and what are present in flood- or drought-prone areas, who live in proximity to dangerous facilities, or who are in a war zone. Unlike a hazards perspective, however, the geography of vulnerability is identified with 'defence-less spaces', with the pattern of frailties and absent protection.

It is unlikely that, anywhere in the world today, local-level or even urban and mass, social vulnerabilities exist, or can be tackled, as isolated problems. They are embedded in, and more or less fully subject to, actions and developments at all levels of governments. They involve multinational commercial enterprises, professional bodies, mass media and religious or other powerful cultural groupings. All these levels have some part in the risks that may attend, say, housing development, the use of an antibiotic or pesticide in a village house-hold in the High Andes, or mining in the Australian outback. What a house-hold and 'ordinary life' vision of vulnerability and mitigations does is to alter the terms upon which we examine and decide what is important. It leads to a view of spatial organisation and its consequences as seen *on the ground* and, societally, *from below*. This contrasts with the abstract or 'modelled' logic, as if gazing upon reality from above or on behalf of states and other large forma-tions. This becomes especially relevant where unjust and brutal regimes or political economy forcibly endanger the disenfranchised or powerless (Table 6.4).

Table 6.3   The forms of geographical vulnerability: vulnerable places and people

Exposure to hazards through location, as in:
• flood- or drought-prone areas
• proximity to dangerous facilities
• war zones

Impoverished and impaired habitats:
• regions with 'mined out' soils, ground water, forests, endangered wildlife
• pest- or disease-infected areas
• polluted areas
• environment devastated by warfare and weapons testing
• environments destabilised by human activity (deforestation, desertification, radioactive fallout)
• adverse natural impacts (climate change; volcanic activity)

'Defenceless spaces' resulting from:
• lack of physical protection such as flood works or sea walls, wind-breaks, fire walls, chemical or thermal pollution controls, air-raid and fallout shelters
• absence of material, design and environmental quality standards or their effective monitoring and enforcement. This applies to areas lacking physical safety regulations and building codes; standards applying to water, air, food, medicines, etc.; or effluent and other waste disposal from industries and municipalities
• absence of community organisation and practices of looking out for and caring for inherently weaker members; of public spaces that can be effectively policed against acts that jeopardise safety
• collapse of community responsibility and care for 'the commons'
• lack of warning and emergency preparedness in the event of damaging events

Locational disadvantage:
• slums and impoverished ghettos; 'wrong side of the tracks'
• peripheries

Enforced geographic vulnerability

Where large numbers of people within a given population are unusually vulnerable this commonly reflects a political space in which corrupt, unjust and illegitimate governance has more or less free rein. The forcible imposition of demands and goals of illegitimate or despotic governments operates to increase the vulnerability of those under their control. It is something associated especially with forces of occupation, colonial rule and 'government by terror'. Such places or countries are singled out as *regions of misrule*, usually the main source of exceptional and unfair relative vulnerability, and the disasters that accompany it.

Dispossession is forced geographical change that commonly results in severe endangerment for the dispossessed. If long-time residents and other users of given areas are deprived of rights of access to resources they previously enjoyed, their vulnerability must increase. If those lands or resources are taken over for other use – perhaps mining, logging, or plantations using dangerous pesticides – this may expose the dispossessed to new hazards. These

Table 6.4 Enforced geographical vulnerability or endangerment through geography

Regions of misrule:
corrupt, unjust and illegitimate governance; forces of occupation; colonial rule; 'government by terror'; support for terrorist organisations and state terror.

Exploitation and dispossession:
forced labour; seizure of traditional lands and leasing to outside users; reneging on traditional obligations between land-owning classes, peasants and other landless residents.

Forced resettlement, uprootings and expulsions:
'ghettoisation', 'reservations', 'pacification' by resettlement; removal to less safe or less known environments; expelling people from their long-time homes and countries; exile in foreign areas and uncertain futures elsewhere.

Ostracised and excluded states, regions and peoples:
economic blockade or sanctions; political manipulation of international food and other assistance.

have become a global source of impoverishment and disaster for indigenous peoples, notably as a result of European imperial rule and post-colonial state formation in newly independent countries.

Deprivation of rights to land and resources is often a prelude to, or accompanied by, actual *uprooting*, *expulsions* and *resettlement*. The 'enclosures' and 'Highland clearances' in early modern Britain provide examples of peoples dispossessed, uprooted and impoverished. Afterwards they were subjected to a range of further misfortunes and displacements. In such cases, vulnerability arises from enforced removal to less known, if not less safe and more impoverished, environments, from rural to urban settings, if not between countries and cultures.

In recent state-building, in civil and international wars, large numbers of people have been subjected to forced resettlement as a 'pacification' measure. They have been placed on 'reservations' or in so-called 'strategic hamlets' such as were developed in Vietnam and other counter-insurgency wars (see Chapter 10). Throughout Latin America rural folk have been forced into urban squatter settlements and ghettos by terror in the countryside from both insurgent and state forces. They may be driven from known lands to unknown and uncertain futures elsewhere. Temporarily or permanently they become more vulnerable as a direct result and integral part of their changed geography.

Similar results may come about when a given territory and people are cut off from established sources of wealth, exchanges of goods, services and benefits. Groups and even whole states may be made more vulnerable by being ostracised, subjected to economic blockade or sanctions, or excluded from programmes to provide, say, emergency or relief food and other assistance.

Every one of these geographic vulnerabilities was criminally imposed upon the Jews of Europe in areas controlled by National Socialist Germany and some other fascist regimes in the 1930s and Second World War (Chapter 12). In greater or lesser degree, such privations and vulnerabilities have been imposed upon the indigenous peoples of North America for five hundred years. And still today, these and many other minorities, impoverished classes and disenfranchised peoples are subject to some or all of these positive endangerments 'by geography'.

Concluding remarks: problems with a vulnerability paradigm

To summarise, vulnerability is a product of the circumstances that put people and property on a collision course with given dangers, or that make them less able to withstand or cope with disaster. It depends, in large measure, upon ongoing conditions of material and social life, or their transformations. For that reason, this perspective also draws attention to cultural and ecological contexts that constrain or enhance people's abilities to respond and cope.

The broadened awareness and analyses of a vulnerability perspective have had a salutary impact upon our ideas of risk and disaster. Nevertheless, 'vulnerability' may prove to be an unfortunate term. Unlike much of the work that it labels, the word emphasises a 'condition' and encourages a sense of societies or 'people' as passive. Indeed, as happens with the hazards paradigm, vulnerability can treat human individuals, the public or communities simply as pathetic and weak.

Latterly, vulnerability is being recast as yet another of the 'social pathologies' like, or derived from, poverty, underdevelopment and overpopulation (cf. Table 6.1). Vulnerability to major disasters is increasingly seen in much the same way as aging, mental instability or unemployment have been viewed in modern societies, that is, as 'social problems' posed by 'victims', to be dealt with through professional treatment. This is to ignore how the elderly or unemployed, or those going through mental torments, often have enormous capabilities; that their problems derive largely from the social context rather than their inherent qualities. Much the same expert systems and public 'minders' vision may emerge as the dominant view of a new 'pathology' called vulnerability. Problems and settings that lack 'advanced' technologies and governmental intervention, or a strong consumer economy and especially professional expertise, are likely to be seen as vulnerable for those reasons.

In fairness, much of the actual literature of vulnerability emphasises human resilience and adaptability as well as the presence of effective risk aversion arrangements in all sorts of societies. Persons and groups are shown to be anything but mere playthings of nature, fate or even governments. This leads on to critiques of national and global strategies that *generate* vulnerability, or undermine people's capabilities to avoid or recover from disaster.

Vulnerability perspectives

Suggested reading

Berer, M. and Sunanda Ray (1990) *Women and AIDS*. Pandora, London.
Blaikie, P. *et al.* (1994) *At risk: natural hazards, people's vulnerability, and disasters.* Routledge, London.
Bohle, H. G. (ed.) (1993a) *Worlds of pain and hunger: geographical perspectives on disaster vulnerability and food security.* Freiburg Studies in Development Geography no. 5. Verlag Breitenbach, Saarbrücken.
Chambers, R. (1989) *Vulnerability: how the poor cope.* IDS Bulletin, Special Issues.
Mann, J. M. *et al.* (eds) (1992) *AIDS in the world.* Harvard University Press, Cambridge, Mass.
Watts, M. J. and H. G. Bohle (1993) The space of vulnerability: the causal structure of hunger and famine, *Progress in Human Geography*, **17** (1), 43–67.
Wilson, M. E., R. Levins, A. Spielman and I. Eckardt (eds) (1994) Diseases in evolution: global changes and emergence of infectious diseases. *Annals, New York Academy of Sciences*, **740**, 15 December.

CHAPTER 7

Active perspectives: responses to disaster and adjustments to risk

> Let it not be said that geographers have become so habituated to talking about the world that they are reluctant to make themselves a vital instrument for changing the world . . . Each of us should ask what in his teaching and research is helping our fellow men to strengthen their capacity to survive in a peaceful world.
> Gilbert F. White, *Geography and public policy* (1972, 104)

Introduction: questions of response

Mitigation

Work in the field of risk and disasters is justified largely in terms of the desire to reduce human misery and material losses, and to maintain control over other activities. It is essentially an 'applied' field. Professionals and formal organisations need, or are forced by events, to work together in terms of public dangers and common goals. Investigations are most often in reaction to actual disasters, or involve research and practices relevant to places where they have occurred. Geographers have also looked mainly at the nature or assessment of practical needs and public policy. This is what gives our field a positive purpose, a responsible and a hopeful orientation, even in the midst of disaster.

Scientific and official actions have usually been directed at improving technical measures and official responses. These are well-entrenched in the urban-industrial heartlands of today's world, and the arena in which those who read books like this are likely to work. Yet they are not the only areas of response that need to be recognised. Rather, they are just one expression of the adaptable nature of human societies, and the struggles necessary for more secure living in the face of changeable and sometimes life-threatening conditions. Human capacities and responsibility for safer practices apply in many different contexts and at many different scales. Households and neighbourhoods, villages and small businesses are as involved in risk-reducing measures as state and international agencies, if in differing terms. The everyday life of many traditional societies seems to be organised around risk-reducing practices. This might include, say, crop types and patterns that are less than optimal for production in most years but reduce losses in adverse ones, and the chances of calamitous failures. A comparative, geographical study of human responses must take such variety into account.

In general, a vision of complete safety is not a realistic one. Rather, our work is sustained by evidence of differences between places and people that are more secure and have developed safer practices, and those lacking them. And if there are safety systems that lie beyond the skill or resources of a given community, there are invariably others within its reach. Since risk and the severity of disasters have this relative, but improvable, quality, caution is needed in how we define useful actions. The idea of disaster *prevention*, for example, has been criticised for giving an exaggerated sense of what is possible. Terms such as risk *reduction*, hazard *abatement* and vulnerability *mitigation* are more in keeping with what has been done, and can be expected. 'Mitigate' means to abate, moderate, alleviate or make less severe. In the context of social hazards it seems to be an appropriate word, since it can mean to render something less violent, to appease or mollify.

Because we address urgent and often dismaying questions of human need, many of those involved are obliged, above all, to act. Often, they are expected to do so quickly. This encourages what is called a 'mission-oriented' approach. Even when exploring and working towards new possibilities, there is a sense of urgency. Other dangers or goals compete for public attention and the same limited resources. For all that, there is a need – some would argue, under present world conditions, an *urgent* need – to step back from immediate problems and take stock. Possibilities and performance need to be reviewed without pressure to fulfil some particular mission or appeal to a particular constituency. Such an approach is at least desirable in an introductory overview.

The context and scope of action

> Disasters, however, are never unique. They are recurrent social situations, and even the dramatic should be firmly rooted in the continuity of social life.
> Russell R. Dynes and E. L. Quarantelli, in Kreps and Bosworth, (1994, 13)

What is and can be done to reduce dangers or respond to disaster depends upon two broad considerations. On the one hand, there are the conditions and circumstances that constrain action. This relates to the combination of resources, skills and organisation for action, and the values that motivate it. Pressures of time and competing concerns, of jurisdiction and the relations between administrative areas, may be critical. Each of these constraints forms the *context* in which actions take place.

On the other hand, there is the question of the kinds of action that people or organisations are prepared to undertake. There may be well-established procedures, customary or legislated responses. These in turn relate to societal priorities and historical precedent, occasionally to a culture of innovative or experimental approaches. In general, we can approach this by considering the scope or *range of actions* that are known or that might be possible.

Context and action

The most obvious differences are between measures against possible or future dangers, and responses in disaster. Although interdependent, these are very different contexts in terms of what can be done. Crises and disasters drastically curtail possible actions. When no crisis is present, systematic planning and improvement of public safety are possible. Measures to reduce vulnerability and mitigate hazards can be taken. Resources and knowledge can be used to make life more secure and head off or avoid disaster. Communities can improve building quality or relocate activities, set aside emergency food supplies or plan evacuation routes. It is also when training and preparedness against emergencies can be carried out. There is time to put warning and safety systems in place. Many organisations and persons that we might not think of as keys to public safety and security or expect to find in the front line of disaster response are then engaged in risk-altering activities. Meanwhile, how people view and react to hazards depends, in part, upon the history of disasters in their community, or personal experience of the dangers involved.

The contexts of action are further complicated and constrained by risk or disaster geography. Few societies or organisations are equally well-prepared and resourceful over all the territory, or all eventualities for which they have responsibility. Decision making, or key aspects of it, may be centralised in some areas, decentralised in others. Past decisions may leave highly uneven patterns of capability, expectations and risk in different places. In disaster zones, what can be done, as well as what needs to be done, varies with the social geography of capacities that survive or are destroyed. Regional and local knowledge, and site- or society-specific understanding, are always relevant to putting viable safety measures in place.

Since in some sense context refers to a society's 'world', its full extent exceeds what we would usually expect to deal with or to understand. This is one argument for recognising and always trying to incorporate knowledge of those 'on the ground' and most familiar with how their world works. Meanwhile, it invites us to approach the problem of action in terms of broad, flexible options, at least until some specific response has been decided upon. In technical risk management we find an increasing emphasis upon multiple back-up measures, and the call for open and flexible learning within a 'safety culture'. More generally, there is a growing advocacy of flexible and context-sensitive approaches. This makes for common ground between recognising the role of context and thinking in terms of the range and choice of possible actions.

171

'Alternative adjustments' and human choice as a framework

Over the past several decades, perhaps the most influential notion in geographers' work on hazards has been that of *a range of adjustments*. It was first proposed by Gilbert F. White in the context of flood hazards, and summed up in this way:

> It is a view which considers all possible alternatives for reducing or preventing flood losses . . . takes account of all relevant benefits and costs . . . analyses the factors affecting the success of possible uses of a floodplain . . . [and] seeks to find a use of the floodplain which yields maximum returns to society with minimum social costs, and it promotes that use.
> Gilbert F. White, 1945 (quoted in Kates and Burton (eds) 1986, 16)

This view considers the alternative actions that people take in different times and places, but especially the different possible kinds of option in any given place. These may include ways to control or prevent a dangerous process, change human land uses and practices so as to avoid it, insurance against damage, and emergency measures.

The alternative adjustments idea was soon applied to other natural hazards, and in other parts of the world (Table 7.1). Later, some of us found this concept useful in examining responses to technological, disease and even war hazards (Table 7.2). And while it favoured the hazards perspective, the 'classes of adjustment' include measures that would reduce vulnerability and improve the context or general security of human groups.

In addition to the classes of adjustment in these tables, other kinds of option were recognised from the beginning. Responses might seek to avoid or prevent damage, or might react to that which occurs. 'Preventive' measures are generally those that try to control or alter the dangerous physical process, or reduce human exposure to it. Options include building safer structures or changing human behaviour or land occupancy. A case may be made for fiscal incentives or penalties, or for the promotion of scientific research and public education. Actions may be directed at certain parts or sectors of a society, perhaps the risks faced by women or the elderly, transportation or public health. They might also favour, overtly or not, influential economic and political groups.

Perhaps the idea that, for any given danger, a variety of responses exist is rather obvious, but one must understand the context in which it was first proposed, and its larger implications. Societies may apply only a few of the adjustments possible. They may favour one type of response over others. White's argument began as a challenge to the 'technological fixes' he found dominated organised responses to flood hazards in the United States. Engineering control works, instrumental monitoring and warning systems seemed to be the main preoccupations of responsible agencies. He showed that the result was seriously to limit what was done to reduce hazard. Worse, such a narrow approach could promote developments that increased risks. Greater

Table 7.1 An early version setting out alternative adjustments to natural hazards, or the 'theoretical range of adjustments to geophysical events' (from Burton *et al.* 1968)

Class of adjustments	Event		
	Earthquakes	*Floods*	*Snow*
Affect the cause	no known way of altering the earthquake mechanism	reduce flood flows by: land use treatment; cloud seeding	change geographical distribution by cloud seeding
Modify the hazard	stable site selection: soil and slope stabilisation; sea wave barriers; fire protection	control flood flows by: reservoir storage; levees; channel improvement; flood fighting	reduce impact by snow fences; snow removal; salting and sanding of highways
Modify loss potential	warning systems; emergency evacuation and preparation; building design; land use change; permanent evacuation	warning systems; emergency evacuation and preparation; building design; land use change; permanent evacuation	forecasting; rescheduling; inventory control; building design; seasonal adjustments (snow tyres, chains); seasonal migration; designation of snow emergency routes
Adjust to losses			
Spread the losses	public relief; subsidised insurance	public relief; subsidised insurance	public relief; subsidised insurance
Plan for losses	insurance and reserve funds	insurance and reserve funds	insurance and reserve funds
Bear the losses	individual loss bearing	individual loss bearing	individual loss bearing

occupancy of flood plains was sometimes encouraged by a false sense of the degree of security provided by large control works and forecasting systems. To be eligible for government assistance for flood-control measures, municipalities had to reach a certain 'threshold' in, say, numbers of flood-endangered homes. In growth-oriented communities, that could be a disincentive to limit or preclude development in a flood plain. In this way, the alternative adjustments approach also showed the need to examine the mix and relations among actions.

Here, a second element in White's thought gave the alternative adjustments idea its most distinctive and innovative significance. This was the idea that measures taken represent a *choice of adjustments*. He proposed that, in some sense, the adjustments that people make represent personal or collective decisions. They reflect how people, or those with the authority, view dangers: the

173

Table 7.2 The range of adjustments to war hazards. An attempt to view social dangers in terms of the alternative adjustments (after Hewitt, 1987)

Class of adjustment	Range of adjustments	Present status		
		major	limited	minor or absent
1. 'Affect the causes of war'	*Military (offensive)*			
	Overwhelming superiority	×		
	Deterrent force	×		
	Alliance	×		
	Arming allies/'friends'	×		
	Composition/roles of armed forces	×		
	Rapid deployment forces	×		
	Compulsory military service	×		
	Nuclear freeze		×	
	No weapons testing		×	
	Non-proliferation	×		
	Unilateral disarmament			×
	Political–diplomatic			
	Arms limitation/disarmament	×		
	Propaganda (a) sabre-rattling	×		
	(b) peaceful		×	
	Conflict-reducing initiatives		×	
	Military/economic aid	×		
	Global/national referenda			×
	International peace-keeping		×	
	Goodwill exchange/ties with potential enemies		×	
	Civil disobedience			×
	Hostage-taking		×	
	Economic–industrial			
	Conversion of military industries		×	
	Oppose arms traffic			×
2. 'Modify the war hazard	*Tension-reducing*			
	'No first use'	×		
	Realignment/non-alignment			×
	Nuclear weapons-free zone(s)		×	
	'Greening of policy'			×
	Peace movement			×
	Vulnerability-reducing (military)			
	Defensive (home) forces only		×	
	Monitoring and warning systems	×		
	Active civil defence		×	

Table 7.2   contd.

Class of adjustment	Range of adjustments	Present status major	limited	minor or absent
	Bomb-/fallout-proof shelters		×	
	'War chest'		×	
	Protect leadership	×		
	Industry/resource protection		×	
	Decontamination systems		×	
	Build adequate, safe refuges			
	a) Leadership	×		
	b) Military		×	
	c) Civil population			×
	Emergency education			×
	'Non-violent' preparedness			×
	Prayer	×		
	Emigrate	×		
3. 'Modify war loss potential'	*Emergency measures* ('passive' civil defence)			
	Mass evacuation		×	
	Planned national/regional dispersal (industry, population, etc.)	×		
	Civil warning systems	×		
	Decommission/disperse dangerous installations		×	
	Home/municipal radiation-detection systems			×
	Emergency decontamination centres (mass)			×
	Retrofitting existing health utilities, etc. for war casualties centres			×
	Extensive national refuge			×
	Army/police control systems	×		
	Relief and rehabilitation plans			×
	Other			
	'Civilian defence'			×
4. 'Adjust to expected losses'	War insurance			×
	Individual loss-bearing	×		
	Fatalism	×		

priorities, policies and values upon which they act. This was to introduce, in the strict sense of the word, an *active* principle into the heart of the field.

It was not argued that all, or even a broad selection of, 'theoretical' adjustments are equally possible. Some are much more costly in resources or human effort. Some will violate social priorities, or favour one type or part of society over another. They may benefit urban areas more than rural, farmers or motorists at the expense of suburban householders or pedestrians. Entrenched ways of responding, or the history of how things are done, may close off options. People may plead 'necessity' in order to further narrow interests. Then again, greater efficiency or cheapness may be worth sacrificing to achieve more pleasing or equitable results.

The alternative adjustments idea is also a valuable teaching and learning tool. It helps us to review risks, damages and contexts with action in mind and an awareness of practical frameworks. What I will do here is to adapt this approach to represent, more fully, the importance of reducing vulnerability, problems of intervening conditions and risk context, as well as the hazards that were the original focus. This serves to elaborate and redefine some of the classes of adjustment and range of choice, but without abandoning the original idea.

Hazards reduction

> Individuals may be quite knowledgeable about the existence of flood hazards or seismic hazards in the region, and may indeed have experienced flooding and earthquakes, but the significant issue is the translation of this knowledge and experience into attention to the hazard.
>
> Palm (1990, 90)

Hazards reduction deals with the range of actions that may be taken to affect or counteract dangerous agents, or to modify human behaviour in direct relation to them. For example, coastal flood dangers may be countered through protective dikes, 'flood-proofing' structures or shelters, encouraging or legislating vulnerable people to leave the shoreline and low-lying areas, and flood warning systems. In each case, the guiding consideration is a high water hazard. What impacts can it have? Where and when may it occur? Likewise, containment structures and monitoring devices for radioactive materials, or excluding settlement from the zone surrounding where they are concentrated, are hazard-reducing actions. So too are investigations, experiments and design work intended to improve performance in relation to, say, pests, fires or pollutants.

Some, if not all, such actions will relate to vulnerability-reducing goals as well. However, the nature of the hazardous agent is the focus of the adjustments. As stated earlier, this has been the main preoccupation of our field, and likewise of responses. The literature on structural safety of buildings under seismic and high-wind stresses, for example, would fill a sizable library. The same is true for fire prevention in built-up areas, toxic and radiation hazards, forest and grassland fires, insect plagues, flood-control works, 'terrorism', or efforts to control 'dubious weapons' and indiscriminate warfare. These and most other areas of hazards reduction have become highly specialised and complex sub-fields. Often they are the responsibility of particular agencies and professional bodies with their own journals and meetings, standards and leading figures.

Reducing vulnerability

It is vital that . . . those involved in disasters work accept that the reduction of disasters is about reducing vulnerability, and that this involves changing the processes that put people at risk.

Blaikie *et al.* (1993, 219)

The focus of actions to mitigate vulnerability, while not ignoring dangerous agents, is upon fragile and defenceless persons, property and communities. It is designed to respond in terms of the exposure and needs of those at risk. We look at ways of protecting what cannot be strengthened or made more resilient, or the provision of safer alternatives or places to occupy (Table 7.3).

To grasp how this differs from a focus upon hazardous agents, more familiar, everyday examples may help. 'Childproofing' a home certainly includes keeping dangerous substances out of reach and closing off potentially lethal spots such as electrical sockets or steep stairs, but the rooms and facilities also need to be arranged in terms of the likely behaviour, needs, special weaknesses and inexperience of the small child. Otherwise, accidents will happen even without specific or obviously hazardous items.

Essentially similar considerations apply to making airports or amusement parks safe for young and old, or in making public spaces available to the handicapped. Safety measures must grasp and be directed at the way in which the person at risk is vulnerable or can become so, rather than only the conditions that might trigger a crisis. The problems may or may not arise from external dangers. People may be put at risk through certain personal behaviours, or regardless of how they behave. The main source of trouble may be neglect or separation of dependent members of families or groups. A crisis may arise from neglecting or forgetting medication or the frequency or length of exposure to conditions that are tolerable most of the time. Mitigating public dangers and disaster reduction require a similar vision of the way families or age groups, employers or users of facilities, communities and settlements are vulnerable to natural, technological or social dangers.

Not everyone or everything in given places or societies is equally vulnerable. Hence, the distribution of vulnerabilities or their social geography must be taken into account in efforts to reduce risks. People at greatest risk from,

Active perspectives

Table 7.3 Extending the notion of alternative adjustments to include actions to mitigate vulnerability and adverse intervening and societal contexts of risk

Focus of actions	General forms of action and practices	Examples
Dangerous agents and conditions (hazards)	Affect generative process(es)	fog dispersal; cloud seeding; slope stabilisation; watershed improvement.
		eradicate disease vector; biological control of pests.
		structural safety.
		safety standards for workplace, installations, food additives, birth control devices, toxic waste disposal.
		ban CFCs (ozone layer threat).
		benign substitutes for dangerous pesticides, carcinogens, radiation technology.
		outlaw anti-personnel and CBT weapons; comprehensive nuclear test ban; disarmament.
	Modify agent behaviour/impact	snow fences; fire breaks; flood control works; lower traffic speed limits/emissions.
		standards; toxic waste disposal practices.
		rules of war, tension-reducing acts, conscientious objectors.
Vulnerability (of persons, property, places in danger)	Avoidance (reduce exposure)	land use planning and exclusion zones (e.g. for flood plains, steep slopes, active volcanic terrain, dangerous facilities, etc.).
		protective perimeters; 'cordon sanitaire'; nuclear-free zones.
		strategic dispersal of vulnerable persons/industries.

Table 7.3   contd.

Focus of actions	General forms of action and practices	Examples
	Protection:	
	general public	flood-proofing; wind breaks; fire walls; storm sewers; sprinkler systems; immunisation; sunblock; food banks.
		arms trade protocol (e.g. none to regions of conflict, repressive or aggressive regimes); peacekeeping forces.
	specific	for children, the elderly, art treasures, the infirm, minorities.
		humanitarian assistance of all kinds.
	strengthening	'Health and Safety Act'; Mine Safety Act – inspections and enforcement; Environmental Protection Act.
Intervening conditions (societal, environmental contexts)	Material uplift	political and military alliances. improve diet; education and training; investment in infrastructure.
		sustainability agendas for environment, resources; full-employment legislation; equalisation payments to depressed regions; overseas aid; 'save the children'.
	Public security	social welfare; insurance; gun ownership laws.
		economic sanctions against human rights violators, etc.; peace and security alliances.

Table 7.3   contd.

Focus of actions	General forms of action and practices	Examples
	Social justice (empowerment)	equitable safety standards/ measures; equitable entitlements; affirmative action; civil and minority rights legislation, 'rule of law'; reduce (or penalise beneficiaries of) 'structural violence' and unsafe practices.
	Constitutional and policy frameworks	public accountability in dangerous practices and involuntary risks; 'sustainability' legislation; victims' rights; constitutional restraints on state uses of violence; non-aligned, non-violent posture.
Disaster response		protocols for disaster relief, epidemic disease, permanent agencies, war measures act.
	Emergency preparedness: warning systems	extreme event watch/warnings (e.g. severe weather, flood, forest fire, etc.) SOS protocol (shipping, fire, crime, etc.).
	Emergency/disaster agencies (permament): national civil defence agency	emergency evacuation plans; emergency communications; air raid shelters, gas mask programme.
	international agencies	UNDRO, USAID.
	(non-governmental)	Red Cross, Medecins Sans Frontières, Mennonite Central Committee, Red Crescent, Oxfam, etc.
	Coping	self/family/informal community responses of sheltering, flight, rescue, first aid, etc.
	(Organised) relief and rescue	food and medicine air drops; field hospitals; temporary shelters; emergency evacuations; disease control.

Table 7.3   contd.

Focus of actions	*General forms of action and practices*	*Examples*
	Rehabilitation	clearing rubble and other debris; restoring communications and other life-supporting infrastructure; temporary housing.
	Reconstruction	planning for greater public safety and reduced vulnerability; establish and apply safety codes; improve land use planning; respond to knowledge/needs/ concerns of victims, residents and others at risk.
	Establish cause/ blame	courts/commissions of enquiry; institutional and governmental accountability; compensation of innocent victims, war crimes tribunals.
Improve knowledge and information	Monitoring and detection	seismic network; extreme weather watch; world data centres and networks; International Search and Rescue Advisory Group.
		building/siting/product/worker safety inspections; Nuclear Safety Inspectorate; environmental risk assessments.
		medicines/drugs screening; reports on social indicators.
		'watchdog' reports: consumer, nuclear industry, genetic engineering, arms trade, minority rights, human rights abuses.
		referenda on pressing public issues.
	Research	disaster research centres; 'disaster scenario' planning.
		geotechnical, medical, accident research; peace research; conflict resolution.
		priority research for disadvantaged and vulnerable groups (women, children, minorities, etc.).

Table 7.3   contd.

Focus of actions	General forms of action and practices	Examples
	Education, training, and awareness	
	improve public access to information	public hearings and workshops; outreach programmes; freedom of information.
	Official/professional training	for police, military, government officials, corporate executives, social workers, academics, NGOs, etc.
		training in search and rescue, disaster communications, first aid, disaster medicine, post-disaster trauma and counselling; building standards; non-violent and peacekeeping practices.
		training and 'sensitivity sessions' to reduce racial-, gender-, religion-, class-biased attitudes and behaviours.
		international exchange and understanding programmes, courses, workshops.
Do nothing (deliberate/ inadvertent)	Risk taking/ accepting	voluntary, involuntary; acceptance of a technological or weapon imperative (i.e. 'if it exists it must/will be used'); 'survival-of-the-fittest' approach'; Total War; ('War is W-A-R'!); victor's justice.
	Loss bearing	by victims; communal; state level; 'winners' and 'losers' philosophy.

say, famine or flood are likely to be weakened in many other ways, and to lack protection against a range of other hazards. To rescue them from one danger may even increase or make more agonising their exposure to others. Persons saved from hunger may, for example, face acute unemployment or displacement.

It was pointed out earlier that vulnerability may, and in many modern dangers does, arise from conditions unrelated to particular hazardous agents. People's vulnerability may be decided by commonplace, everyday arrangements or some sudden change in them, or by modernisation directed at quite other goals or problems. If this is so, then measures to reduce risk will fail unless they involve the transformation of everyday vulnerabilities, or are incorporated into development programmes. In this sense, vulnerability reduction for those most at risk is hardly separable from general socioeconomic uplift or empowerment (see page 186).

Emergency measures and coping

> All definitions of disaster, whatever their comprehensiveness, suffer from the same insufficiency; they rarely consider the response to disaster impacts as part of the same event – as if society functioned without in-built reactive mechanisms and the external, especially international, response was the only means of recovery.
>
> Alabala-Bertrand (1992, 10)

There is little or no prospect of preventing all, or perhaps most, disasters from occurring, although much could be done to reduce their severity. Hence, preparations for emergencies, and questions of how people respond in them, are of major concern. Disaster preparedness also turns upon what is done to reduce dangers or vulnerability and is not, therefore, completely separate from the previous concerns. However, *emergency measures* and civil defence are widely treated as distinct problems.

In crises, swift action may be critical to saving lives, reducing distress and aiding recovery. As noted earlier, disaster raises novel organisational and practical dilemmas. Responses in disaster involve unique conditions for the victims, and even for unharmed parts of society that try to help. If there is indeed a disaster, some or all of the usual means to save lives, treat the injured and protect property are overwhelmed. The actions of unimpaired survivors and outside agencies are frustrated by destroyed infrastructure, by the scale of damage and need. They are rendered physically or organisationally incapable of doing what we might expect or wish of them. Time to consider, let alone to try out, new or alternative actions is absent or severely limited. Institutions and customary arrangements break down.

The condition of survivors, often isolated, severely impaired, and surrounded by injured and dead family members or acquaintances, creates a world of special stresses and needs. These are the concerns of a special part of disaster research, often identified with the idea of *coping*, as distinct from the responses of organised emergency measures. Important contributions to their understanding come from psychology of stress and disaster sociology, but they must also be seen as an integral part of the general problem of disaster.

In areas of heavy damage, there will usually be a certain time before individuals or families, neighbourhoods or institutions, link up with other survivors. They may be isolated and trying to deal with continually life-threatening conditions for days, weeks and, in the worst of epidemics and famines, for months or years. They must often take on tasks they would not do at other times – digging out and burying the dead, rebuilding their homes or makeshift shelters, evacuating the injured or vulnerable persons, and seeking alternative sources of food. Following the devastation, survivors must adapt to a radically trans-formed, grief-stricken social milieu. They will continue to exist in, and be set apart as, a distinct 'disaster community'. Few of them will ever be quite as they were before. They will be a generation sharing something that even their chil-dren, or the outsiders who come to help, cannot. During and after the worst calamities, there is a special world of 'survivors', in the sense and examples that Robert J. Lifton has done so much to explore and try to understand.

Organisations that remain viable through a disaster may be given different powers or take on different roles. Army units are brought in to assist or per-form the tasks of civilians. Extraordinary powers and measures are invoked. The usual system of legal recourse and social values may be suspended. In traditional societies, customary divisions of labour between men and women, or different trades and classes, are temporarily ignored. In a modern city, the designated disaster area or damaged buildings may be cordoned off. Even those who live there may be kept out. Functions normally left to local govern-ment or the domestic sphere are taken over by agencies of the central govern-ment, international aid agencies or Non-governmental Organisations (NGOs). These functions might include housing and feeding, child care, traffic and public health control. Some people see their influence soar at these times, not least the (police/military) security forces and disaster 'experts'. All of these possibilities are also problematic.

Alabala-Bertrand, in the quotation at the beginning of this section, properly questions the tendency to see disaster response only in terms of official mea-sures and outside emergency agencies. These are usually needed and often life-saving in major disasters, and may be essential to recovery. Nevertheless, over-emphasis upon them is typical of modern approaches and accords closely with the hazards paradigm. Seeing the problem as one of *external, unexpected hazards* and out-of-control events, it also focuses upon *external, planned emer-gency assistance*. However, most disasters involve critical combinations of the response of survivors and some or all parts of the larger society affected, not merely the emergency or civil defence and humanitarian organisations.

Emergency measures are sometimes too concerned with imposing order and control by state agencies at the expense of the immediate needs of affected communities, which must look elsewhere or take care of their own. We have many examples of heavy-handed or corrupt emergency measures that worsen conditions and losses for victims, or cause dismay and anger in local societies more aware of, and often better able to treat, the needs and concerns of survivors. Indeed, in those cases where emergency and humanitarian assistance is of foremost importance, the victims have usually abandoned, or have been abandoned by, their society. In practice, successful disaster responses require a balance between, and mutually reinforcing mix of, survivor and society-wide coping, and emergency assistance.

Responses to loss and damage

Since failures occur in all systems, means to recovery are critical.

Charles Perrow, *High-risk systems* (1984, 92)

Once destruction has taken place and the crisis period is over, other distinctive problems arise. There are stages and opportunities associated with post-disaster conditions, the plight of survivors, and the form and effectiveness of restoration. Reconstruction, in particular, is the subject of a considerable research effort. Examples range from the rebuilding of cities destroyed by bombing, and places recovering from flood or industrial explosions, to the clean-up and rehabilitation of wilderness areas contaminated by oil spills or weapons testing. However, before we get to this point, and in order to do so, there are questions of how to and who will bear the costs. In White's original scheme it includes the alternatives of 'loss sharing' and 'loss bearing'.

'Loss sharing' refers to all the ways in which those who have not suffered disaster losses, or not the most crippling ones, try to share the burden of those who have. It might be through charitable donations, support of disaster relief and reconstruction work, or help to restore the sources of livelihood of the victims. With the declaration of a disaster, governments have powers to dip into the general tax base to defray rehabilitation costs and compensate victims. Whether they do so, and on what terms, is also a subject of research and some disagreement. Private and traditional social arrangements commonly involve more direct loss sharing. Family members, neighbours or more distant, friendly communities take survivors under their wing. They offer shelter, food, work and amenities that may cut into their own resources. They may also benefit from an influx of cheap and willing labour.

'Loss bearing' conveys the idea of people accepting, or being left to carry, the burden of loss. This may be, in effect, a continuation of 'coping'. Survivors must make do, remaining more or less deprived and burdened by the disaster. They creep slowly back to some sort of 'normality' through their own efforts, or abandon their old ways of life and perhaps their homes. Even where there is some compensation, a critical aspect of vulnerability for some survivors is simply lack of the means or assistance to recover readily, or at all.

Equally, loss bearing may follow from pre-disaster or long-term decisions by a society, or for parts of it, to 'take their chances'. A group may decide that the wealth or benefits to be obtained in a certain dangerous place or activity offset possible losses. The decision may be made to, in effect, 'gamble' with disaster rather than forgo the activity or potential benefits. This sort of risk taking, sometimes voluntary, often enforced, may be widespread, especially in a changeable or competitive environment.

Insurance schemes are, perhaps, the most important ways in which loss bearing can be planned for in advance, and perhaps made relatively painless. There have been repeated efforts in our field to develop insurance as both a risk-reducing technique and a way to exert leverage to make risky undertakings less so. If it can be handled well, payments into an insurance fund may even be used to generate further wealth against possible loss. As happens with life, accident or motor vehicle insurance, those at risk may be rewarded or penalised according to how far they act to lower their risk. Vulnerability-reducing actions, in particular, seem an attractive way to use insurance premiums to reward improved safety.

Some major impediments to this form of loss bearing, or sharing, must be mentioned. Since premiums are relative to the risk, poor people living in vulnerable areas, but unable to move to safer areas, may be unable to afford insurance. Meanwhile, few private, cooperative or even government insurance schemes have, or feel they can, insure against catastrophic losses. These, often referred to as 'Acts of God', are explicitly excluded from insurable risks, like acts of war or some fatal medical conditions. Thus the persons most at risk or the most severe disasters are most likely to be uninsured.

Another problem is how insurance may encourage *more* risky behaviour. This is illustrated by some forms or examples of crop insurance for farmers. These can be a major cushion against drought-, pest-, storm- or flood-induced losses, but they may also tempt farmers to plant crop varieties that give higher yields *if successful* but that are more likely to fail; or to open up more marginal and hazardous lands.

Modifying the context: improved well-being and general safety

Societies incapable of meeting their citizens' needs are the most vulnerable to breakdown and conflict; conflict, in turn, does lasting damage to the political, social, and economic foundations of stable and prosperous societies.

Project Ploughshares (1995, 3)

Vulnerability is primarily a sociopolitical issue rather than a question of protective technology or engineering works.

Alabala-Bertrand (1993, 204)

If risk and disaster arise from the everyday life and the development of societies, rather than just from natural 'extremes' and technological 'accident', then safety for citizens must be rooted in their general well-being. Security turns upon the allocation of risk and protection within the social order, and the guiding principles of the administrative order. This line of argument has received its greatest impetus from the vulnerability perspective. However, it takes us beyond the particular ways in which people and property are placed at risk, or lack specific defences. Rather, we return to the discussion of people's status in and ability to influence society (see Chapter 6). It directs attention to those persons and activities that are unusually vulnerable within their societal context and most likely to suffer great harm in disasters. This may be rooted in perennially adverse or deteriorating features of ecological, economic, social, legal, technological and political conditions.

If we accept Sen's (1971) views on how entitlements affect *whether* a famine occurs or *who* suffers from it, then effective measures will have to deal with much more than hazardous processes or specific vulnerability to them. Watts and Bohle (1993a,b) developed this argument further, suggesting that groups singled out as unusually vulnerable are made so in a 'tripartite structure . . . [involving] entitlement, empowerment and political economy' (p.52). They argue that the prevailing social order, or the structures and (ab)uses of power that maintain it, actively create – in their terms, 'reproduce' – vulnerability.

In such circumstances, robust and enduring reductions of vulnerability are unlikely without changes in the social order or the situation of more vulnerable groups in it. Indeed, as others have argued, temporary measures in crisis, such as famine relief, may tend to exacerbate their longer-term predicament. 'Development' that reinforces the security and power of groups, or functions that ignore or add to the vulnerability of others, increases risk. This is not to argue against humanitarian measures, or the responsibility of governments to feed the victims of local food shortages or provide aid for specific short-term needs. However, unless followed up by more permanent and reliable measures, and guided by the principle of improving, say, the food security of the most vulnerable, the net effect must be to magnify the risk of future disasters.

These are concerns that relate to, if they do not stem from, various other late twentieth century crises. Paul Elkins (1992, 1), for example, reviewing 'global change' literature, conferences and declarations, suggested that it identifies four great threats to the future of humanity:

- 'weapons of mass destruction and the overall level of military expenditure;
- hunger and absolute poverty of some 20 per cent of the human race;
- environmental pollution and ecosystem and species destruction;
- intensifying human repression resulting from increasing denial by governments of the most fundamental human rights.'

We may disagree about the relative importance of each of these, just how serious they are, and which deserves the more urgent attention. Nevertheless, they are far from being separate problems. A major effort to reduce any one of them seems unlikely to succeed without proportionate improvements in the others. However, what needs emphasis here is how each of these forms of general crisis is intimately related to disaster risks and the unusual vulnerability of those most likely to suffer in them.

Here, I refer back to the discussion of chronic and routinely treated dangers in Chapter 2. While it may be true that disaster risks break out of routine arrangements and specialised fields, the capability to withstand or recover from disaster depends upon them. Even emergency plans and services have, as their main role, to inform, mobilise and link up the existing permanent facilities, and public and professional organisations, available outside the disaster zone or that have survived within it. Vital contributions are drawn from agencies and professions permanently at work dealing with chronic dangers, whether injury and disease, public utility breakdowns, crime, military readiness or fire-fighting. As a rule, the better equipped and trained these services are and the higher their morale and performance of public service generally, the more reliable and effective is their role in disasters. However, that is a function of good government and the quality of such services at all times. Moreover, the permanent network of services, care and protection in a society plays a decisive role in whether the disasters will happen, and to whom they happen. Accessibility and geographical distribution or availability of routine care and protection are the best indicators of public safety and who is, or is not, vulnerable to extreme losses.

However, the quality of permanent, routine systems in matters of public safety and security of persons is less a matter of the total wealth of states or their most advanced facilities and more about commitment and equitable access. In particular, the existence of sectors or groups, excluded or poorly served, identifies the most likely sites of disaster. This has been shown to be more relevant, because large parts of a country or economic activity within which disasters recur actually show little or no adverse consequences. The losses are highly concentrated and borne by only certain sectors, while there are often compensating and even beneficial consequences for others. It is not at all unusual for economic conditions and improvements for some to accelerate in a country or region in the aftermath of disaster (Alabala-Bertrand, 1993). Meanwhile, everything we have seen so far indicates that disasters will only be reduced, in scope and numbers, by paying particular attention to services and improvement of life for the most vulnerable, whether they are the already hungry in famine-prone lands, or ordinary civilians in war zones. This is unlikely to occur without the same people having a greater say in political and economic affairs affecting their daily lives. In the last analysis, the four great crises identified above are each associated with *an absence* of good government, the everyday safety and security of citizens, sustainable relations to the environment, and social justice.

Organisation, practice and public policy

Action, public or private, is not simply a 'natural' step from identifying appropriate adjustments to applying them. Even with the necessary resources and publicly accepted priorities, performance depends on the arrangements to put adjustments into effect. In any practical field, what is done is critically dependent upon the social organisation for carrying it out. Whether exactly intended or not, purely scientific explanations, engineering designs, planning documents and political platforms tend to appear complete in themselves, or to be presented as such. This is to ignore the relations of goals or plans (what we intend or say we do) to organisation and practice (how, and what we actually do). No administrator, or anyone reflecting upon an organisation to which they belong, or with which they have had dealings, would think its mode of operation unimportant to its performance. A large concern in any practical field is how people or institutions carry out their tasks.

The organisational basis of disaster reduction

The challenge is to look at both organisation and practices. Here, 'organisation' is used in the widest sense of any activity of two or more persons sharing, or believing they share, common goals and practices. No organisation is merely a 'transparent' or formal order for a certain purpose. If set up to reflect what people want to do, it will be a factor, sometimes the most decisive one, constraining what they do do. And that is not to consider how institutions tend to take on a life of their own, and to be promoted and protected by their members, sometimes in ways that undermine, or conflict with, what is assumed to be their mandate.

'Practice', as its other meaning conveys, identifies how things are done more or less repeatedly. In the early stages of an institution, and in periods of great change, there may be a struggle to accommodate questions of principle and mandate to an administrative order and explore alternative ways of problem solving. But there is usually a 'learning curve' in which organisations settle down into routines, chains of command, customs and expectations: the 'culture' of the group or institution. This will tend to determine who does what, the resources they have available and the standards applied.

Flexibility and innovation may be of key value in disaster. The ability to function or bounce back, despite damaged items and loss of members, may be at a premium, yet what is done largely follows from established arrangements and responsibilities. The tasks might range from a programme of vaccination or restoring water supply to evacuating survivors or giving haven to refugees. How these are done will be shaped by more or less standardised practices.

Some preparations and responses to dangers are, or are best, taken at the domestic and local community level. Others require and depend upon the actions of various levels of government. Some may call upon international organisations, or transfers of assistance and trained persons between states. Institutions skilled in resolving conflicts among different organisations may have a key part to play. There are risks left largely to the internal safety systems and expertise of an industry. Others depend upon safe management in facilities open to the general public, such as ports and recreational areas, subway stations and arenas.

Of course, practical aspects of responses differ quite radically among kinds of organisational units. For families and face-to-face groups, the concerns and constraints are rarely the same as in formal, technical organisations. Local government or small private and professional groups may operate on very different terms from the hierarchical, bureaucratically controlled agencies of government in so-called mass society. For some humanitarian organisations, how they conduct themselves and whom they help may be more important than exactly what they do. For professional organisations, technical skill and standardised practices may be of foremost concern. In others, the chain of command may get most attention, hoping to ensure that what is carried out on the ground will reflect as exactly as possible instructions from distant centres. A prevailing concern with 'command and control' is most evident, for example, in military and police organisations, which play a major part in disaster response.

Organisations and their performance are subjects of specialised and extensive study. There are literatures and numerous case studies on fire-fighting and police units, aid agencies, social services, and the role of the military in disaster relief and civil defence. Some highly specialised work has been done on, for instance, delivery of 'disaster medicine', or the organisational role of telecommunications in crises. The personal and interpersonal problems of those involved in search and rescue, or peacekeeping in highly hazardous situations, are of concern. So too are debates on how best to provide care for, and help restore the lives of, bereaved survivors. Less often looked at, but not less important, is how partially impaired and disorganised institutions, or those formed spontaneously by groups of survivors, act in disasters. Anthropologists and some others have investigated 'traditional' modes of preparedness and coping with disasters. They highlight the risk-averting strategies of hunter–gatherer peoples, nomadic herders and peasant agriculturalists.

Some of the most important and vexed questions of organised response to public dangers involve relations among different organisations, places and people. Disasters break down, as well as overlap, barriers between institutions and jurisdictions, when they do not actually destroy the organisations involved. We saw how technological disasters will often arise in the ill-defined or unexamined areas between organisations. They may hinge upon places of ill-defined jurisdiction, or where they come into uncertain contact with other institutions and the general public. Disasters challenge or destroy the familiar, everyday phenomena of 'in-house' operations and professional 'turf'.

The emergence of the non-governmental organisation

Geographers and other professionals have given most attention to governmental responsibilities. However, a striking development in the late twentieth century is the growth and influence of Non-governmental Organisations (NGOs). Many of them are involved, directly or indirectly, with the dangers and crises that concern us. Some have local and quite particular concerns, like those who run 'food banks' for the impoverished in North American cities, or clinics and self-help projects in the cities of South Asia. Others, like Oxfam, Amnesty International and Medecins Sans Frontières, have a global presence. Small or large, however, most have sprung up around pressing issues of human suffering, the protection of endangered places and peoples, or particular crises.

Many are not merely 'non-governmental' but, in one way or another, work for solutions that governments are not delivering, or not adequately and justly. Some, like Amnesty International, peace movements and champions of the poor and oppressed, spend much of their efforts monitoring or confronting the actions of governments, challenging their policies and lobbying for change. Others, like the Mennonite Central Committee and the Society of Friends, go quietly about the business of bringing relief and assistance directly to the victims of disaster, and try to stay out of 'politics' and high-profile public debates. A few, like Greenpeace, are involved in confronting corporate enterprises that they see as causing or risking great damage to land and life.

There are observers who see NGOs, or some of them, as unsatisfactory or overly biased in the way they deal with these problems. Whatever one thinks of them, today they are a major part of the organised response to risk and disaster, sometimes the most influential or exemplary part. They now include a number whose primary interest is in hazards and vulnerability mitigation, and disaster preparedness and recovery. Some of them argue strongly that their type of operation is more sensitive, and much more effective, in these situations.

Choices, values and powers

> The Red Cross, born of a desire to bring assistance without discrimination to the wounded on the battlefield, endeavours . . . to prevent and alleviate human suffering wherever it may be found. Its purpose is to protect life and health and to ensure respect for the human being. It promotes mutual understanding, friendship, co-operation and lasting peace amongst peoples.
>
> *International Red Cross Handbook* (1983)

> The best protection for the victims of plague will be a community that already has taken social justice to heart.
>
> Mary Douglas (1992, 120)

Ethics and disaster

Humanitarian concern and philanthropic gestures, expressions of deep concern and unselfish helping, are an integral part of responses to disaster. But are they a necessary part? They may seem to take us to the opposite extreme from practicalities and operations discussed in the previous section. The view developed here is that practices not informed by principle are meaningless; principles without action, worthless. Nevertheless, just as it seems convenient to look at kinds of adjustment, so the ethical principles that should inform carrying them out are the focus of other discussions. In modern settings, some people or institutions seem much more, if not largely, concerned with principles, others

191

with practice. The General Assembly of the United Nations or the supreme courts of Canada and the United States, and many religious teachers, would be examples of those mainly concerned with principles in the broadest sense.

As shown in the 'mission statement' of the League of Red Cross Societies, quoted above, many of those involved in practical, organised response to dangers and disaster define their work in terms of morality and justice. Nationally or internationally, it is widely agreed that the relief of suffering should not be contingent upon who is a victim, or the preferences of the people providing assistance. This has seemed fundamental to professionals and philanthropists, as well as to some religious groups, and on practical as well as ethical grounds. Assistance can only be effective if rooted in sympathetic and friendly attitudes toward, and peaceful, even-handed relations with, any who may need and request assistance. Thus, in addition to the humanitarian ethic quoted above, the International Red Cross also binds itself to applying several other fundamental principles: impartiality, neutrality, independence (in the sense of not being beholden to governments or other clients), and universality.

Such principles and others like them are elaborated and entrenched in a range of international conventions, protocols and declarations agreed to by most or all state governments, many with a bearing on our concerns. Often these reflect, more than anything else, how the modern history and lessons of public safety and security involve excessive state or authoritarian control as much as too little. For most people, risks from dangerous enterprises turn out to arise from the 'efficiency' and power of technocratic, impersonal institutions more often than from their failures. There is no indication that such grave dangers arise from too much civil freedom, high-quality and equitably distributed social services, devolution of powers, or 'traditional' arrangements for social security and disaster response.

Meanwhile, as we saw in Chapters 4 and 5, scientific discoveries, or their exploitation by narrow interests, have produced some of the gravest dangers and disasters we have to confront. 'Social theories', such as those that informed the ideology and practices of Nazism, Stalinism or the Pol Pot regime in Cambodia, have been used to justify some of the greatest disasters of the century. The gravest developments in insecurity, untimely death, and the uprooting and impoverishment of peoples have derived from totalitarian violence, monopolistic deployments of science and technology, and a 'war paradigm'. There is no basis for believing that technical work, how ever sophisticated, will avoid unjust allocation of risks, or be prevented from creating a more dangerous world, unless subject to ethical and humanitarian constraints. Ethics, and the responsible use of powers, are at the heart of the problem.

In any particular situation or field, as Mitchell and Draper (1983) pointed out, ethics may involve several different levels of conduct and regulation. There may be 'individual self regulation' by the researcher, consultant or practitioner. There are 'disciplinary responses' or 'institutional controls', perhaps in the form of review procedures or codes of professional conduct. 'External controls' refer mainly to governmental guidelines or legislation that might apply to conflicts of interest, safety standards, or the rights of citizens to protect themselves against unwanted intrusions or secrecy in matters affecting their well-being and safety.

Action, choices and powers

In her study *The Human Condition*, Hannah Arendt (1958) described 'activity' as the defining characteristic of our life in the world. Following Aristotle, she discussed this in terms of the combined activities of labour, work and 'action'. Labour refers to the necessities of biological existence by which life itself is sustained. Work is what we do to fabricate and maintain the artificial world of things made with, but distinctly different from, natural phenomena. This provides for a material culture transcending individual and biological existence. 'Action', however, she treats as something more than and distinct from physical change or observed 'behaviour'. It is about intelligent, motivated deeds, 'the capacity of beginning something anew' (p.9). Acts of communication, in particular, allow a public life and 'politics', in the original sense of the practice of debating and choosing ways to order and direct that life, to arise. In particular, Arendt emphasises the irreducible links between speech and action as the defining and realising arena of collective human life. 'Webs' of dialogue and shared acts 'empower' what is done, and define the distinctive nature of political power as something quite distinct from any measures of wealth, although it may be used to amass or defend these. To be powerless in something is to be incapable of sharing and realising one's thoughts in action. This is a way of describing the terms in which people differ from things, and government differs from mechanical order.

Among other things, Arendt set out to challenge and refute the rather common view in the sciences that human life is ultimately a function of, or only explained by, the material conditions of our existence. As we have seen, nowhere is this more prevalent than in studies of risk and crisis, and in the 'hazards paradigm' in particular. However, in rejecting a deterministic view of natural hazards, Gilbert F. White situated the practical issues of hazards within problems of human choice and public policy. In Arendt's terms, that is to see it as a problem of priorities and action, not merely of knowledge and technique.

Conscious investigation of hazards and disasters is surely concerned, above all, with deliberate activity to avoid or cope with dangers: to mitigate, respond to and recover from disaster. Following Arendt and the spirit of White's 'choice of adjustments' approach, humans appear, for good or ill, as active participants in all aspects of private and public risks. Unless seriously impaired, or deprived of their humanity by external forces, they continually converse about, and act to moderate or exaggerate, dangers. In general, people's treatment of hazards and disaster will be as wide and creative, or as narrow and suppressed, as their social context allows. If creative living and works are our unique achievements, survival of self and people in the face of

193

continually changing, often threatening, developments are the decisive contexts of human ingenuity and adaptability.

However, what can be hoped for or done by human action can also be denied or undone by other humans. Some voices can be much louder than others. The words guiding action may derive not from dialogue but from dictates. Without the power to choose, or to have one's concerns listened to and weighed justly, knowing and not knowing the possible adjustments are of little use. To be powerless in something is to lack the capacity for action. That is the most serious relation between greater vulnerability and the inability to influence public affairs. For this reason some say that it is never enough, nor a likely way to reduce danger, to import and impose technically more sophisticated protection from one society or context to another. More important is to empower the vulnerable, so that they can develop or demand safer living for themselves. This may apply to women or peasant farmers, to dependent countries or ethnic groups. It requires us to look to legal and other instruments to ensure social justice.

Suggested reading

Alabala-Bertrand, J. M. (1993) *Political economy of large natural disasters: with special reference to developing countries.* Clarendon Press, Oxford.

Arney, W. R. (1986) *Experts in the age of systems.* University of New Mexico Press, Albuquerque.

Hass, J. E. *et al.* (eds) (1977) *Reconstruction following disaster.* MIT Press, Cambridge, Mass.

International Federation of Red Cross and Red Crescent Societies (IFRCRCS) 1994 *World Disasters Report*, 1994, Martinus Niihoff, Dordrecht.

Kates, R. W. and I. Burton (eds) (1986) *Geography, resources, and environment*, 2 vols. University of Chicago Press.

Kent, R. C. (1987) *Anatomy of disaster relief: the international network in action.* Pinter, London.

Maskrey, A. (1989) *Disaster mitigation: A community based approach.* Oxfam, Oxford.

Palm, R. I. and M. E. Hodgson (1992) *After a California earthquake: attitude and behavior change.* Geography Research Paper no. 233, University of Chicago.

White, G. F. (1964) The choice of use in resource management. *Natural Resources Journal*, **1**, 254.

Part 2

Communities at risk, places of disaster

CHAPTER 8

'Unnatural' disasters: the case of earthquake hazards

First the earthquake, then the disaster.

Doughty (1986, 48)

Today, humans are playing too large a role in natural disasters for us to go on calling them 'natural'.

Wijkman and Timberlake (1984, 11)

General remarks

During this century, some 1.5 million people have been killed in earthquakes, more than half of them since 1950. Perhaps ten times as many have suffered serious injury, and more than 100 million have had their homes destroyed. Since the Second World War, earthquakes have initiated over 200 major disasters (Table 8.1, Fig. 8.1). Many more have caused some destruction.

The disasters follow directly from the occurrence of large earthquake events, and most of them have occurred in regions with relatively high levels of earthquake activity. Nevertheless, a great many more earthquakes of potentially damaging size occur without giving rise to disasters. Actual destruction is always dependent upon the presence and character of human settlement and land uses. The worst catastrophes are generally those in which severe shaking affects large, densely occupied urban areas. However, this is less frequent than disasters affecting rural and small settlements, partly because these involve much larger areas within earthquake-prone regions. In this case, too, destruction tends to be proportional to the number, size and density of affected settlements. This, in turn, will be shown to reflect their vulnerability as a function of human activity and socioeconomic conditions that may or may not show regard for earthquake dangers and risk-reducing practices.

197

Table 8.1 Earthquake disasters. Selected examples of catastrophic earthquakes. 1945–1995

Place	Date	Disaster features	Deaths
China, Shensi Prov.	2 Feb., 1556	Most lethal earthquake disaster known, affecting an immense area of central China. Details few but death toll from imperial records accepted.	820 000
India, Calcutta	11 Oct., 1737	Most lethal recorded in subcontinent.	300 000
Portugal, Lisbon	1 Nov., 1755	Catastrophe damaging many other centres in SW Iberian peninsula and Morocco. In Lisbon, Portuguese capital, 30 000 + deaths, 80% of housing destroyed and most buildings of religious, commercial and cultural value. Much of damage due to fires raging through city for six days, helped by strong winds. Widespread damage to port and waterfront areas due to tsunamis and, inland, to landslides. Singular effects on European religious and scientific thought.	50 000 + (?)
Italy, Calabria	5 Feb.– 27 March, 1783	Series of large shocks destroyed 181 towns and villages. Countless landslides, some very large, often burying people and structures, and changing the landscape. Landslide-generated coastal waves drowned 1000 + persons. Later, 10 000 + survivors died in epidemics.	30 000 +
Italy, Sicily and Calabria	28 Dec., 1908	Worst twentieth century disaster in Italy. Greatest devastation in Messina with 98% homes wrecked, most public and commercial buildings collapsed, and at least 45 200 dead, possibly 58 000. Almost as severe in Reggio di Calabria with about 10 000 deaths. Great fires caused much damage, and tsunamis along coasts drowned 220 persons.	47 000– 82 000(?)
China, Kansu	16 Dec., 1920	One of most severe this century, and the second most lethal. Ten ancient cities razed and thousands of villages. Immense landslides buried whole communities and dammed rivers. Many thousands perished from exposure to freezing temperatures.	180 000

Table 8.1   contd.

Place	Date	Disaster features	Deaths
Japan, Kanto Plain	1–2 Sept., 1923	Earthquake and strong winds fanning mass fires in congested areas of capital city, and in port city of Yokohama. Most lethal Japanese earthquake, injured 20 000 and destroyed homes of 500 000. 44% Tokyo and 28% Yokohama burnt out. 'Firestorm' engulfed and burned to death 40 000 sheltering in military clothing depot. Thousands also died seeking safety from fires in canals and waterfront when oil storage depot exploded and spilled flaming oil into bay. 1.7 million had homes destroyed or severely damaged. Inland great devastation from ground failure and landslides, and from tsunamis along shorelines. 4000 Koreans, blamed for 'offending the spirits' and causing the disaster, were beheaded!	143 000
Pakistan, Quetta (British India)	30 May, 1935	Great devastation to capital of Baluchistan.	30 000
Chile, Chillan	25 Jan., 1939	Second most lethal in Western Hemisphere. *70% of the dead were children*, while 60,000 persons were injured, 700,000 lost their homes. Five major centres devastated. Chillan, closest to epicentre, razed except for three buildings, and Concepcion 70% destroyed.	50 000
Turkey, Erzincan	27 Dec., 1930	Most lethal in modern Turkey, with hundreds of thousands injured and made homeless. Erzincan city razed to ground, all its doctors and nurses killed. Large number of deaths due to severe winter weather and blizzard immediately following earthquake.	50 000
Iran/USSR, Ashkhabad	5 Oct. 1948	City of Ashkahabad in Turkmen Republic wrecked and heavy damage in Iranian holy city of Meshed.	19 800

199

Table 8.1   contd.

Place	Date	Disaster features	Deaths
Iran (NE), Dasht-e Bayaz	31 Aug., 1968	Some 200 towns and villages damaged or razed, leaving 75 000 + homeless. In Dasht-e Bayaz, 3000 of 3500 people killed, and 4000 of 6500 in Kalkk, which was totally destroyed. Damage attributed partly to poor adobe construction but also to location and sitting on alluvium with high water tables.	10 000
Peru, Yungay	31 May, 1970	Most lethal in Western Hemisphere. Also 50 000 injured and 200 000 left homeless. Major losses in port cities, especially Chimbote, but greatest casualties, disruption and devastation due to countless landslides in mountainous interior. Includes the most lethal example known, a catastrophic avalanche–debris flow that totally annihilated and buried the Andean town of Yungay and killed about 18 000 persons. Cold, destroyed communications and inadequate relief measures added to plight of survivors.	70 000
Iran, Ghir	10 April, 1972	Lethal and destructive throughout Fars Province. Many landslides. Levelled nearly 45 villages, Ghir and Karzin totally devastated.	17 000
Nicaragua, Managua	23 Dec., 1972	Most of capital city (405 000 pop.) razed leaving 70% homeless and half without employment.	10 000
Guatemala, Guatemala City	4 Feb., 1976	More than 1 million people made homeless, 70 000 injured. Majority poor residents of squatter areas of Guatemala City, but also poorer rural settlements. Note that Guatemala City underwent near total devastation in 1586, 1773, 1874, 1902 and 1918.	22 000

Table 8.1   contd.

Place	Date	Disaster features	Deaths
China, N. China Plain	28 July, 1976	Most lethal of the century. In and around the coal mining city of Tangshan, some 650 000 homes destroyed or seriously damaged, and 78% of industrial structures. 10 000 coal miners trapped underground, their life supports destroyed. Some destruction in Beijing and severe damage along Bohai coast. Official death toll given here is disputed, many reports saying 650 000 or more.	270 000
Romania, Bucharest	4 March, 1977	Worst in country's history, felt throughout but worst damage in capital, Bucharest. 10 500 injured and tens of thousands left homeless. Collapse of poorly constructed residential buildings large factor. Major damage in oil town of Ploetsi.	15 000
Mexico, Mexico City	1985	Widespread damage along Pacific coast but most and worst in Mexico City (see Figs 8.3 and 8.4). Injured 40 000, left 30 000 + homeless in capital. Collapse of one apartment building, Nuevo Leon, killed nearly everyone in 200 families. The General Hospital collapsed burying 600 staff and patients in rubble. Losses US$4 billion.	7000 +
USSR, Armenia	12 Dec., 1988	Major devastation and most casualties in towns of Leninakan, Spitak and Kirovakan. Dozens of villages badly damaged. 130 000 injured, 250 000 + homeless. Poor official emergency response. US$11 billion in losses.	55 000
Iran (NW), Caspian Sea area	21 June, 1990	105 000 injured and 500 000 made homeless. Devastated 100 + towns and villages, major landslide damage. Dam burst near hardest-hit town of Rasht causing more deaths and destruction.	40 000
India, Latur–Osmanabad	30 Sept., 1993	Losses US$80 million (cf. Tables 8.2, 8.9).	7600

Table 8.1 contd.

Place	Date	Disaster features	Deaths
California, Northridge	17 Jan., 1994	Thousands injured, 25 000 homeless, six important freeways and highway intersections closed. Tens of thousands without water or electricity. Estimated losses US$17 billion. Federal emergency fund requested US$6.6 billion.	56
Japan, Kobe	17 Jan., 1995	Massive devastation to port city incl. homes, railways, freeways, factories, port buildings. Major fires raged in residential areas. 50 000 injured, 300 000 made homeless. Damage US$100 billion + .	6000 +

Suggested reading[1]

Coburn, A., A. Pomonis, and S. Sakia (1989) Assessing strategies to reduce casualties in earthquakes. In *International workshop on earthquake injury epidemiology for mitigation and response.* Johns Hopkins University Press, Baltimore, Md.

Alexander, D. (1993) *Natural disasters.* Chapman and Hall, New York, chapters 2, 5 and 7. (An excellent overview of the geophysical background and impacts of earthquake.)

Earthquake damage

The built environment and domestic life

The direct threat of earthquake is, overwhelmingly, to built structures of all kinds. Residental buildings are nearly always damaged in disasters in the greatest numbers. Deaths and injuries are mainly due to collapse of buildings, most often people's homes. The largest numbers of more seriously affected survivors are those left homeless. Provision of shelter creates pressing humanitarian problems for relief and reconstruction. Thus, in general it is *domestic space* – the losses and needs of families, especially dependent members, and residential neighbourhoods – that lies at the core of earthquake risk.

The Friuli, Italy, earthquake of 6 May, 1976, provides a typical example in this regard, if studied more than most. Damage to homes was overwhelmingly the largest loss. The homes destroyed were neither built nor maintained to

---

[1]Throughout the remainder of the text, at the end of many sections, some references will be suggested to follow up the points being made. In part this is to recommend worthwhile or influential items, but also to help the reader find a way through voluminous literature.

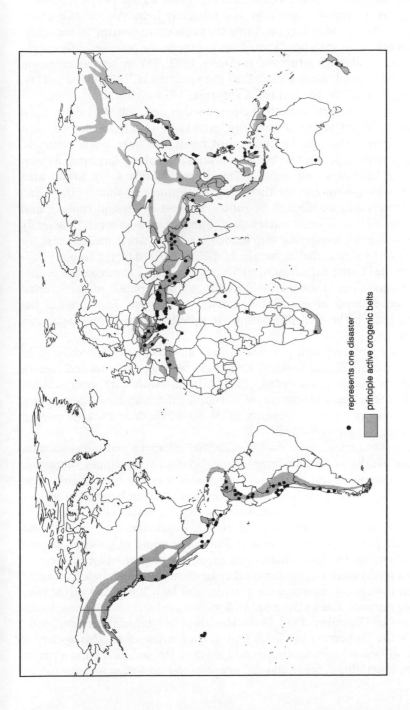

Fig. 8.1 Global distribution of major damaging earthquakes, 1950–1979 (after Hewitt, 1982)

represents one disaster

principle active orogenic belts

resist earthquakes. This was the major source of casualties. In all, about 1000 persons died and 2400 were injured. Some 32 000 homes were totally destroyed and 157 000 damaged to some extent, leaving about 89 000 people homeless. Yet damage to the regional economy was relatively light. As one assessment put it, 'typological vulnerability [meaning the inability to control "a particular sort of environmental variation"] *was bounded inside the housing system,* leaving whole the productive structure' (Pelanda 1982, 75). A similar comment could fairly be applied, among others, to the Guatemala, 1976; Mexico City, 1985; Armenia, 1987; or Northridge, California, 1994 disasters.

In many cases, however, domestic harm is further aggravated by destruction and disruption of public services and infrastructure, usually the second most widespread form of damage. Destruction of roads, rail lines, water channels, pipelines and reservoirs, and fuel and power lines can subject large populations to continuing hardships and danger. This is particularly so in urban area disasters, in which there may be thousands of ruptures to water, power and fuel lines. Streets may be blocked by rubble, severed by ground rupture, and flooded by subsidence or burst water mains. Public transit lines can be severely affected or destroyed, interfering with immediate rescue work, medical care for the injured, or food and shelter for the homeless. Such damage to built 'lifelines' contributes to the hardships of survivors, if not to a continuation of life-threatening conditions. Damage to hospitals, schools, public markets, retail stores and government offices, although fewer in number, generally has the worst impact upon the civil population. It is felt the most, once again, in domestic life.

Death to the ordinary civil population may also come through damage to commercial, industrial and cultural buildings, together with related loss of employment. Meanwhile, the many other ways in which built environments are central to the life and culture of a people (the historical associations, gathering points and sense of identity in places of residence) are threatened by this built-environment hazard.

Damage to industries, major communications networks and infrastructure, and harbour works and other large installations can sometimes be severe. Potentially widespread and disastrous consequences in the future could stem from destruction of modern facilities such as those storing large quantities of toxic, radioactive, inflammable or explosive materials; industrial plants producing hazardous materials; and, in particular, hydroelectric megaprojects and nuclear power plants. However, these receive much more of government and corporate attention, technical studies and more thorough safety measures.

There is a tendency to underestimate the significance of economic losses and wider social disruptions affecting the poorest and least influential earthquake victims. For instance, one of the more widespread and serious economic losses in the disaster of December, 1974, in the Himalaya of Pakistan was destruction of terrace walls supporting the fields and irrigation channels. This society is largely one of peasant agriculturalists and herders. As well as leaving a major task of rebuilding, this damage allowed precious topsoil to cascade downslope and be lost into the rivers. Such losses for more traditional, rural economies or low-energy technologies – a feature of the majority of disasters that occur in rural areas and less wealthy nations – rarely appear large in monetary terms. Often they are not accounted at all. Yet they may have more severe human consequences than apparently more costly examples in wealthier societies (Table 8.2).

However, this primary threat and harm to domestic life has generally been taken for granted or rarely given much attention in professional research and official actions. In keeping with the hazards paradigm, it has been treated as an unfortunate side-effect of the nature of seismic events and the performance of built structures, rather than the central reality of these disasters. Interpretations often take it as something that varies with, or follows from, national building styles or densities of living, or demographic and economic profiles. The social geography of the risks and the social implications they identify are commonly ignored. In many parts of the world, this is further aggravated by the predominance of impoverished classes and groups having little or no voice in economic and political life among earthquake victims. We will see that the association between (low) socioeconomic status, sometimes gender or age disadvantages, and losses in earthquake is often the strongest aspect of the social geography of harm. These are of central concern for a human ecology of

Table 8.2   Economic devastation in agricultural village communities, Maharashtra, India, 1993. The survey covered 69 villages, with a total of 34 446 households and 170 954 persons, in the Latur and Osmanabad districts of the Deccan region. The data show the impact upon domestic and agricultural assets, whose real severity for those affected is not at all conveyed by comparisons with disasters in wealthier regions or equivalent monetary losses (after Parasuraman, 1995, 159, 167–8)

Houses completely destroyed:	
In 32 villages	> 90%
In 37 villages	50–90%
Livestock losses:	*% killed attributed to households*
Cattle	19.1
Buffaloes	25.9
Goats and sheep	62.6
Donkeys	37.9
Bullocks	12.3
Poultry	69.5
Destroyed agricultural assets:	*%*
Bullock carts	36.6
Tractors	48.9
Ploughs	50.2
Pumps	47.8
Cattle sheds	67.2
Sprayers	62.3

earthquake risk and, arguably, should be the main focus of risk reduction and disaster relief measures.

Suggested reading

Geipel, R. (1982) *Disaster and reconstruction: the Friuli (Italy) earthquakes of 1976* (trans. P. Wagner). Allen and Unwin, London.
Hewitt, K. (1976) Earthquake hazards in the mountains. *Natural History*, **85**(5), 30–7.
Parasuraman, S. (1995) The impact of the 1993 Latur–Osmanabad (Maharashtra) earthquake on lives, livelihoods and property. *Disasters*, **14**(2), 156–69.

Damage forms

The *primary* or direct damage due to earthquake is physical or mechanical and, as noted, is largely to built structures. A visit to an earthquake disaster area is an object lesson in the forms of structural weakness and failure, but also, in buildings that survive, of what provides strength. Damage ranges from cracking and flaking of plaster or toppling of loose, unstable objects on shelves to the total collapse of buildings. Other direct impacts affect the ground surface through the loss of bearing strength and slope failure. The range and scale of such primary damage phenomena are invoked in various intensity scales, ranking observable impacts. The most widely used is the Modified Mercalli Intensity Scale (Table 8.3).

The importance of these damage phenomena is further reflected in the forms of death and injury. In most cases these are due mainly to crushing and burial in buildings, or fatal blows from falling masonry and other materials broken or shaken loose. The evidence is not well-documented, but it seems clear that most injuries have the same causes. After a disaster, survivors in hospitals and the streets commonly have head injuries, broken bones, ruptured soft organs, cuts and severe bruising. A survey of 4803 persons hospitalised after the 1993 earthquake in India showed 18.4 per cent with injuries to the head, 24 per cent with injuries to the upper body and 14 per cent with injuries to the lower body (Parasuraman, *op. cit.*, 165).

Most people buried but not immediately killed in the rubble of ruined buildings also have more or less severe injuries of these sorts. The seemingly miraculous survival of those buried for a week and more is a major beacon for rescue efforts and anxious families. Unfortunately, very few people survive more than 12 hours. This is why rescue efforts by family members and other local, unharmed survivors usually save the most lives.

Suggested reading

Glass, R. I. *et al.* (1977) Earthquake injuries related to housing in a Guatemalan village. *Science*, **197**, 638–43.

Table 8.3 Earthquake intensity: the Modified Mercalli Scale (after Wood and Neumann, 1931)

I   Not felt except by a very few under especially favourable circumstances.

II   Felt only by a few persons at rest, especially on upper floors of buildings. Delicately suspended objects may swing.

III   Felt quite noticeably indoors, especially on upper floors of buildings, but many people do not recognise it as an earthquake. Standing motor cars may rock slightly. Vibration like passing of truck. Duration estimated.

IV   During the day felt indoors by many, outdoors by few. At night some awakened. Dishes, windows, doors disturbed, walls make cracking sound. Sensation like heavy truck striking building. Standing motor cars rock noticeably.

V   Felt by nearly everyone, many awakened. Some dishes, windows, etc., broken; a few instances of cracked plaster; unstable objects overturned. Disturbances of trees, poles and other tall objects sometimes noticed. Pendulum clocks may stop.

VI   Felt by all, many frightened and run outdoors. Some heavy furniture moved; a few instances of fallen plaster or damaged chimneys. Damage slight.

VII   Everybody runs outdoors. Damage negligible in buildings of good design and construction; slight to moderate in well-built ordinary structures; considerable in poorly built or badly designed structures; some chimneys broken. Noticed by persons driving motor cars.

VIII Damage slight in specially designed structures; considerable in ordinary substantial buildings, with partial collapse; great in poorly built structures. Panel walls thrown out of frame structures. Fall of chimneys, factory stacks, columns, monuments, walls. Heavy furniture overturned. Sand and mud ejected in small amounts. Changes in well water. Persons driving motor cars disturbed.

IX   Damage considerable in specially designed structures; well-designed frame structures thrown out of plumb; great in substantial buildings, with partial collapse. Buildings shifted off foundations. Ground cracked conspicuously. Underground pipes broken.

X   Some well-built wooden structures destroyed; most masonry and frame structures destroyed with foundations; ground badly cracked. Rails bent. Landslides considerable from river banks and steep slopes. Shifted sand and mud. Water splashed (slopped) over banks.

XI   Few, if any, (masonry) structures remain standing. Bridges destroyed. Broad fissures in ground. Underground pipelines completely out of service. Earth slumps and landslips in soft ground. Rails bent greatly.

XII   Damage total. Practically all works of construction are greatly damaged or destroyed. Waves seen on ground surface. Lines of sight and level are distorted. Objects are thrown upward into the air.

Secondary damage

Calling an earthquake a geological hazard not only identifies its source in the solid Earth but also defines it as a risk that affects the stability of the land surface. Here damage arises from the indirect or derived impact of seismic shaking. In a large fraction of earthquake disasters, damage and death come from the collapse or shifting of weak, unconsolidated materials under buildings and landslides triggered on unstable slopes (Table 8.4). While ground and slope failures are direct effects of earthquakes, they in turn may do great secondary damage. Large numbers of landslides are commonly reported, and also locally disastrous ground failures (Table 8.5).

The larger part of the devastation and some 180 000 deaths in the great Kansu, China, catastrophe of December 1920 came from slope failures in weak loess and alluvial deposits, and some huge landslips. Many destructive individual landslides have occurred in earthquakes. The most devastating example in this century in the Western Hemisphere occurred in the May 1970 Peruvian disaster. The earthquake triggered an avalanche of ice and entrained debris from Nevados Huascaran (6665 m) in the Cordillera Blanca. The avalanche was 50 million to 100 million cubic meters in volume, travelled about 14.5 km and attained speeds in excess of 300 km/hour. A tongue of the debris avalanche killed some 18 000 persons in and around the town of Yungay. All of that lovely town was annihilated, and barely 300 of its resident population of 4500 survived. It may be added that thousands of landslides occurred elsewhere in the Cordillera Blanca. In many other areas too, especially coastal settlements, destructive ground failure occurred by subsidence, fissuring and liquefying (see page 200). Together, these secondary effects were responsible for major losses over an area of some 103 000 sq. km.

In coastal zones in active seismic regions, another secondary hazard is the seismic sea wave or *tsunami*. This is a water wave generated by sudden shifts in the sea bed. The greatest risk is usually to nearby shores. Undersea earthquakes may be accompanied by large changes in offshore topography, causing the sea to retreat suddenly but soon afterwards to rush back in as a huge, devastating wave. Some of the worst damage has accompanied such waves, together with high death tolls where people remained beside or even ran down to the exposed shore areas, only to be engulfed by the returning tsunami. Submarine earthquakes, even those that are not destructive where they occur, may generate tsunamis that cross thousands of kilometres of ocean to wreak havoc on a distant shore (Table 8.6).

Tertiary damage

Severe physical damage and harm may arise from processes only indirectly related to shaking, or to ground or slope failures. Fires started in collapsed buildings, perhaps from household fireplaces, broken gas mains or other fuel or inflammable items, can spread to cause damage, occasionally of the worst type. About four-fifths of the property damage in San Francisco in the 1906 disaster was attributed to the great fire. Tokyo suffered a similar fate in the 1923 Kanto Plain earthquake, the fires also being the major factor in some 110 000 deaths. In both cases, the wind acted as an important intervening factor to fan and spread the flames. Damage to water supply systems hampered fire-fighting efforts.

Table 8.4 Examples of earthquake disasters in which loss of strength and collapse of 'susceptible' soils or alluvial deposits resulted in major devastation

Date	Location	Damage
22 May, 1950	Cuzco, Peru	Worst damage on thickest part of Rio Huatanay alluvial fan, water saturated to within 1 m of surface.
21 July, 1952	Kern County, California, USA	Subsidence, fissuring slumps in alluvium of San Joaquin Valley.
21 July, 1956	Rann of Cutch, India	Widespread fissuring, subsidence of alluvial plains associated with main damage.
28 June, 1957	Mexico City, Mexico	96% houses damaged built on old Lake Texcoco floor (cf. 1985 disaster).
4 Dec., 1957	E. Altai Mts., Mongolia	Break up, fissuring, outbursts of ground water on alluvial fans and flats associated with main damage.
29 Feb., 1960	Agadir, Morocco	Port facilities on artificial fill severely damaged by subsidence, fissuring.
21 May, 1960	Concepcion, Chile	Soil liquefaction, fissuring, subsidence, major source damage at Valdivia, Puerto Montt and other port towns. Major damage at sites of artificial fill.
26 July, 1963	Skopje, Yugoslavia	Great destruction almost wholly confined to area of thick alluvium on flood plain of Vardar R. in downtown Skopje.
27 Mar., 1964	Alaska, USA	Innumerable examples of fissuring, subsidence, slumps, sand boils, etc. on coastal and riverine alluvial sites.
16 June, 1964	Niigata, Japan	Spectacular damage to high-rise buildings, port, and transport facilities due to soil liquefaction and subsidence in sand of coastal and deltaic alluvium.
29 July, 1967	Caracas, Venezuela	Particularly destructive to buildings on alluvial sites.
1 Aug., 1968	Luzon, Philippines	Damage, well outside epicentral area, due to alluvial subsoil conditions, especially in Manila.
28 Mar., 1970	Gediz, Turkey	Much of damage associated with poorly consolidated valley floor sediments and high water table.
10 Apr., 1972	Ghir, Iran	Intensity of shaking and damage greatest along alluvial tracts beside rivers.
1 Feb., 1974	Izmir, Turkey	Nearly all heavy damage, regardless of built structure, on alluvial valley fill.
19 Mar., 1976	Khulm, Afghanistan	Damage in Khulm concentrated in lower, finer-grained sediments of alluvial fan site. Damage due to subsidence in alluvium.

Table 8.4   contd.

Date	Location	Damage
27–28 July, 1976	Tangshan, China	Parts of Tangshan on unconsolidated alluvium that fissured and settled, assisting in the great damage.
12 June, 1978	Miyagi-ken-Oki, Japan	Soil liquefaction resulted in damage to concrete structures, bridges, and dikes around Bay of Sendai and port of Ishinomaki.
19 Sept., 1985	Mexico City, Mexico	Most of catastrophic damage on susceptible old lake bed alluvium.
1989	Lomo Prieta, California	Areas of construction on saturated fill and bay mud had major damage and loss of life.
1995	Kobe, Japan	Devastation and building collapses on coastal alluvium and artificial fill.

Dangers from landslides do not end with the moving mass of earth or rock. The most disruptive and widespread damage in many recent earthquakes has been from landslides burying or carrying away roads. Small and large slides may also dam rivers. The rapid failure or breaching of such dams can generate large flood waves, with calamitous results (Table 8.7).

Indirect threats include the exposure of the homeless to inclement weather, a particular threat for injured survivors, small children and the elderly. The severe weather may prevent victims being taken to safety and medical care,

Table 8.5   Large landslides or large numbers of landslides in earthquakes: selected examples of earthquake disasters, 1950–1995, in which a major part of the damage involved landsliding

North America		East, Central and Southwest Asian mainland	
1959	Hebgen Lake, Montana	1950	Assam, India
1964	Prince William Sound, Alaska	1953	Turud, Iran
1971	San Fernando Valley, Calif.	1956	Hindu Kush, Afghanistan
		1957	Altai Mountains, Mongolia
Central and South America		1957	Elburz Mts., Iran
1960	Valdivia, Chile	1962	Elburz Mts., Iran
1970	Peru	1966	Varto, Turkey
1973	Orizuba, Mexico	1970	Gediz, Turkey
1974	Arequipa Province, Peru	1972	Zagros Mts., Iran
1976	Guatemala City, Guatemala	1974	Pattan, Pakistan
1992	Colombia	1975	Kinnaur, India
1994	Paez River, Colombia	1988	Armenia
Southeast Asia and Australasia			
1964	Niigata, Coastal Mts., Japan		
1970	Madang, New Guinea		
1993	Goroka, Papua New Guinea		

Table 8.6 Selected earthquakes with severe, sometimes mainly, tsunami damage

Date	Place	Damage
1964	Prince William Sound, Alaska	Major coastal damage.
1965	Celebes Sea, Indonesia	Extensive damage along coastlines.
1966	Lima and Callao, Peru	Extensive waterfront damage.
1969	Celebes, Indonesia	Extensive shoreline damage, 600 drowned.
1973	Hilo, Hawaii	Shoreline damage to piers, roads.
1973	Hokkaido, Japan	Main losses in 2 m tsunami.
1975	Kilauea, Hawaii	Widespread devastation from several waves (2–4 m) in quick succession.
1976	Mindanao and Salu Is., Philippines	Enormous shoreline devastation and deaths (4.5 m waves).
1977	Sumbawa, Bali and Lombok Is., Indonesia	Great damage.
1978	Thessaloniki, Greece	Damage.
1979	Dalmatian Coast, Yugoslavia	Extensive damage to port facilities and docks.
1979	Tomaco, Columbia	Heavy damage (3 m waves), 10 villages swept away.
1992	Flores Is., Indonesia	Some 2000 people killed, 250 000 homeless.
1994	S. Java, Indonesia	200 killed, major damage to houses, fishing boats, about US $2.2 million damage.

or rescue teams from reaching them. Such problems, magnified by cold and snow under winter conditions in the mountains, aggravated the disasters at Pattan, Pakistan, in December 1974, Italian Campagna in November 1980, and Armenia in 1987.

There is risk of disease outbreaks where survivors are crowded into makeshift refuges lacking adequate sanitation, food, fuel and safe water. However, the consensus is that this danger has been exaggerated. Chronic physical ailments and psychological distress affect people in such situations too, sometimes aggravated by delayed or inappropriate government and international relief efforts. These are not, of course, problems peculiar to earthquakes.

Suggested reading

Reed, S. A. (1906) *The San Francisco conflagration of April, 1906: special report.* National Board of Fire Underwriters, New York.

Pflaker, G. *et al.* (1971) Geological aspects of the May 31, 1970, Peru earthquake. *Bulletin Seismological Society of America*, **61**(3), 543–57.

Keefer, D. K. and N. E. Tannaci (1984) *Bibliography on landslides, soil liquefaction, and related ground failure in selected historic earthquakes.* US Geological Survey, Open File Report 81-572, Menlo Park, Calif.

Table 8.7 Earthquake disasters in which natural damming occurred, and where the bursting of these or artificial dams caused major damage

Date	Place	Event
15 Aug., 1950	Assam, India–Tibet	Large landslide blocks tributaries of Brahmaputra. Devastating outburst floods.
3–6 Aug., 1951	Potosi, Nicaragua	Crater Lake burst flooding towns.
9 Sept., 1954	Orleansville, Algeria	Lamartine Dam collapsed; 200+ drowned in flood; damage to Ponte-la Dam.
8 June, 1956	Hindu Kush, Afghanistan	Main cause of death and destruction (few details).
2 July, 1957	Elburz Mts., Iran	Landslides dam streams. Largest 20 m high: reservoir 7 km long, 100 m wide.
17 Aug., 1957	Hebgen Lake, USA	a) Madison River Canyon dammed to form 'Earthquake Lake' (drained artificially) b) Hebgen Dam spillway and casting cracked.
21 May, 1960	Concepcion, Chile	a) Many landslide dams on Andean streams. b) Large slide 27 m high and 1500 m long raised level of Rinihue Lake.
24 May, 1968	Inangahua, New Zealand	Buller River dammed by slide. 40 ft high, 4 mile long lake. Later raised by second slump.
3 Sept., 1968	Bartin-Amasra, Turkey	Buyuk Dere R. dammed by large landslide forming lake.
8 July, 1971	Valparaiso, Chile	a) Cerro Negro tailings dams collapsed and four others damaged. b) Embankment of water dam collapsed.
10 May, 1974	Yunnan Prov., China	Landslides blocked stream valley. Reservoir 30 m deep at dam.
19 Jan., 1975	Kinnaur, India	Parachu valley dammed by major slide. Reservoir 60 m deep at dam. Overflow changed river course.
21 Mar., 1977	Bandar Abbas, Iran	Huge rockfall blocked Bohregh Canyon.

Earthquake risk

Having reviewed the kinds of damage earthquakes may involve, the conditions or elements of risk that contribute to them can now be examined. Discussion will be confined to some of the more salient ones.

Seismic dangers

The role of the earthquake itself is governed by geophysical and geological conditions. Earthquakes originate more or less far down – tens to hundreds of kilometres deep – in the Earth's crust. Their size and frequency depend upon stresses and fractures in crustal rocks. Those that may do damage involve sudden release of large forces and transmission as seismic waves through the Earth. Actual danger arises as these waves reach and interact with the Earth's surface.

The place within the crust or lithosphere where an earthquake is generated is referred to as the source, focus or hypocentre. The epicentre is the spot on the Earth's surface immediately above. Properties of the source or 'focal' mechanism commonly identified with risk are depth and magnitude. There is a rapid decline in earthquake numbers below about 200 km and though some of the largest are deep-focus, damage is mostly from shallow ones. Regions with more frequent, large, shallow earthquakes, such as southern California, feature prominently in the disaster count.

'Magnitude' refers to the energy release or, at least, to some measure of the relative size of the seismic event. It should be distinguished from 'intensity' which, as shown earlier, refers to observed effects at the surface. The 'Richter Magnitude' is widely quoted and is often assumed to bracket risk. The Richter Scale used today – open-ended, but generally presented as running from 1 to 9 – derives from the maximum displacement of the trace on a seismograph, corrected for distance from the source. The value reported is actually the logarithm of the displacement, so that each unit increase of magnitude is ten times greater than the one below. Few destructive earthquakes have a magnitude of less than 5. The majority of disasters involve magnitudes between about 6 and 7.5. High magnitudes are much rarer. Very few in this century are thought to have reached 9.

Special interest in geological faults or ruptures relates to whether they are active, hence guides to possible seismic sources, and whether they reach to and undergo displacement at the surface. A compelling image is of rupture at the surface, perhaps slicing roads or buildings in half.

It is easy to see why, from a seismic point of view, magnitude and other properties of seismic waves appear significant, since geophysical behaviour follows largely from the initial energy release and wave propagation. However, the relationship between seismic magnitude and the severity of, or damage in, earthquake disasters is quite weak (Table 8.8).

Again, fault-related surface changes are invaluable guides to geological interpretation. They are rarely significant in terms of overall damage, mostly causing localised if unusual losses.

Suggested reading

Reiter, L. (1990) *Earthquake hazard analysis: issues and insights.* Columbia University Press, New York.

Intervening variables

The depth, magnitude, distance and direction of the earthquake source provides the original seismic energy. The seismicity that affects society essentially
involves acceleration and amplitude of ground motion at the surface, especially
so-called strong motion. In this, the intensity or effects of seismic shaking can
be greatly modified by near-surface rock and soil; by ground slope and topography; by surface, soil and ground water conditions; and by vegetation cover.
The greater the depth of, or distance from, the source the more critical these
can become. These conditions act as 'intervening variables' of earthquake
danger. Severe damage within disaster zones is commonly associated with
locally unstable surface conditions.

Seismic waves may be amplified on a steepening slope and certain forms of
salient or cliff, and at the interfaces of bedrock and unconsolidated earth
materials. The amplification of earthquake effects and ground failure in 'susceptible' alluvial deposits has been a serious cause of building collapses, often
in buildings of otherwise favourable earthquake-resisting properties. Dramatic
effects are associated with the passage of seismic waves into areas of deep
alluvium, especially where the water table is close to the surface (Fig. 8.2).
The safest foundations are usually on intact, unweathered bedrock; well-
drained, coarse soils are the next best. Moist clay soils, certain sandy deposits,
especially if wet, and artificial fill are likely to prove unstable foundations. The
most dangerous sites are on deep, fine-grained alluvial materials with high

Table 8.8 Selected earthquake disasters of 'moderate' magnitudes (less than Richter
M = 7), showing a large range of damage and loss of life

Date	Place	Richter	Deaths	US dollar loss estimates
1949	Amatao, Ecuador	6.8	6000	66 million
1954	El Asnam, Algeria	6.8	1250	–
1960	Agadir, Morocco	5.6	12 000	500 million
1963	Skopje, Yugoslavia	5.8	1200	1.5 billion
1967	Caracas, Venezuela	6.5	277	150 million
1969	Ceres, South Africa	6.3	9	24 million
1971	San Fernando, California	6.6	65	750 million
1971	Tuscany, Italy	5.0	18	42 million
1972	Managua, Nicaragua	5.6	5000	1 billion
1976	Gemona del Friuli, Italy	6.5	965	2.5 billion
1988	India/Nepal	6.7	1100+	25 million
1988	Armenia	6.7	35 000	14–15 billion
1989	Newcastle, Australia	5.5	10	792 million
1990	Sicily, Italy	4.7	20+	Total economic and physical damage 10–15 billion
1994	Northridge, California	6.6	57	30 billion

water tables. Some types undergo *liquefaction* with seismic shaking – in effect, the foundation changes from a solid to a suspension of soil particles in liquid. Buildings have been observed to sink into such liquefied materials as if they had suddenly been placed in deep water. This is a particular danger of coastal and flood plain areas (see Table 8.4).

Because so much depends upon the stability of near-surface materials and slopes, vegetation cover or its removal can play an important role. Thus a large part of the damage in a Himalayan earthquake in Pakistan in 1974 was due to landslides (Hewitt, 1972), but I found no evidence of landslides where there was a forest cover. Photographs by a European visitor to the area in the 1920s indicated much more forest cover than today. The many buildings that fell down may also relate to a timber shortage. Traditional structures with generous use of timber generally stood up well. Most newer structures, without any timber in the walls, performed badly, many collapsing on their occupants. It seemed to me this was as much a 'deforestation' as an earthquake disaster.

Human vulnerability

Since the greatest impacts of earthquakes upon society occur through built structures, their vulnerability is a paramount concern. It depends upon:

- construction materials, their quality, availability and cost;
- building design, including key items in structural integrity such as doorways, windows and roof lines, and layout of rooms;
- conditions that affect secondary and non-structural damage, notably water mains, fuel, fireplaces and chimneys, furnishings, ornamentation, storage of loose items;
- upkeep and renovation or their absence, since, over time, structural strength tends to decline through deterioration of all of the above items;
- site and situation, especially the stability of foundations, and proximity to slopes and other structures that may fail.

Where earthquake-resistant building design and land use practices have received concerted effort, as in parts of Japan, southern California and Greece over recent decades, they help explain there being fewer or less lethal events than exposure to large earthquakes would indicate. Unfortunately, vigorous application of earthquake-resistant building standards and land use planning is the exception. Domestic architecture, and the conditions under which it may be made safer, has received the smaller share of attention in work on building safety by modern researchers, even though, as we saw, it involves the greatest dangers.

As a result, vulnerability tends to reflect the more or less local 'building culture'. This involves available construction materials and their costs, economic activity, social and political organisation, the history and modern transformations of construction technique, and customary or fashion preferences. These can involve large differences within, not merely between, different socie-

215

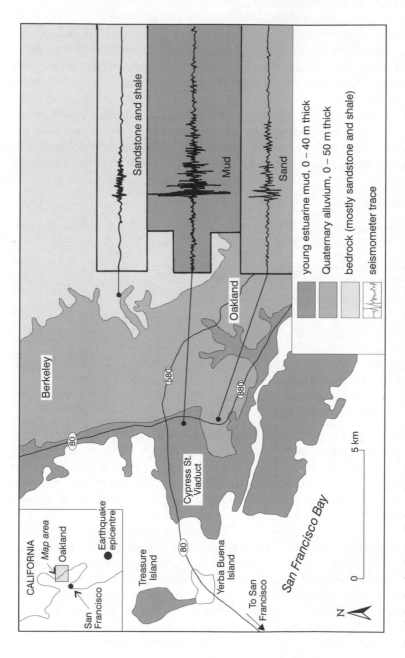

Fig. 8.2 In the Lomo Prieta, San Francisco, earthquake of 17 October, 1985, much of the damage was in areas of weak bay alluvium and soggy fill, in places with liquefaction. The seismograph traces show the dramatically greater amplitudes and more intensive shaking in the estuarine mud. This was a major factor in the collapse of the Cypress Street Viaduct in Oakland, which crushed 42 people and many motor vehicles (after US Geological Survey, 1989; and Coch, 1995, 134)

ties. In any case, people can build structures only with locally available construction materials that they can afford and with which they, or the society's professional builders, have experience.

Even more important may be choice of or access to sites, apparent from the social and land use geography of communities involved. The problem was, perhaps, most starkly evident in the February, 1978, Guatemala catastrophe, in which almost 1.2 million people lost their homes. In Guatemala City, nearly all of some 59 000 destroyed homes were in urban slums built in ravines, above and below steep, unstable bluffs, or on poorly consolidated young fluvio-volcanic sediments. Losses to the rest of the city, and among more expensive homes, were negligible, since they occupied much more stable sites. Most of the remaining loss of homes was in poor, rural communities in the more rugged interior.

Relations to social class were more complex in the Managua, Nicaragua, disaster of 1972, where 10 000 died, 300 000 lost their homes, and almost the entire inner city area was razed. By numbers, the poorest 70 per cent of the city's population suffered more than the rest of society and lost about 26 per cent of their homes. However, Haas, et al. (1977, 133–7) point out that 'lower middle-class' people suffered a 60 per cent loss of housing. Their homes were exceptionally concentrated in the core area, where almost total destruction occurred, but extreme overcrowding there had led many of the poor to 'leap-frog' to squatter settlements or barrios around the periphery of the city, which was not as severely affected. Middle- and upper-income housing loss was about 34 per cent, but there were some conspicuous examples of housing and housing complexes owned by these classes that suffered no damage.

Less wealthy people are inevitably exposed in greater numbers, since they form the vast majority of most societies. However, it does not necessarily follow that families or communities with low status, according to official economic measures, will be unusually vulnerable or even the most vulnerable members of a society. Nor does it necessarily follow from material conditions that may seem wretched compared with the furnishings and appliances in, say, an average, comfortable, Western home. There are plenty of examples of traditional subsistence communities whose buildings perform well in earthquakes. Indeed, in a great many recent earthquake disasters, unusual vulnerability and often the main destruction and loss of life are associated with new developments, new structures, neglected older buildings or, more generally, environmental and social change.

Suggested reading

Davis, I. (ed.) (1981) *Disasters and the small dwelling.* Pergamon Press, Oxford.
Jones, B. G. and M. Tomazevic (eds) (1982) *Social and economic aspects of earthquakes.* Proceedings of the Third International Conference, Bled, Yugoslavia, 1981. Cornell University, Ithaca, NY.

Who dies?

Only rarely do studies pay attention to the social and demographic distribution of earthquake casualties. The most readily available national statistics rarely provide such data, yet the problem was again starkly evident in the February, 1978, Guatemala catastrophe. Some 23 000 were killed, largely in the urban slums of Guatemala City, and poor rural communities. Losses among better-off citizens were negligible, leading some to call this a 'classquake'.

If the commonest place of death in an earthquake is the home, it could be expected that women and, perhaps, dependent children and the elderly would often be disproportionately killed and injured. Again the data are rarely made available, but the few examples reveal the problem. In the Friuli disaster in 1976, this seems to have been reflected in the almost 20 per cent greater loss of life among women than men, partly due to the timing of the earthquakes. Women were more likely to be closer to, or inside, the home early in that warm summer's evening when the earthquake struck.

In a singularly detailed survey of the 1993 disaster in Maharashtra, India, Parasuraman (1995) revealed the extraordinary vulnerability of children, women and the elderly to disaster striking homes and domestic space (Table 8.9). Deaths were disproportionately among women and children. The only age group in which more males were killed was under 15 – the age when boys are most often in the home. Whereas 445 women were left as widows by the disaster, there were 932 widowers. Whereas 7.2 per cent of widows lost both husband and children, 22.4 per cent of widowers lost both. It is striking how similar their distributions are to mortality in the urban bombing attacks described in Chapter 11. It suggests that the same persons tend to be most vulnerable to the sudden destruction of homes in an earthquake or a bombing raid, and perhaps most other devastating forces. However, in several villages in the 1974 Pakistan disaster, casualties among men were disproportionate. The main shock occurred at the time of prayer, when most men were in the mosques. Where these collapsed, casualties to men were high. In one small village all but one adult male was killed. At the same time, many women were indoors, so their losses were high too as homes collapsed. Children were outside, so their losses were fewer. As a result, the number of orphans was high. If this is an exceptional case, it reinforces the point that social conditions and practices are closely involved with vulnerability and loss in these built environment and domestic disasters.

Suggested reading

Seaman, J., S. Leivesey and C. Hogg (1984) *Epidemiology of natural disasters.* Karger, Basle.
Alexander, D. (1985) Death and injury in earthquakes. *Disasters,* **9** (1), 57–60.

Table 8.9 The social composition of earthquake deaths in agricultural village communities, Maharashtra, India 1993. As was shown in Table 8.2, the deaths were due almost wholly to the collapse of homes and reflect who was more likely to be in them when the earthquake struck (after Parasuraman, 1995, 159, 167–8)

	Age group					
	0–14	*15–24*	*25–44*	*45–59*	*60+*	*Total*
Female deaths	1833	615	959	422	478	4307
% pre-disaster female population	5.9	4.8	4.3	5.2	6.3	5.2
% dead females	42.6	14.3	22.3	9.8	11.0	100.0
% all deaths this age	48.7	61.6	63.2	60.5	58.0	58.4
Male deaths	1927	383	558	275	347	3490
% pre-disaster male population	5.8	2.4	2.4	3.3	4.2	3.9
% dead males	55.2	11.0	16.0	7.9	9.9	100.0
All deaths	3760	998	1517	697	825	7797
% all dead	48.2	12.8	19.4	8.9	10.7	100.0

	Age group		
	0–9	*10–18*	*Total*
Orphans (all)	651	861	1512
One parent dead	585	686	1271
Both parents dead	66	175	241
Female child	304	342	646
Male child	347	519	866

Total households with deaths		3465
Households with:	one death	45.0%
	two deaths	24.0%
	three deaths	15.0%
	four deaths	8.0%
	five + deaths	8.4%
	72 households lost all members	
Total deaths:	in district 8311 (including 222 visitors)	

The space of disaster events

Questions of what influences damage, and what and whom it affects, also direct attention to *where* it occurs, both in a general geographical sense, and in terms of the patterns or social geography of impacts in a disaster zone. We will

219

consider the latter first, and then broader questions of the global geography of these disasters.

The 'isoseismal assumption' and its limitations

Other things being equal, the larger and closer an earthquake is to a human settlement, the greater and more varied the damage it is likely to cause. However, 'other things' – that is, the other elements entering into risk – are rarely equal. There is little or no correspondence between the distribution of human settlement over the Earth and of earthquakes or their magnitudes, depth of focus and other properties. There should be, but rarely is, a relationship between the socioeconomic factors that decide human or structural vulnerability, and seismic conditions. Distance to sites of human occupancy is obviously critical. The smaller the earthquake or the greater the distance, the more critical are surface conditions and human vulnerability to whether it will cause damage. In effect, surface conditions and vulnerability act as 'spatial filters' of the risk, and are always crucial to damage.

It has become a commonplace of disaster surveys to plot damage or the intensity of effects as isoseismal lines of impact that fall off radially from the earthquake epicentre. Yet damage patterns are rarely, if ever, of this radial kind, and are poorly predicted by the radial attenuation or dissipation of seismic energy away from the source. The worst damage commonly occurs next to much lesser damage throughout most severely affected areas. In general, damage is highly uneven in distribution and heterogeneous in any given locality.

The appalling losses in a not exceptionally strong (M = 6.7) Armenian earthquake of 7 December, 1988, were concentrated fairly near the epicentre. This was 20 km from Spitak, a town of about 25 000 people, which suffered almost total annihilation, and 23 km from heavily damaged Leninakan, the second largest city in Armenia (pop. 300 000). Most other devastation was within about 70 km of the epicentre, yet it was equally clear that the scale of devastation and concentrated loss of life reflected recent, rapid population influx into these towns, and the poorly designed and sited residential structures in which they were housed. In Leninakan, losses came mostly from the collapse of some 120 modern high-rise structures, recently built of prefabricated concrete blocks. In all, about 35 000 persons died, and the estimated economic loss was US$14 billion to $15 billion. The death toll would have been higher but for prompt action by survivors. In the first few hours, they managed to dig out most of some 15 000 persons eventually pulled from wrecked buildings. Of these, 12 000 were injured and would surely have died without such early rescue. The toll might have been *lower*, however, if the authorities had provided tools promptly for handling the heavy rubble of concrete buildings, a major impediment to local efforts. Bad weather, poor transportation and inadequate medical facilities further aggravated the plight of survivors, and increased the death toll. Even here, then, the actual damage pattern depended critically upon the distribution of vulnerable structures or sites, and recent social change. However, such problems and damage may be concentrated far from the earth-quake source.

Enormous damage in the 1970 Peruvian earthquake was due to landslides triggered 150–200 km away from the epicentre, which was out in the Pacific. Yungay, buried in the avalanche with the most concentrated loss of life, lay 120 km away. The 1985 Mexico City disaster provides even more striking evidence, with the major damage and loss of life occurring 370 km from the epicentre (Fig. 8.3). It is not that no damage occurred closer to it and in the intervening areas. What is remarkable is that the heaviest damage was so far away. Most of the 10 000 death toll and costs were concentrated in one area of the capital city, while damage nearby was generally much smaller, if not negligible. The deci-sive factors were a combination of very unfavourable intervening conditions and vulnerable structures housing large numbers of people (Fig. 8.4).

Such events indicate that the isoseismal picture of damage as a function of nearness to the earthquake source is a theoretical notion: an abstract model that defines only one variable in a multivariate problem. Actual damage, and especially the spatial pattern and social geography of harm, more directly and closely reflects differences in human vulnerability.

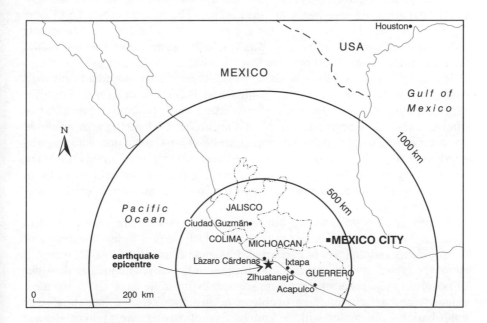

Fig. 8.3 The location of the 1985 earthquake, Mexico City, and distances from the epicentre (after Degg, 1992, 105)

Suggested reading

Degg, M. R. (1992) Some implications of the 1985 Mexican earthquake for hazard assessment. In McCall, G. J. H., D. J. C. Laming and S. C. Scott (eds) (1992).

Worst-case events: urban disasters

The scale and scope of human vulnerability means that the greatest threats or absolute risks tend to be in cities, where the density of human occupancy and dependence on built structures is the greatest. They happen more rarely than disasters for smaller settlements and rural populations, but most of the great earthquake calamities of history have involved large cities, and recent decades are no exception (cf. Table 8.8). In a survey that identified 650 modern earthquakes causing some deaths, barely 20 events involved over 10 000 deaths (Coburn *et al.*, 1989). In each case, however, high and concentrated casualties occurred in cities. The most extreme example in recent decades was in Tangshan, China, in 1976, with over 250 000 persons killed (some argue over 600 000 and perhaps as many as 800 000).

Similar arguments apply to the scale of economic losses. The greatest also tend to occur when the damage zone involves, if it is not exclusively in, urban areas. In the September 1985 disaster in Mexico City, property damage was estimated at US$4 billion, the most costly in Latin America. The death toll in the 1989 Loma Prieta, California, earthquake was 63 persons, with the economic loss also estimated at about US$4 billion. The disaster centred on Kobe, Japan, in January 1995 was probably the most costly where modern economic values apply. Property destruction was variously estimated at US$100 billion to $150 billion. Some 6000 persons lost their lives.

Incidentally, in a rapidly urbanising world, these events carried a heavy load of disappointment, if not dismay. China had become a beacon of hope for more traditional, social methods of predicting and preparing against earthquakes. Japan had been considered at the forefront of earthquake research, earthquake-resistant building design and other risk-reduction efforts, and nowhere has more funding been spent or research reported than California. Mexico City, of course, represents the emerging class of 'megacities' in Latin America, South and Southeast Asia, and the fact that most are in earthquake-prone areas.

As ever more people, wealth and large built structures are concentrated in towns and cities, these disasters in places in which there is a clear awareness of the problems and dangers of earthquakes is some cause for alarm. However, it is not the growth itself so much as the placing of far more and greater densities of people in vulnerable structures on seismically unstable land. Often the most extensive and serious of these problems lie in major centres, including large conurbations. Examples with a known risk of strong earthquakes include Mexico City, Caracas, Lima, the Los Angeles conurbation, Cairo, Istanbul, Athens, the Haifa–Tel Aviv region, Tokyo and Manila. Only major invest-

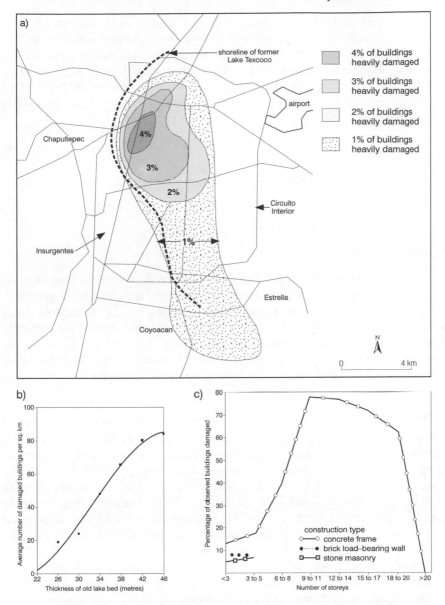

Fig. 8.4 Intervening conditions, vulnerability and major destruction in Mexico City from the 1985 earthquake (cf. Fig. 8.3): (a) The relationship between the percentages of buildings heavily damaged and the area of the bed of dry Lake Texcoco (after Blaikie *et al.* 1993, 176); (b) The relationship between the density of damage in the western part of the Lake Texcoco area of Mexico City and the thickness of the susceptible clays deposited in the lake (after Degg, 1992, 110); (c) The relationship between height (number of storeys) and type of building in the Lake Texcoco area of Mexico City and the percentage destroyed (after Degg, 1992, 110)

223

ments in earthquake-resistant structures and land uses, avoidance of areas unstable in earthquakes, and efficient disaster preparedness can counter this growing risk. In a great many cities, however, the evidence is rather to the contrary. Vulnerable structures proliferate and crowd onto unstable sites. Meanwhile, few governments and municipalities are willing to provide the means or give the freedom to local communities to improve their own safety (see Chapter 10).

Suggested reading

Berlin, G. L. (1980) *Earthquakes and the urban environment.* CRC Press, Boca Raton, Fla.
Bilham, R. (1988) Earthquakes and urban growth. *Nature*, **336**, 625–6.

Earthquake country

Turning to the broader and global geography of earthquake risk, it again seems important to work from known events and damage, rather than the geophysical hazard (see Fig. 8.2). A global survey covering the 243 most destructive earthquakes reported from January 1950 to 31 December 1990 was made to help identify conditions repeatedly associated with disasters. The data are certainly incomplete, but such information as is available allows a preliminary assessment and leads to some surprising geographical insights.

Seismic geography

All but a handful of the events were located in the most active seismic zones, notably around the margins of the Pacific Ocean and the series of mountain ranges from the western Mediterranean to China. Strings of disasters have also been shown to relate to some of the most active fault systems, notably the San Andreas and related faults in California, and the Anatolian Fault running roughly east–west through Turkey. More generally, most of the earthquakes record processes of collision and mountain-building disturbances associated with crustal plate margins. According to plate tectonic theory, seismic activity is a result of the large-scale evolution and movements of the Earth's crust. Except in a broad sense, however, the relation of the disasters to seismicity is uneven and inconsistent. There are substantial, highly seismic and tectonically active areas with few or no events, while clusters of disasters appear in areas that are not the most active. Again, in part this must be attributed to areas with greater incidence of strong earthquakes, shallow ones or other tectonic conditions that increase seismic forces at the surface.

However, there are clear indications that other factors play a large role, especially intervening surface environments and human vulnerability. This emerges when we look at the incidence of the disasters in terms of conditions that apply or are reported in a relatively large number of cases.

Disaster geography

For the inventory of 243 late twentieth century disasters, 12 attributes or variables, in addition to seismicity, were found to have more or less widespread significance. Three were derived from locational correlates of terrain and climate, and nine were repeatedly mentioned as causes of significant damage in reports (Table 8.10).

After seismicity itself, the most common feature of these events was occurrence in, or partly in, regions of high relief. Taking that to mean an elevation range of at least 1000 m within the damage zone, more than 95 per cent of the disasters involved mountain lands (see Chapter 9). Some of the most destructive examples, or the greatest damage in others, have occurred in areas of low relief. However, they are few in number, and disasters occurring wholly outside the mountains appear very rare (see Fig. 8.1).

Seismic events are, of course, closely related to mountain building, and the association between them may be taken for granted. However, other and distinctive aspects of mountain environments and settlement are critical to this hazard. This appears first in relation to intervening variables. We have seen that topography and steepness of slope are important factors in seismic responses. In 44 per cent of the examples, reports emphasised the destructive role of landslides. Nearly all of these were in rugged terrain. Reports were not always complete, and it is likely that landsliding and slope failure were major sources of destruction in many more of the events.

Some or all destruction in almost *one-third* of the events occurred in settlements in rugged mountainous terrain. In 20 per cent victims and relief efforts were severely affected by bad weather and exposure, usually involving cold and snow. These, and some 13 per cent with earthquake-triggered damming of streams and dam-burst floods, were also in mountain environments.

However, two of the least often noted but most common attributes of these disasters have no obvious relation to seismology, orogeny, ruggedness or seismically triggered surface changes. During surveys in the Mediterranean, and Southwest and South Asia, I was struck by the frequency with which the worst earthquake damage involved settlements in mountain foot and intermontane basin areas. With that in mind, the events in the inventory were examined to determine the locations of severe and concentrated damage, or the place from which the disaster was named. It was found that these included mountain foot settings *in nearly all cases*, and the worst damage in 75 per cent. If their total share of damage cannot be determined, it is certain they prevail in certain regions with the highest incidence of disaster (i.e. Italy, Iran, Chile, Japan, southern California). In these places, virtually all major settlements, including the major cities, are located in such mountain foot locations. The same locations figured prominently in the 20 per cent of cases where major damage was reported due to the failure of susceptible soils, including liquefaction. More detailed examination usually revealed that these were sites on alluvial fans, river terraces, flood plain areas, old lake floors and coastal alluvium in moun-

Table 8.10　Disaster geography: geographical and damage attributes of the 243 most destructive earthquake disasters reported 1 January, 1950, to 31 December, 1990 (after Hewitt, 1992)[1]

Rank	Damage-associated attribute[2]	n (total 243)	% of cases
1	Mountain land	236	97.12
2	Piedmont/intermontane basin	183	75.31
3	'Dry land'	173	71.19
4	'Development' and change stresses	143	58.85
5	Landsliding/slope instability	106	43.62
6	Coastal mountain settlements	73	30.04
7	'Susceptible' regolith	72	29.63
8	Mountainous interior settlements	68	27.98
9	Severe weather/exposure	42	17.28
10	Tsunami damage	26	10.70
11	Damming/dam bursts	20	8.23
12	Fires	12	4.94

Notes

[1] The examples are counted only where reports indicate substantial damage associated with the attribute. In many cases, however, there is only a verbal report, rather than a careful survey. Data must, therefore, be considered incomplete for all attributes except 1, 2 and 3, which are derived from cartographic location.

[2] Attributes:

1. Mountain land: damage occurs partly or wholly in areas with local relief exceeding 1000 m.
2. Piedmont/intermontane basin: substantial and often most damage in settlements located in mountain foot, foothills or intermontane basin sites.
3. 'Dry land' relation: where regional climate of the damage zone has an arid to subhumid climate, with a seasonal (145 events) or perennial (28 events) moisture deficit, and specifically with a Budyko–Lettau Dryness Ratio of 1 or greater (see Fig. 8.6 and Hewitt, 1982).
4. 'Development' stresses: where recent and on-going changes in settlement, economic activity and habitat are prominent features of the damaged area, and cited as causes of vulnerability in reports.
5,7 'Microseismic' responses: slope and earth surface materials that prove unstable under seismic shaking, resulting in substantial damage.
6. Coastal settlements: where substantial or all damage reported is in the mountainous coastline settlements.
8. Mountain interior: where substantial damage is reported in high-relief and steep slope areas.
9. Severe weather: where substantial damage or human casualties and hardship are reported from weather conditions and exposure following the earthquakes.
10. Tsunami damage: substantial damage to coastal settlements from seismic water waves.
11. Damming/dam bursts: reports of devastation due to flooding induced by earthquake-triggered mass movements and dam-break floods.
12. Fires: fire damage reported following earthquakes. Some of the most devastating urban earthquake disasters have involved mass fires, but they are relatively rare in this set.

tain foot, mountainous coastline or intermontane basin settings. The examples at Kobe, Mexico City and Lomo Prieta, described earlier, fit this category.

More surprising is that, although mountains are associated with higher humidity, and although there is no causal relation between climate and seismicity, a majority (71 per cent) of earthquake disaster zones occur where the climate is semi-arid or sub-humid (Fig. 8.5). It was also found that among these, places with a pronounced dry season (58 per cent) figure most prominently. In fact, in comparison with events where seismicity, land area or human populations was the key factor, an exceptional number of these mountain foot disasters occurred in 'mediterranean' and 'sub-mediterranean' climatic regimes (25 per cent). Related to this is another variant of mountain foot settlement: the concentration of damage along mountainous coastlines recorded in 30 per cent of the examples. These included 13 per cent of cases reporting tsunami damage.

In sum, the global geography of earthquake disasters reveals an overwhelming concentration of harm in mountain foot and intermontane basin settlements, and along mountainous coastlines. These, and the special difficulties of wrecked communities and communications in mountainous interiors, appear as key factors in the nature of human vulnerability and the effectiveness of rescue, relief and reconstruction efforts. In terms of the geography of risk, it seems they must reflect, primarily, the relations between habitat conditions in mountainous zones or mountain rim areas and human settlement. But how far is it useful to see conditions in, say, southern California as comparable with those in Peru or eastern Iran? Along with this, there is the seemingly problematic association with drier and drought-prone climates. I suggest, however, that there are important and useful parallels to be drawn here that apply to the historical and ecological contexts of destruction, and questions of resistance to and survival in earthquakes.

Risk and change in 'basal zone' ecotones

The risk context mainly involves the balance between the benefits and attractions of mountain foot settlement, especially in drier lands and coastal mountains or mountainous islands, and the unique constraints they impose. We have to consider settlement patterns in southern California or Chile; around the Mediterranean from Morocco to Turkey; across Iran and the great mountain ranges of Central and South Asia; and in the mountainous islands of Southeast Asia from Japan to New Guinea. Nearly all cities, strings of countless smaller towns and villages, and concentrations of agricultural population are found in mountain foot and smaller intermontane basin settings.

Consider cities like Kobe, San Francisco, Lima, Agadir, Corinth and Antakya – all identified with major historic disasters. Their setting is a mountain foot coastline and harbour. There is generally a limited amount of flat or gentle terrain, much of it marine, river fan or estuarine alluvium, often extended with artificial fill or sediment traps. Alternatively, consider the sites of interior cities

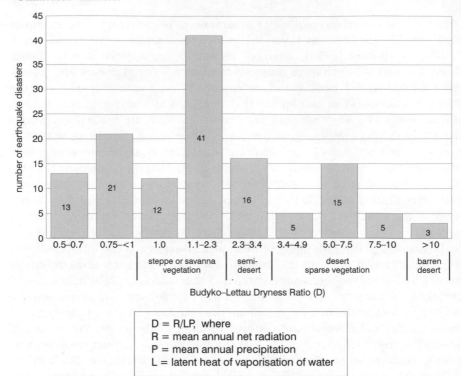

Fig. 8.5 The relations between climatic dryness and earthquake disasters geography. Climate at the locations of the 143 most destructive disasters 1950–1979 is defined by the Budyko–Lettau Dryness Ratio. This depends upon the relation of annual precipitation to net radiation and is considered to be a fair indicator of moisture stress (Hewitt, 1982)

with a record of destructive earthquakes. These include Santiago (Chile), Cuzco (Peru), Mexico City, Guatemala City, Bakersfield (California), Al Asnam (Algeria), Gemona del Friuli (Italy), Erzurum (Turkey), Bucharest (Rumania), Tabas-e-Golshan (Iran), Ashkabad (Turkmenistan) and Quetta (Pakistan). A detailed map shows their cores or older centres of settlement on alluvial fans or other mountain foot depositional aprons. Often, today, they encroach upon river terraces, deltaic or lake bed areas and steep surrounding slopes.

In the disasters associated with each of these coastal and interior centres, the phenomenon of 'susceptible soils' – situations in which the foundations of structures lose strength, break up, undergo liquefaction or become waterlogged – is commonly identified with devastation. Likewise, slope failure and landslides cause major damage. In part this reflects a problem of soil mechanics or slope stability in areas where weak soils, high water tables or steep slopes occur. Mainly it records pressures or incentives in mountain foot areas for human communities to build on them. The occurrence of risky sites identifies common features of the mountain foot microseismic environment. They reflect a complex depositional, topographic, hydrological and, above all, dynamic geomorphological environment.

The relationship of piedmont settlement to dryness or seasonal drought environments reflects major historic patterns of human settlement and exploitation, readily exemplified by the Mediterranean lands. They record the attractions of mountain foot settlement. In dryland areas, especially, these combine better water supply than surrounding drier lowlands and ecotonal variability, with locations where the ruggedness and poor soil problems of the mountainous interior are relaxed. These have been vitally important in the development of many of these agricultural societies. It is no coincidence that a great many of the main domesticated crop plants and pastoral animals originated in or near these places. Important towns and cities developed because such sites provided a strategic location between the resources of mountain and plains, control of the route through the mountains, and access to sea routes. In the modern world, increasingly, the amenities and aesthetic appeal of sunnier, drier locations with mountains and, especially, the sea, draw increasing numbers of residents and tourists to these same places. Southern California is the epitome of this phenomenon in its 'post-industrial' form. Of course, in wholly mountainous regions, islands, or countries, there may be little option but to develop mountain foot and coastal sites for urban if not for other forms of settlement.

Such habitats were recognised as a distinctive type by the plant geographer Schimper (1903), who referred to them as 'basal zone' environments. More importantly, they are areas of rapid transition for virtually all aspects of the habitat – geology and soils, moisture supply, topography, likelihood of inundation, climate, vegetation, and so forth. Environments characterised by strong gradients or transitions are referred to as ecotones or ecotonal. If the basal zones are typified as much by a complex patchwork or mosaic of sites and conditions as by gradients, this at least conveys the sense of the kind of habitat to which human activities must be adapted. This links together the earthquake problems of these otherwise very different cultural and natural settings around the margins of the world's mountains.

This interpretation shifts attention away from the dominant seismic preoccupations in explanation of earthquake risk by drawing attention to comparable ecological features of human settlement in a great many disaster settings. It still sounds like an environmental determinist explanation, but where detailed surveys of damage exist, two crucial qualifications are revealed. First, they show the greater part of the damage to be associated with recent changes in socioeconomic and habitat conditions. For 55 per cent of the disasters, there is independent evidence of rapid social and habitat change in the area of damage. For most of them, observers directly attributed damage to new developments or social and habitat change. The 1976 Guatemala City, 1978 Armenia and 1985 Mexico City disasters, described earlier, are particularly tragic examples of high death tolls for relative newcomers pushed onto unsafe sites or into recently erected but unsafe buildings. Typically, in most detailed reports, damage is to relatively new structures, recently occupied sites, or neglected buildings. They often include relatively earthquake-resistant structures, whose foundations collapsed because of locations on weak materials on alluvial fans, flood plain areas, dry lake beds, shoreline alluvium or artificial fill, perhaps with high water tables.

By contrast, it must be emphasised that a great many structures, often the majority, in the identified disaster zones survived intact, or with minor damage. It is quite fair to say that there are structures, traditional and modern, and an abundance of sites in these areas, that are as resistant to collapse in earthquakes as will be found anywhere. In other words, high risk and disastrous damage are not inevitable consequences of human settlement in basal zone environments.

What is identified as the key to the unusual incidence of disasters in them is a combination of the powerful habitat constraints imposed upon safe land use and building, and *socioeconomic developments that disregard these constraints*. For Japan, this was the essential story in Kobe in 1995 and Niigata in 1964. These are as much disasters of (unsafe) 'development' and urbanisation, or related social and demographic changes, as of earthquake or ecotonal conditions. They typify 'development' focused on amenity, and profit growth in the larger political economy, with little or inadequate attention to the endangerment of local sites and communities. However, because there are indeed safe sites, and methods of land use and construction, risk and damage tend to be highly differentiated within all these places. The least safe sites are often undesirable for other reasons – steep slopes, waterlogging or frequent flooding – and known for weak soils. Those who know about the problems and can afford to avoid them do so. Newcomers, or those who have little choice of where to live, end up on the unsafe sites.

Suggested reading

Hewitt, K. (1984) Ecotonal settlement and natural hazards in mountain regions: the case of earthquake risks. *Mountain Research and Development*, **4** (1), 31–7.

Concluding remarks

Earthquakes are in many ways the archetype of natural hazard. They originate in an uncontrollable, spontaneous process that is part of the Earth's natural development. While we know the places and regions where severe earthquakes are more likely to occur, nearly all actual occurrences are unpredicted. This combines long-term certainty of crisis with a high degree of uncertainty in the short and medium term.

Yet damage is not closely related to the severity of the natural hazard itself. The geography of exposure to given seismic events, and especially the social geography of vulnerability, is of decisive significance. Meanwhile, conditions and developments that increase vulnerability to earthquakes have themselves been increasing in quantity and concentration in recent decades. This is mainly a result of rapid land use change, urbanisation and habitat abuse. These are processes that, to date, have been largely carried forward, regardless of how they affect earthquake risks. One result is dramatic differences in risk among different items, groups and societies. By and large, if the rich are getting richer and in relative numbers, fewer, the 'safe' are getting safer. The vulnerable – like the poor, whom they most often include – are increasing in numbers and becoming more vulnerable.

Thus, earthquakes are associated with typical examples of what are ironically termed '*un*natural disasters'. The reader might well find that a comparable analysis of other natural hazards such as tropical cyclones, droughts or landsliding would reveal different and distinctive geographies and properties, but the same sort of historical and human ecological dilemmas.

CHAPTER 9

Contexts of risk: mountain land hazards and vulnerabilities

A mountain is 'high' according to the human scale, and in relation to human designs. Beyond some reference to a project or a lived experience, these notions of height [etc.] have no meaning. Anthropocentrism, you say! But one must insist upon it; outside an actual or an imagined human presence there can be no geography, not even a physical one, but only a useless science. Anthropocentrism is not a defect, but an inescapable condition.

Eric Dardel (1952, 9–10)

Introduction: regional habitats

Geographical setting is always a presence and a constraint affecting risk. People's relations to the climate, waters, bedrock, soils, topography, wild plants and animals attached to particular places are an integral part of their security and the dangers they face. This applies even to places covered in tall buildings and asphalt, where the noise of traffic, and the smell and smokes of industrial activity, fill the air. We saw it tragically revealed by the 1985 earthquake in Mexico, where the worst destruction was in heavily built-up areas; or again in Kobe in 1995, when many new industrial and port structures located on unstable coastal sediments were destroyed. Yet it was easy, perhaps, to fail to recognise before the disasters the habitat conditions with such fateful consequences.

To speak of these as *geographical* environments highlights the importance of the actual places or regions where people live. It is to emphasise how all places have a distinctive geography, but also a definite situation relative to surrounding places and within modern, globalised systems of exchange and influence. When considering the range of living conditions in different geographical regions and cultures, certain overarching or recurring benefits and dangers appear. They are associated with life in different cultures, and their relations to arid lands, lowland rain forests, sub-arctic permafrost environments, rocky coastlines or small oceanic islands.

Difficulties arise in classifying and distinguishing regional environments. They are often reduced to a few standard measures, with their populations stereotyped in terms of narrow demographic and economic criteria. Details and circumstances of key importance to danger can be overlooked. It will be shown that it is not the general features of desert or mountain environments that place people at risk but specific sites, practices and, in the modern world, outside influences upon them.

The commonest approach looks to particular hazards or damaging conditions as worse in, or unique to, particular regions. This is not without merit. Clearly, tropical cyclone and storm surge risks are strongly concentrated along low-lying or indented ocean coastlines facing eastwards or equatorwards in the tropics. 'Tornado alley' in the United States is, indeed, the region with the highest frequency of tornadoes and recurring destruction from them. Avalanche hazards are virtually confined to high, perennially or seasonally snow-covered mountains. As in the case of earthquake and mountain foot areas, however, it is rarely these natural extremes, or intrinsic properties of the environments where they occur, that lead to disaster. It is a problem of whether and how human settlement observes the constraints in given places. For, while hazards and vulnerability relate closely to the geographical environments in which people live, little about them is inevitable or unavoidable. Hence the geographical environment is seen, here, as drawing attention to the *contexts* of risk and the role of *intervening conditions* (see Chapter 1).

In many respects, hazards and disaster geography comprise a sub-field of environmental studies. This conclusion is suggested by such phrases as 'the environment as hazard' (Burton *et al.*, 1978), 'environmental hazards' (Smith, 1992) and 'environmental risks' (Cutter, 1994). Hazards are commonly treated as a part of ecological or biosphere concerns, especially in endangered and, supposedly, more fragile or harsh environments. The importance of natural hazards, at least, has long been recognised in UNESCO's 'Man and Biosphere (MAB) Programme' which includes a mountain or 'alpine' component, and by the United Nations Environmental Programme (UNEP) in its work on desertification. The Earth Summit in Rio de Janeiro recognised the importance of climate change and natural hazards for sustainable development. 'Sustainable Mountain Development' was one of its components. To illustrate this view we will examine dangers and conditions that shape risk in mountain lands. It is not suggested that they are more important than other regions, in which the reader may wish to make a similar investigation.

The distinctive features of mountain environments tend to become strongly articulated where there are local elevation differences of 1000 m or more, although they may be seen in areas of lower relief, particularly in high latitudes or drier regions. However, if we concentrate on areas where elevation varies at least 1000 m over 25 km horizontally, this would embrace about one-quarter of the world's land areas. Some 10 per cent of the world's population resides in them. Of course, events in the mountains, notably affecting watersheds, can have large consequences far beyond. This is most obvious in the heavily populated foothills and plains below much of the Himalaya, European Alps, American Rocky Mountains and High Andes. It is thought that perhaps 2 billion other people are, or can be placed, at risk from certain kinds of disaster and adverse environmental change in the mountains. In the other direction, the actions of people in lowland areas, including virtually all cities, indirectly affect conditions in the mountains. From them come temporary but influential visitors, mass communications, military campaigns and air pollution.

Suggested reading

'Mountain Agenda' (1992) *The state of the world's mountains – a Global Report.* Zed Books, London.
ICIMOD (1993) *Our mountains: the Hindu Kush–Himalayas – a decade of effort towards integrated mountain development, 1983–1993.* Kathmandu, Nepal.
Denniston, D. (1995) *High priorities: conserving mountain ecosystems and cultures.* World Watch Paper 123, Worldwatch Institute, Washington DC.

Mountain habitats and risks

Mountains are defined, primarily, in terms of topography. They combine relatively high relief, meaning a large elevation range within a given area, and ruggedness, a strongly dissected landscape with steep slopes. There is no infallible measure of how high, rugged or steep the surface must be to qualify. Indeed, even short bluffs or cliffs in otherwise subdued areas exhibit processes and risks found especially in mountainous lands. However, it is the prevalence and extent of these topographic conditions, and how they appear in relation to surrounding lands, that characterise what seems distinctive about mountain habitats.

An elevation change of 1000 m, say, is sufficient to produce marked differences in temperature and pressure conditions. Ambient temperature, sunshine, winds, moisture supply and cloudiness will vary to such an extent that they help create different habitats vertically, somewhat as latitudinal differences do horizontally but with noticeable variations over much shorter distances. The effect tends to be greater as altitude above sea level increases, and to become increasingly critical to living conditions. At high altitudes, say above 3000 m, and in the major mountain ranges, elevation itself becomes a severe constraint, via oxygen deprivation and much greater penetration of solar radiation.

Equally importantly, steep slopes and valleys ensure rapid downslope movement of water and erosional materials. Thus, not only is there usually much heavier precipitation in the higher areas, but it can also be transferred swiftly downslope to lower levels. That is, or can be, an advantage. However, steep slopes, heavy rains or compressed seasonal snowmelt, and large variations in sunshine and temperature, result in large, sudden changes and in a flood-prone hydrology. The moisture may also move downslope as large avalanches, debris torrents or mudflows. These are important constraints in most high mountain areas, and sources of disasters there.

Conditions also vary with the orientation of slopes and valleys, or their *aspect*. This modifies climate, especially, through exposure to the sun and the direction of prevailing winds. From the subtropics to high latitudes, exposure is critical in distinguishing sunnier, often more droughty, slopes facing towards the equator, and those that are more shaded, humid and cooler, facing pole-wards. Relations to atmospheric circulation define windward and lee slopes. They can differ markedly in windiness, precipitation and humidity, and often also in cloudiness. In extensive mountain lands, topography serves to steer regional winds, to create local valley wind systems, and to generate 'chinook' or föhn-type winds. The latter are warm, dry, descending winds that can occur on the lee slopes of mountain ranges. These create further distinctive sub-environments and risks.

Thus, strong vertical and horizontal gradients or sub-environments typify the everyday conditions in the mountains. Seasonal variations are superimposed upon these, and with different strength or severity according to altitude and aspect. 'Seasonality' is as much a spatial as a temporal constraint of mountains. In combination with highly variable geology and geomorphological features, and surface and ground waters, these conditions help to create a mosaic of micro-environments and ecozones. In other words, mountains are characterised by complex and often large local variation in habitat conditions. This leads to very diverse plant and animal populations, or communities adapted to particular ecozones. There are also species whose life cycle is adapted to moving between the different habitats, especially over a range of altitudes seasonally. Examples include many insects, and large mammals such as the North American grizzly bear.

Most indigenous inhabitants of mountain areas, and many visitor activities, also range between several zones, as in the common seasonal pastoralism some-times called transhumance.They may engage in exchanges between different elevational or other ecozones. Sometimes this involves a single, mobile com-munity, or a system of exchange between several occupying different ecological zones, a well-known feature of the cultures of the High Andes.

There is, perhaps, as much diversity within and among mountain lands and their residents as in all other environments together. Mountains are found in every continent, from the equator to polar regions. They occur as mountainous islands and coastal ranges, and in the innermost continental areas. There are isolated massifs or cones, single high ridges or cordilleras, and vast regions of contiguous ranges and networks of intermontane valleys. Not all have high rock and ice summits. Those that do usually have a highly varied set of envir-onments below them: perhaps densely forested slopes or arid valley floors; perhaps glaciers or fiords winding through them.

It may be wondered, therefore, if anything is to be gained from comparing such a varied set of peoples and environments, or whether an outsider like the author and most of his readers could ever know sufficient about even one mountain area to do it justice. In certain respects, these are proper and key questions. Experience on the ground and of the people at risk needs to be given much greater emphasis and authority. However, on the one hand, we will not be making any claims about the 'true nature' of the mountains, or a comprehensive depiction of life in any location. On the other hand, the problem is one in which no such niceties and sensitivities are displayed in what turn out to be the main sources of disaster, or growing dangers, in the mountains. Rather, mountain land risks and disasters, though involving distinctive features of the habitat and vulnerabilities of mountain peoples, increasingly reflect external influences.They appear largely a result of more or less enforced social and environmental changes in response to the latter. Where dangers and disaster are concentrated, recent transformations invite or magnify damaging events. They create unsafe conditions, or render local people and visitors more vulnerable.

Suggested reading

Ives, J. G. and R. Barry (eds) (1974) *Arctic and alpine environments*. Methuen, London.
Barry, R. G. (1981) *Mountain weather and climate*. Methuen, London.
Price, L. W. (1981) *Mountains and man: a study of process and environment*. University of California Press, Berkeley.
Allen, N. J. R., G. W. Knapp and C. Stadel (eds) (1988) *Human impact on mountains*. Rowman and Littlefield, Totowa, NJ.

Mountain hazards and disasters

The socio-cultural impact of tourism may well be the most serious problem at the present time.

di Castri and Glaser (1980, 8)

A few hazards are found only or largely in mountain environments. These include snow avalanches and mountain sickness due to altitude and oxygen deprivation, or hypoxia. Other hazards such as earthquakes, rockfalls and vulcanism are more common or severe in the mountains. Some technological hazards may be more common in the mountains, for example, those associated with large dams or social hazards such as guerrilla, counter-insurgency and drug wars. The context is reflected in the way most disasters involve a magnifying of human or natural forces by the environment. Steep slopes and narrow valleys, severe climate or weather fluctuations, and earthquakes and vulcanism, help to exaggerate specific problems and vulnerabilities. Examples appear in deforestation-induced mass movements; failure and floods from human-made dams; highway landslide problems; or winter sports accidents. In terms of expanding and unusual environmental risks in mountains, perhaps the most widespread are associated with the huge scale of water, minerals and timber resource extraction, and the rapid growth of mountain tourism and recreation. Access roads, resorts, tours and expeditions have increased exponentially in recent decades. Winter sports, mountaineering and trekking can expose people who may be ill-prepared or unusually vulnerable, in places or at times when severe events can occur.

Suggested reading

Ives, J. D. and B. Messerli (1989) *The Himalayan dilemma: reconciling development and conservation.* London.
Hewitt, K. (1992) Mountain hazards. *GeoJournal,* **27,** 47–60.
Ryan, C. (1993) Crime, violence, terrorism and tourism: an accidental or intrinsic relationship? *Tourism Management,* **14**(3), 173–182.

Natural hazards and disasters

Natural hazards have received the greatest attention and are often taken to be definitive of risk in mountain environments. They are an important part of mountain risk and a frequent source of disasters (Table 9.1). Strong forces act frequently or continuously in the mountain environment. Others are ready to be released by extreme events or adverse human actions. Many damaging events reflect the more extreme environmental processes.

Global incidence of lethal and destructive disasters points to the overwhelming roles of earthquakes and floods (Table 9.2). The impact of landslides is under-represented, however, since damage in many earthquakes and floods, sometimes the worst of them, is due to landslides and related processes, reflecting the steepness of slopes. However, slope instability and landsliding become more frequent and very costly problems where modern roads are pushed through the mountains.

The majority of active volcanoes occur in mountain cordilleras or mountainous islands, or as isolated mountains built of eruptive materials. Destructive consequences of eruptions are associated with steep slopes and canyons. The melting of snow and glaciers during eruptions on high volcanic cones may generate extremely destructive lahars (volcanic mudflows). Major flood disasters in mountains have been triggered by bursting of crater lakes and glacier outburst floods from sub-glacial vulcanism. Even more widespread and similar in scale and severity are other natural damburst floods. They arise from failure of ice- and moraine-dammed lakes in glacial terrain, landslide dams and, though rarely as large, bursting of log and river ice jams (Table 9.3).

Forest fires present special problems in mountains due to high winds or valley wind systems, and the greater incidence of severe lightning activity or vulcanism. These are aggravated in seasonally dry areas, for example, of med-

237

Table 9.1 Examples of natural disasters in mountain lands associated with catastrophic Earth surface processes, reflecting the potential for extreme events in mountain environments (after Hewitt, 1992, 49)

Catastrophic phenomena	Disaster events
Rockslides	Elm, Swiss Alps, 1881
	Frank, Canadian Rockies, 1906
	Vaiont, Italian Alps, 1963
Mud and debris flows	Mayunmarca, Peruvian Andes, 1974
	Karakoram, Pakistan, 1980
	European Alps, 1987
	Huanuco Province, Peru, 1989
Debris torrents	Coast Range, British Columbia
	Rio Colorado, Chilean Andes, 1987
(snow and ice) Avalanches	Steven's Pass, Cascade Range, USA, 1910
	Austro-Italian Alps, war front, 1916
	Swiss Alps, 1985
	Korak, Pakistan Himalaya, 1989
	Hakkari, Turkey, 1989
	Western Iran, 1990
Earthquake-triggered mass movements	Hebgen Lake, Montana, USA, 1959
	Huascaran, Peruvian Andes, 1970
	Pattan, Pakistan Himalaya, 1974
	Campagna, Italian Appenines, 1980
	Mount Ontake, Japan, 1984
	Sharora, Tadjikistan, Pamirs, 1989
Vulcanism-triggered mass movements	Mount Tokachi, Japan, 1925
	Ruapehu, New Zealand, 1969
	Villarica, Chilean Andes, 1971
	Mount St Helens, Cascade Range, USA, 1980
	Nevado del Ruiz volcano, Columbia, 1985
Weather-triggered mass movements from volcanoes	Mount Kelut, Indonesia, 1966
	Mount Semeru, Java, 1981
Natural dams and dam-break floods	
• Landslide dams	Indus Gorge, Western Himalaya, 1841
	Lake Rinihue, Chilean Andes, 1960
	Zerafshan, Pamirs, 1963
	Ecuadorean Andes, 1987
• Glacier dams	Gietroz Glacier, Swiss Alps, 1818
	Chong Khumdan Glacier, Karakoram, 1926, 1929
	Rio Plomo, Argentine Andes, 1930s
	'Ape Lake', British Columbia, 1984
• Moraine dams	Khumbu, Nepal Himalaya, 1985
• Avalanche dams	Santa River, Peruvian Andes, 1962
• Vegetation dams	New Guinea Highlands, 1970

Table 9.1   contd.

Catastrophic phenomena	Disaster events
• Artificial dam failures	Buffalo Creek, Appalachians, USA, 1972
	Teton Dam, Idaho, USA, 1976
	Arandas, Mexico, 1980
	Java, 1989
	Shanxi Province, China, 1989

iterranean- and monsoon-type climates. Destructive fires in the hills of southern California in October–November 1993 illustrate the problems. During the 'Santa Ana', föhn-type winds, downed power lines and camp fires started huge, uncontrollable fires that swept through the tinder-dry chaparral. More than 100 000 hectares were burned and over 1100 homes destroyed, including some expensive ones in the canyons of Malibu. In the summer of 1988, vast fires burned almost half of Yellowstone National Park, Wyoming. The extensive forests in the mountain parkland were made more vulnerable by drought and by extensive areas in which trees had been weakened or killed off by insect infestations. Lightning started the fires. High winds helped overwhelm fire-fighting efforts. Some, however, blame the extent of the fires on decades of efforts to suppress and contain previous fires, leaving excessive amounts of inflammable timber.

There is a growing incidence of skiing, trekking and mountaineering accidents with multiple deaths, most often due to avalanches, and sometimes from other extreme events. In November 1995 landslides and floods from extreme snow and rains in the Nepalese foothills killed over 60 persons. Almost half were foreign trekkers. The Nepalese killed were mainly guides and others serving the trekking industry. About 550 persons had to be rescued by helicopter, most of them also foreign trekkers. A similar disaster occurred in the Northwest and Karakoram Himalaya in September 1992 (see page 260)

Avalanches are probably better researched, understood, routinely monitored and actively modified than any other mountain hazard. Presumably this is because they threaten some of the most high-profile and influential of mountain visitors and mountain land uses. The fact that the toll of deaths and costs from avalanches continue to rise, however, is not so much a reflec-

Table 9.2 Distribution of major natural disasters affecting mountain regions, 1953–1988 (after Hewitt, 1992, 50)

Region	Natural agent						TOTAL
	Earthquake	Volcano	Landslide	Avalanche	Flood	Storm	
Mediterranean basin (excl. Turkey)	21	1	4	–	13	1	40
Southwest and South Asia	49	–	15	2	44	3	113
East Asia	11	2	19	2	14	10	58
Southeast Asia, Australia and Oceania	21	8	7	1	9	18	64
Africa	3	2	1	–	5	2	13
Europe (excl. Medit.)	2	–	4	6	5	2	19
South and Central America	33	5	16	7	22	4	87
North America	14	1	4	–	6	1	26
TOTALS	154	19	70	18	118	41	

Table 9.3 Examples of natural dam-break or outburst floods in mountain regions with disastrous consequences

Dam/Location	Date	Comments
Giétroz Glacier, W. Alps, Switzerland	1818	Glacier dammed Drance de Bagnes R., followed by sudden outbreak of 3 km long lake, devastating valleys towards Rhône, 50 deaths.
Indus R./Nanga Parbat (present day Pakistan)	1841	Massive earthquake-triggered landslide dam failed after 6 months, creating largest and most destructive historic flood on Indus with thousands of deaths.
Kokuzou Mt., Sai R., Japan	1847	Earthquake-triggered landslide dam drowned 30 villages, then failed, flood wave killing 110+.
Shyok River, Karakoram Himalaya	1929	Ice dam of Chong Khumdan Glacier failed. Major devastation to Indus plains.
Deixi landslide, Min R., Sichuan, China	1933	Three earthquake-triggered landslides dammed river, the Deixi, to depth of 255 m. This dam failed suddenly after 43 days, the resulting flood wave killing 2423+ persons.
Rio Plomo, Mendoza, Argentina	1934	Glacier dammed river followed by destructive outburst flood.
Chin-Sui-Chi R., Taiwan	1942	Failure of large, earthquake-triggered dam following heavy rainfall. Major devastation and 154 killed.
Cerro Condor, Mantaro R., Peru	1945	Landslide blocked canyon to depth of 100 m, creating lake 21 km long in 73 days. Overtopped and drained in 7 hours, outburst flood destroying numerous bridges, farm land, but few casualties in sparsely inhabited area.
Lake Rinihue, Valdivia, Chile	1960	Earthquake-triggered landslides impounded lake. Outburst flooded much of Valdivia city, but loss of life averted by control of works and warnings.
Khumbu Himal, Nepal	1985	Moraine-dammed lake breached by wave triggered by ice avalanche. Severe flood erosion and destruction in the valley.
Colorado R., Andes, near Santiago, Chile	1987	Landslide dammed main river, followed by outburst and debris flow, killing 37, destroying power plant.

tion of these hazards of the mountain environment or lack of understanding. It records the ever-increasing pressures of exploitation. Ultimately, the sheer scale and extent of mountain tourism overwhelms the considerable research into the natural processes involved, and efforts to monitor and prevent such disasters.

There is a common perception that disasters in the mountains generally reflect the inherent severity and scale of natural processes, or lack of development among their populations. However, this is an outsider's perspective, and a version of the now familiar technocratic 'hazards paradigm', exaggerated by environmental and cultural stereotypes. The diversity and strong energy exchanges in the mountains can be dangerous, but they are also valuable resources, offering a generous mix of options for the well-adapted society. The dangers are usually clear, even starkly evident in the landscape. Most are readily recognised in hazards mapping projects (Keinholz, *et al.*, 1983). Yet the mountains do present complex problems for social change and new developments. It is mainly in relation to modern technologies like all-weather highways and large-scale resource extraction or tourism, and their specialised and intensive land uses, that the mountains appear more hazardous than other lands.

Suggested reading

Sheets, P. D. and D. K. Grayson (eds) (1979) *Volcanic activity and human ecology*. Academic Press, New York.
Price, L. W. (1981) *Mountains and man: a study of process and environment*. University of California Press, Berkeley.
Tufnell, L. (1984) *Glacier hazards*. Longman, London.
Schuster, R. L. (ed.) (1986) *Landslide dams: processes, risk, and mitigation*. Geotechnical Special Publication no. 3. American Society of Civil Engineers, New York.
Romme, W. H. and D. G. Despain (1989) The Yellowstone Fires. *Scientific American*, **261**, 37–46.

Technological hazards in the mountains

All of the ways in which structures, industrial processes, modern transportation and other infrastructure may fail or cause harm can occur when they are used in mountains. However, certain dangers of failure or destruction are more likely or more threatening there. Some technologies, such as large dams, tunnels, high bridges, ski lifts and funiculars, are more common. In rugged terrain with large climate variations, mines, forest industries and tourism entail more concentrated exposure of, and potentially more damaging forces from, modern technology.

Perhaps the most widespread, frequent and, in total, expensive damage is that involving mountain roads. Mountain conditions pose severe risks for all-weather transport and communications, but modern administrative, military, commercial and recreational interests tend to expect that highways be built, and be open at all times. The main results are expensive but increasingly routine monitoring and warning, and rapid reaction to treacherous conditions and restoration of damaged communications. These are put in place because they have the attention of governments, mining or forestry and other resource interests, and the tourism business. In most countries the physical hazards are in the hands of army engineers and pioneers, amply provided with heavy equipment, explosives and human resources. Ski patrols, professional avalanche monitoring, controls and artificial release with artillery are standard in most winter resort areas and in high mountain road passes. But there are multiple closures, and sometimes serious casualties, in every major mountain area during most years.

One of the worst mountain transportation disasters occurred on November 2, 1982, in the Salang Tunnel (3370 m elevation), Hindu Kush, Afghanistan. A vehicle in a Soviet military convoy struck a fuel truck near one entrance to the tunnel. The explosion and burning fuel killed dozens of soldiers and civilian occupants of other vehicles instantly. Also in the tunnel were hundreds of private motor vehicles, and buses packed with passengers. Believing the explosion to be from an attack by the Afghan rebel forces, the Soviet army sealed both ends. However, it was extremely cold and people in vehicles distant from the explosion sat with their engines on to keep warm, adding carbon monoxide to the air. Unfortunately, the tunnel's ventilation system was not working. Many civilians not burned by the spreading flames died of asphyxiation and carbon monoxide poisoning. Estimates of the death toll range from 2000 to 3000.

This was an appalling instance of the compounding of extreme natural, technological and war hazards. However, it is rarely possible or helpful to separate natural mountain environment risks from technological hazards. Conditions or processes in one trigger or magnify dangers of the other. For example, in the 1963 Vaiont Dam disaster in the Italian Alps, the immediate cause was a large rockslide that swept into the reservoir, but the high death toll of about 2500 persons and devastation to communities in the region were due to the flood wave from water forced over the top of the dam. As it swept down the narrow Piave valley, in places it reached a height of nearly 70 m. The arch dam was strong enough to be little damaged by the event. Nevertheless, it had to be written off, partly because the space behind it was largely filled with rock and earth. The disaster raised serious questions about the role of the project in triggering the slide, and of poor planning and design.

Dam projects are increasingly attracted by favourable sites or untapped water and power potential in the mountains. The larger fraction of all dam failures have been to mountain reservoirs, sometimes with calamitous results (see Table 9.3). Some particularly nasty disasters due to failure of mine tailings ponds are scattered through the history of mountain areas. Survivors of one of them, at Buffalo Creek, West Virginia, in 1972, were quoted in Chapter 2.

Suggested reading

Müller, L. (1968) New Considerations on the Vaiont slide. *Rock Mechanics and Engineering Geology*, **6**, 1–91.

Jansen, R. B. (1980) *Dams and public safety: a water resources technical publication*. US Department of the Interior, Washington DC.

Jones, D. K. C. (1992) Landslide hazard assessment in the context of development. Chapter 12 in McCall *et al.* (eds) (1992) 117–41.

Social violence: disasters of armed violence

Armed violence appears to be the principal source of social disasters in mountain lands. Of 34 armed conflicts involving regular state forces in 1991, one-third occurred largely in mountain areas, and nearly all involved them to some extent (Table 9.4). Moreover, 30 of the 33 wars in, or affecting, mountain regions had been going on for ten years or more. Several of these intensified in the early 1990s, including those in Indian Kashmir, Sri Lanka, Afghanistan, Guatemala, Turkish and Iraqi Kurdistan, and Somalia. A rash of new conflicts also engulfed mountain areas in the same years, notably in Bosnia and Croatia, Georgia and Chechnya on the northeast slopes of the Caucasus, Tadjikistan, Chiapas in Mexico, and between Peru and Ecuador on the east side of the Andes. There have been large civilian casualties and uprootings, and destruction of settlements and habitat. Some involved claims of war crimes against civilians and prisoners of war, and widespread raping of women. In some cases, such as Bosnia, the crimes have been confirmed by international observers.

For the twentieth century as a whole, armed conflicts involving state forces and mountain lands form a permanent feature of armed violence (Table 9.5). Some wars involved mountain areas to a limited degree, but others were fought entirely within them. And this evidence is probably just the tip of the iceberg. The Second World War involved as many mountain regions, perhaps as much devastation and death to their peoples, as all the other wars together. Because it was a 'world war', its many mountain campaigns can disappear in the larger view of the conflict, but no one could count the mountain towns and villages bombed, strafed, shelled and torched in the land campaigns. There was destruction from the mountainous islands of the western Pacific to the highlands of Sicily and North Africa; from Norway to the uplands of East Africa.

These wars do not include countless other uses of armed force against or affecting mountain peoples and habitats. They have involved policing and punitive expeditions, acts of military occupation, weapons testing, and military training activities by home forces. A great many of the victims of violence in the so-called 'Arc of Crisis', from the Horn of Africa to the Western Himalaya, have been mountain peoples caught up in regional disputes, civil and guerrilla wars, or state actions underwritten by Cold War and regional strategies. Colonial and counter-insurgency warfare has included experiments or regular use of modern and 'dubious' weaponry, including the machine gun, poison gas and napalm (Chapter 5, Table 5.4). This weaponry has been used against less well-

Table 9.4 Examples of the failure of artificial dams in mountainous terrain with catastrophic outbreak floods (after Jansen, 1980; cf. Table 8.7)

Dam location	Date	Dam height (m)	Attendant conditions	Deaths
Dale Dykes, near Sheffield, UK	11 March, 1864	29.0	Heavy rains, strong winds, wave action on dam structure, sudden outbreak flood.	250
South Fork, near Johnstown, Pennsylvania, USA	31 May, 1874	14.2	Heavy rains, prior flooding, outbreak flood steepened in canyon.	2200
Gleno, near Bergamo, Italy	1 Dec., 1923	43.6	Heavy rains and floods, sudden break.	600
St Francis, near Los Angeles, USA	12 March, 1928	62.5	Reservoir full at time of collapse. No warning. Flood wave 28 m high.	450
Alla Sella Zerbino, Appenines, near Genoa, Italy	13 Aug., 1935	42.0	Heavy rains and floods into dam. Dam overtopped, full-height failure.	100 +
Möhne, Ruhr, Germany	17 May, 1943	40.3	'Dambusters' bombing raid released 10 m + flood wave.	1200 +
Malpasset, near Fréjus, France	2 Dec., 1959	61.0	Heavy rains. Weak rock layers in foundations. No warning for public.	421
Hyokiri, Namwon, South Korea	12 July, 1961	?	Sudden failure. No details.	114
Vaiont, near Belluno, Italy	9 Oct., 1963	265.0	Catastrophic rockslide into reservoir. Water forced over dam. Flood wave below caused deaths, most destruction.	2600
Sempor, near Mangebong, Java, Indonesia	1 Dec., 1967	54.0	Heavy monsoon rains, construction delays, part-finished dam failed.	2000 +
Frias, near Mendoza, Argentina	4 Jan., 1970	35.0	Torrential rains, torrents, flood wave into town full of tourists.	42 +

Table 9.4 contd.

Dam location	Date	Dam height (m)	Attendant conditions	Deaths
Buffalo Creek, W. Virginia, USA	26 Feb., 1972	46.0	Mine tailings dam, burning embankment burst after torrential rains, 6 m flood wave.	125
Stava, Dolomites, Italy	19 July, 1985	?	Earth walls of fluorite mine settlement tank failed. Flood wave into resort full of tourists.	250+
Rimac R., near Lima, Peru	9 March, 1987	?	Several earthen dams failed. Torrents carried boulders into narrow valleys, devastating small towns.	100+
Shanxi Province, China	20 June, 1989	?	Dam burst in mountainous area.	38

Table 9.5 Armed conflicts taking place in 1991 by major world regions, identifying those involving mountain lands. The wars each had at least 1000 deaths and involved the regular forces of at least one state (after Project Ploughshares, 1991; cf. Fig. 5.1)

Region	All conflicts	Involving mountain lands	Wholly in mountains	Duration 10+ years
S. & C. America	5	5	5	5
Europe	2	1	1	1
Africa	10	6	3	9
S. & SW. Asia*	8	8	6	7
S. & SE. Asia	9	9	7	9
Totals	34	29	22	31

*Region from Middle East to Indus Valley

armed people, who sometimes have only spears or bows and arrows, and often against undefended settlements and unarmed civilian populations. Today, bombing raids are standard practice for state forces in their troubles with mountain peoples.

Of particular concern are the numbers of struggles between the state and indigenous peoples, in which the latter and their habitats are seriously harmed. Many of John Bodley's (1975) 'victims of progress' are indigenous mountain peoples, used for forced labour, exterminated or uprooted under the guise of national security measures. Such is reported from the Chittagong Hills of Bangladesh, mountain areas of the Philippines and New Guinea, East Africa, the east slopes of the Andes, and Central America. There is continuing use of regular military forces against the Kurds, a mountain people, in Iran, Iraq and Turkey; by Indonesia in East Timor and Irian Jaya; and in Chiapas, Mexico, to name a few. In the longer historical view, this is a continuation of damaging contact with Europeans, usually involving unrestrained armed assaults, throughout the Americas, North, East and southern Africa, the Southeast Asian mainland and mountainous islands, Australasia, and Siberia. In these events, going back more than five centuries, untold numbers of indigenous mountain peoples have been victims of exotic diseases, superior fire power,

Table 9.6 Armed conflicts involving mountains lands, 1900–1980 (after Singer and Small, 1982)

	conflicts	years of war
Wars with over 1000 battle deaths		
Interstate and colonial or imperial wars	33	c. 133
Civil wars	42	c. 111
Wars with less than 1000 battle deaths	88	c. 182
Total mountain land wars	163	406

deliberate extermination, enslavement and so-called 'development'. For them and their habitats, these have been catastrophes of unprecedented scale and scope.

Devastating, long-term consequences of these wars have come from the use of modern weaponry to destroy villages, irrigation systems, orchards, roads and bridges, resulting in both enormous civilian casualties and laying waste their habitats. In the Afghanistan conflict between 1978 and 1987, nearly 10 per cent of the population were killed by war-related actions (between 1 million and 1.5 million people). The largest fraction were civilians. A majority came from mountain towns, villages and farmsteads of the Hindu Kush. Most fled to valley and foothill settlements or camps in adjacent countries, sometimes bringing social upheaval, violence and environmental devastation there too. In 1977, 85 per cent of the Afghan population was agricultural. In ten years *it fell to 23 per cent* (Fig. 9.1). As they have tried to return to their old homes since the Soviet troops withdrew, apart from continuing warfare, they have found the land, farmsteads and infrastructure devastated, the best fields often sown with land mines. All this, of course, is the opposite of a common view of the mountains as refuges from the misfortunes and powers of history. It reflects how modern surveillance and mobility, the weapons and the sheer scale of military force available can render the mountains uninhabitable for any but warring groups.

Suggested reading

Asprey, R. (1974) *War in the shadows: the guerilla in history*, 2 vols. Doubleday, Garden City, New York.

Crosby, A. W. (1986) *Ecological imperialism. The biological expansion of Europe, 900–1900*. Cambridge University Press, Cambridge.

Amnesty International (1986) *Unlawful killings and torture in the Chittagong Hill tracts*. London, September.

Nietschmann, B. (1987) Economic development by invasion of indigenous nations: cases from Indonesia and Bangladesh. *Cultural Survival Quarterly,* **10** (2), 2–12.

Lovell, W. G. (1988) Surviving conquest: the Maya of Guatemala in historical perspective. *Latin American Research Review*, **23** (2), 25–57.

Enforced uprooting, 'resource wars' and genocide

The dangers of war in mountain areas are not confined to warfare itself. Often that leads to secondary and tertiary harm to land uses and livelihood, social breakdown, and famine. Perhaps the largest of the tragedies involve the human and geographical calamity of forcible uprooting (see Chapter 5). The millions of war refugees and other forced migrants in camps from Sudan to Somalia, from Kurdistan to the Indian Himalaya, in Bosnia and surrounding territories of former Yugoslavia, are largely mountain folk.

As we saw, half the population of Afghanistan was displaced within or outside the country. Some 95 per cent of the more than 3 million who fled

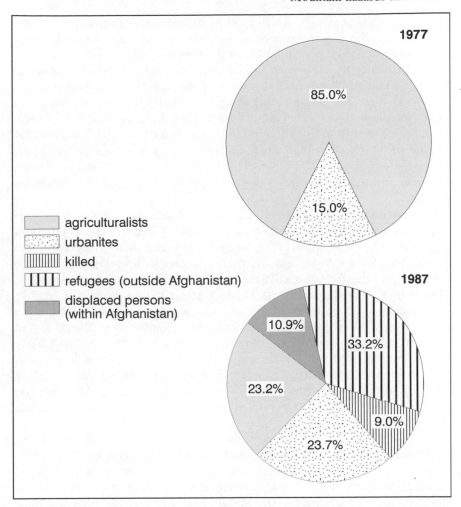

Fig. 9.1 The violent transformation of the people of Afghanistan. The graphs compare
the proportions of rural and urban populations in 1977, before the war, and
after ten years of fighting (1987), and the huge 'new' populations of the killed
(approx. 1.25 million) and displaced generated by war (after Sliwinski, 1989,
54)

to Pakistan were of rural origin. In the refugee camps, almost half were children under 15 years, and about one-third were adult women. The remainder
were mostly old men. Some 35 per cent of women refugees who had been
married were widowed, and 19 per cent of children were orphaned (Dupree,
1988; Sliwinski, 1989, 44).

The slow but steady repatriation in the early 1990s of the more than 5
million Afghan refugees in Pakistan and Iran has been disrupted and even
reversed by further warring. Yet in 1993–94, Afghanistan became a major

refugee *reception* country, to which most of some 500 000 Tadjik civilians fled from civil war. Similar, massive refugee movements are reported from other mountainous Central Asian republics of the former Soviet Union.

Armed violence has also been used against mountain people in genocidal campaigns. The expulsions and massacres of the Armenians of Turkey in the First World War were, among other things, a calamity for a mainly mountain people. Some 1.8 million inhabitants were uprooted. More than half of them were killed or died from privation in forced marches. It is widely believed that genocide, or 'ethnic cleansing', a new euphemism for crimes against humanity, has been occurring in the Chittagong Hill Tracts of Bangladesh, in East Timor, in Bosnia, and against various indigenous peoples in the Central American highlands.

The 'developed nation' mountains are not without their problems of war-related violence. Nuclear and other weapons testing in the United States and Soviet Union include massive explosions near, and fallout in, mountain regions. The Rocky Mountain arsenal in Colorado and Dyshtym in the Urals are notorious for radioactive contamination from weapons programmes. Semipalatinsk, near the main Kazakhstan nuclear testing grounds, is said to have exceptionally high levels of cancers, genetic disease and infant mortality explicable only by excessive radiation exposure. Conditions in the Altai and Tien Shan, which lie downwind of the test grounds and form the most likely areas of concentrated fall- or rainout, have yet to be adequately assessed. Conditions in the Central Asian ranges surrounding the Chinese nuclear testing grounds have also not been revealed, although the fallout from these tests has been considered dangerous to children's health, at least, in North America, 15 000 km or more away.

Finally, conflict over resources brings armed forces into the mountains, usually where minerals and timber, water and power supplies, or trade routes are involved. The desire to capture or control strategic water supplies in the headwaters of important rivers is becoming a major factor in a host of potentially or actually lethal actions. The Middle East is recognised as having some of the more explosive competing demands and conflicts over water, but army occupations and actions in the headwaters of the Nile and in Indian-held Kashmir, and the long India–Pakistan war on the glaciers of the Karakoram Himalaya, are largely about control of water resources. In the process, the conflicts magnify hatreds and political instability, often alienating the peoples in the headwater areas. They also threaten the waters themselves through devastated habitats and toxic war wastes.

Suggested reading

Brogan, P. (1989) *The fighting never stopped: a comprehensive guide to world conflict since 1945*. Random House, New York.
Tobias, M. (ed.) (1986) *Mountain people*. University of Oklahoma Press, Norman.

Allen, N. J. R. (1987) Impact of Afghan refugees on the vegetation resources of Pakistan's Hindukush-Himalaya. *Mountain Research and Development*, **7**, 200–4.
Bryant, B. and P. Mohai (eds) (1992) *Race and the incidence of environmental hazards*. Westview, Boulder, Colo.

'Green mines' and 'white plagues'

In North America and Western Europe, the outstanding late twentieth century disaster involving mountain regions might seem to be international drug trafficking, especially in cocaine and heroin. The raw plant materials come entirely from a few mountain areas (Fig. 9.2). The poppies for the illicit opium/heroin trade have been grown in a series of mountain areas between eastern Turkey and northern Laos. The largest suppliers and most notorious areas in the 1980s and 1990s have been the 'Golden Triangle' and the 'Golden Crescent' of Southeast and Southwest Asia respectively. On the other side of the world, the coca leaves from which cocaine and its derivative 'crack' are made grow in the equatorial Andes, or 'White Triangle' of South America. A substantial part of the illegal trade in marijuana also involves high mountain areas of Latin America and Eurasia.

Ecologically, these are 'montane' crops. The coca belt is concentrated in the *montana* of Peru and the *yungas* of Bolivia. It belongs to the high *selva* (rain forest), or 'eyebrows of the jungle', between the Amazon lowlands and high, rugged crests of the Andes. The opium farmers of northern Laos clear fields from forested slopes below the rugged Annamite mountain chain. Cultivation is in the hands of mountain peasant families who, for the most part, have grown the white poppy or the coca plant for centuries and, perhaps, millennia. Knowledgeable observers argue that, where they are grown, these are traditionally sanctioned crops of choice rather than 'hazards' – often among the few chances of economic advancement and even survival for these folk.

The main human tragedy of the 'white plagues' occurs through sale and addiction in the cities of the plains, especially in North America and Western Europe, but increasingly worldwide. It involves drug sales and use in the destinations of tourists, mainly from wealthier nations, but with ever greater numbers of urbanites from all countries. Nevertheless, there are risks in the mountains and for the growers, and to which they are very vulnerable. Pressures to over-expand production bring ecological and social damage. There are cycles of 'boom and bust', as in all such international commodity trade. The 'bust' hits hardest, while the 'boom' is smallest, for the farmer. Violent rivalries over the spoils spread into the source areas. So-called 'narcoterrorism' has cut a path of blood and fear from the main cities of Colombia, Peru and Bolivia to the mountain villages and fields. Drug profits underwrite the armies and control by 'drug lords' and factions in the Golden Triangle. They helped fund the arms used to fight the war against the government in Afghanistan.

There is a backlash of violence from the centres of power and the countries, notably the United States, in which the white plague has its greatest rewards,

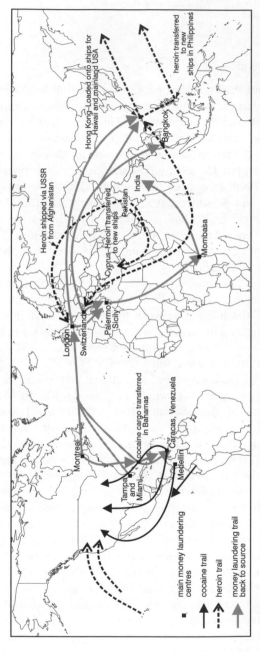

Fig. 9.2 The global hard drugs traffic. The map shows some of the main routes by which cocaine and heroin move from their high mountain sources to the cities of Europe and North America. It also suggests some of the routes by which the earnings of the illegal trade are 'laundered', to be introduced back into the global economy. Of course, if the money does not flow, the drugs do not. As noted in the text, another essential 'flow', missing from the map, is that of arms. These move, directly or indirectly, from the rich nations, where the drugs are mainly used, to the mountainous source regions, where they create other disasters (after Evans, 1989)

and its worst casualties and political fallout. The US 'war on drugs' has had a budget approaching US$2 billion a year. While largely unsuccessful in stemming the tide of drugs entering and being used in North America, it has been carried to fields and processing sites within the high mountains. 'Search and destroy' missions by military forces and narcotics agents use herbicidal sprays and incendiaries, with disastrous consequences for mountain peasants and their habitat.

The links and 'wheels within wheels' of this disaster make it one of the most serious and intractable. They include wars and arms trafficking; drug taking in the armed forces and drug peddling; governments and criminal drug rings; the global network of relations of production, manufacturing, transport and use; the international financial impacts of drug profits, money laundering, and huge funds available for bribery; and violent crime and violent policing. Hard drug users are especially vulnerable to serious, transmissible diseases (see Chapters 6 and 10).

Suggested reading

McCoy, A. W. (1972) *The politics of heroin in Southeast Asia.* Harper and Row, New York.
Westermeyer, J. (1982) *Poppies, pipes and people: opium and its use in Laos.* University of California Press, Berkeley.
McNicoll, A. (1983) *Drug trafficking: A North–South Perspective.* North–South Institute, Ottawa.
Evans, R. (1989) The death industry: world drug economies. *Geographical Magazine*, **LXI** (5), 10–14.
Encyclopaedia Britannica Yearbook (1989) 243.
Lintner, B. (1993) A fatal overdose: civilians butchered in fighting between drug gangs [Thai-Burmese Border]. *Far Eastern Economic Review*, **156** (22), 26–7.

Mountain land vulnerabilities

Like all environments, the sense of what is distinctive and separate about mountains is rooted in human experience and ways of constructing a world view. Among the many people involved, there are clearly two distinct kinds of perspective: those of outsiders, visitors or generalists, and those of mountain land residents and their particular circumstances. The most widely recognised hazards tend to be those that affect visitors, or extensions of the state or international economy into the mountains. These can be acute and costly problems, needing modern technical risk assessment and reduction measures. However, no existing technologies or arrangements can avoid very high costs and recurring damage, interruptions and loss of life. At best, one can look for marginal savings and trying to save lives.

The drug 'hazard' and wars make clear how serious risks in the mountains depend largely upon outside developments and influences. The idea of 'highland–lowland' interaction suggests a framework for recognising these relations. Hitherto, it has been used mainly to highlight environmental abuses in the mountains that may affect the more heavily populated lowlands, for instance where watershed damage increases flooding and sediment yields downstream. The so-called 'Himalayan–Ganga problem' has been seen mainly in this sense, despite the burden of evidence that adverse environmental, social and violent impacts, including those leading to watershed abuses, have moved from the lowlands into the Himalaya. A similar interpretation has been put on the drug trade, or refugee camps in Sudan or Pakistan – that they reflect dangers moving from the mountains to the plains. And the plains or lowland countries increasingly provide refuges for those dispossessed, impoverished or terrorised by events in the mountains. This contrasts with an old stereotype of the mountains being refuges for history's unfortunate folk. However, the refugees from the mountains and others in their endangered refuges are victims of initiatives that *begin* in the 'cities of the plains', and in national or international struggles. Similar geographies of influence and interaction are relevant to other hazards, including natural ones.

Most of the deforested or 'overgrazed' hillslopes identified with accelerated erosion also relate to more or less recent economic and social change. The disastrous floods and mass movements in Thailand in November, 1988, and in the Philippines in November, 1991, though triggered by intense rainstorms, have been blamed primarily on logging of hillslopes, said to be mostly illegal. The greater risks of increased mass movements and flooding also come from 'cleared land' (Hamilton, 1991). In each case, the removal of stabilising trees reflects wider socioeconomic changes. The disaster in the Pakistan Himalaya in 1974, although triggered by an earthquake, appeared to me to be primarily a 'deforestation' or 'development' disaster. Logging and penetration by the timber merchants were combined with clearing of more and steeper hillsides, with less well-made terracing, larger goat herds to produce meat for markets, and – the main cause of death to people and animals – a crisis of building materials for local housing. Failure to replace old, rotted timbers, and the spread of homes built with rubble and mud, were common to most structures that collapsed on their occupants. Photography from early in the century showed much more old growth, heavy forest cover and more timber in built structures (cf. Chapter 8).

In any area in which once extensive, especially mature, forest is logged, the process is invariably accompanied or preceded by drastic socioeconomic changes for the resident populations, changes that may or may not be linked directly to the timber trade itself. It is but one, albeit striking, indicator that hazard and disaster impacts virtually everywhere are closely associated with rapid social, economic and habitat change. In such cases, we see that modern developments and the problems of permanent residents, if profoundly different in human consequences, are closely interwoven. A dissenting but growing view of hazards and disaster sees risk for mountain peoples as well as modern projects as being largely due to influences from beyond the mountains, rather than from the mountain habitat or indigenous practices.

Suggested reading

Davis, I. (1984) A critical review of the work method and findings of the Housing and Natural Hazards Group. In Miller, K. J. (ed.) 200–27.

Hewitt, K. *et al.* (1994) *Mountain hazards in the Americas: a selected bibliography.* Working Paper 3, Cold Regions Research Centre, Wilfrid Laurier University, Waterloo, Ontario.

Disasters and development: two mountain land examples

Natural hazards in the Karakoram Himalaya

The Karakoram is a high, rugged mountain region in Central Asia, its main part lying in the Upper Indus basin of northern Pakistan (Fig. 9.3). Here is some of the most extreme relief on Earth. Valley slopes adjacent to most human settlements rise 3000 to 6000 m. The pace of erosion is high, as is the variety and incidence of large, potentially destructive geomorphological processes. Great relief and ruggedness are combined with extreme variations of climate. The lower settled areas are in arid or semi-arid valleys. The mountain ridges are snow-bound for most or all of the year, and the higher valleys contain the largest area of perennial snow and ice in the subtropics. This includes several thousand individual glaciers but is dominated by some 20–30 ice masses of large size. Despite the extreme continental location, heavy snowfall at high altitudes and steep slopes leads to vigorous ice streams.

The inhabitants live in hundreds of villages, scattered along the valleys up to elevations of about 3000 m above sea level, and in a few small, but rapidly growing towns. The villages, situated in dry, rain-shadowed valleys, depend upon meltwater descending from higher, more humid elevations. Appropriately, they have been called 'mountain oases'. The region is home to more than a dozen, widely different, linguistic groups with distinctive histories and cultures. Nearly all are Muslim, however, and the men generally speak Urdu, the common language of Pakistan. A large part of the population is Shia, and Ismailis are an important group.

They have had more or less strong ties with surrounding lands, through trade, religion and employment and cultural exchanges, going back at least two millennia. In the last century, and early decades of this century, their situation and security were influenced by political rivalries and military interventions from Kashmir, British India, China and Russia. There are many indications that these were years of drastic decline in living standards and depopulation in many of the valleys. This made people more vulnerable to natural hazards, diseases and social violence. Epidemics, natural disasters, wars, taxes and conscriptions of menfolk all served to decimate and impoverish them. British imperial rule imposed a welcome period of peace for about 60 years, but the people remained at the mercy of outsiders and the, often despotic, local rulers upon whom the colonial authorities relied.

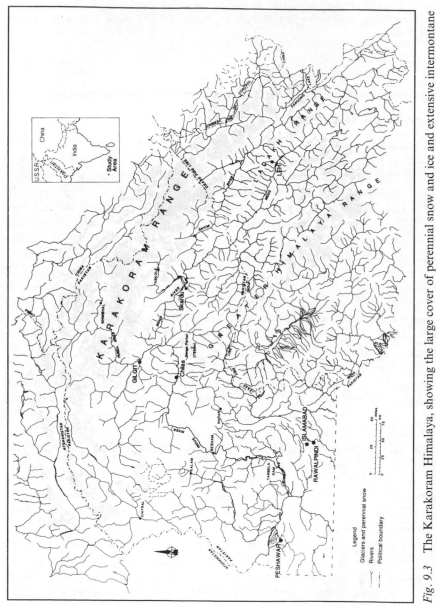

Fig. 9.3 The Karakoram Himalaya, showing the large cover of perennial snow and ice and extensive intermontane river basins, where settlement occurs

Since 1947 most of the area has been administered by Pakistan. there has been improved prosperity and growing numbers in most valleys. Outside towns such as Gilgit and Skardu, the economy is still primarily one of subsistence from irrigation agriculture and animal herding. Little of that is mechanised and depends instead upon human and animal muscle power. Small-scale water mills grind grain, while firewood and dung are used for heat and cooking, and sun and wind to dry crops. However, electricity is increasingly widespread. Unsurfaced 'jeep' roads extend to all but a few villages. Portering and guiding for expeditions and trekkers has been a boom industry. Social and economic change are also transforming the nature or meaning of risk even from natural hazards. Before discussing this further let us review some of the major hazards.

Natural hazards and disasters

Snow avalanches

These occur at all elevations in the Karakoram. They are generally confined to the winter and spring in the zone of permanent habitation, but at high elevations are a year-round danger. A large fraction of the quite high proportion of mountaineers killed in the Karakoram have been swept away by avalanches. In part, this reflects the attractions of the peaks over 8000 m, notably K2 (8611 m), the second highest in the world, and countless, spectacular, unclimbed peaks. In part, it seems to reflect the extremely competitive and aggressively risk-taking nature of modern mountaineering.

Most long-time settlements are well away from avalanche run-out areas, but recently expanded settlement areas may be at risk. The Karakoram Highway and most other link roads are subject to avalanche damage and blockage every year. These usually occur between March and June and may close the roads for several days.

Avalanches tend to have the worst impacts upon the traditional economy indirectly, through erosion damage and temporary damming of streams followed by destructive floods. A disastrous example occurred at Ratul village, Hopar, the highest part of Nagyr, in the spring of 1978. The stream suddenly dried up and was found to be blocked by an avalanche high up. Men of the village went up to try to free the stream before the lake became too large. They were unsuccessful. The dam burst and drowned many of them and, without warning, many in the village below. Much arable land and many homes were destroyed.

Landslides

The region is notorious for large and potentially destructive debris and mud flows, and occasional large rockslides. Since rockwalls make up the single most prevalent landforms, countless small and moderate rockfalls occur at all altitudes, but especially in the spring in the settled areas. The Karakoram Highway is repeatedly blocked by these mass movements, and at dozens of places in a given year. It is usually only the rare, high-magnitude debris-flows that reach and destroy settled areas and agricultural land. A number of these occur every decade somewhere in the region as a result of exceptional rainstorms or spring snowmelt, sometimes triggered by the bursting of small glacier lakes when floodwater picks up large quantities of debris.

Landslide dams

The largest flood disaster known on the Indus occurred in 1841 as the result of an earthquake-triggered rock avalanche. The debris blocked the Indus River below Nanga Parbat. The dam survived six months before failing catastrophically and releasing a huge flood wave, which caused damage and drownings all the way to the Indian Ocean. The second largest flood wave on record came from a landslide dam in Hunza in 1858, with similar if not as destructive results. There is evidence of dozens of landslide dams that blocked the Indus streams in prehistoric times and represent rare, catastrophic events that could occur again. Many small blockages by springtime and summer debris-flows occur that may cause local inundations and hardships.

Glacier hazards

Extreme dangers from glaciers relate mainly to glacier outburst floods and sudden, rapid and large advances known as surges. Unless surges also form glacier dams, they tend to endanger local mountain areas only. This they do through increased erosion, sudden, local floods and preventing movement across the glaciers, which are often important routes. These lead to high pastures, firewood sources and hunting areas and may be followed or crossed by tourists. At least 14 glacier surges have been recorded in the last 100 years, several of them causing devastation and problems for villages in the vicinity.

Glacier dams spread disaster far downstream if the dam fails suddenly, causing a destructive outburst flood. There have been at least 35 of these events in the region over the last 200 years. Some had flows and reached heights not possible in rainfall- or meltwater-generated floods. They travelled as destructive waves hundreds of kilometres through the mountain canyons, and even reached the Indus plains. Every major tributary of the Indus draining the high Karakoram ranges has a history of glacier damming, and rarer but catastrophic floods (Fig. 9.4). The last disastrous cases derived from the Khumdan ice dams on the upper Shyok River in 1926 and 1929. However, an outburst causing some destruction to communities in Sinkiang Province, China, came from an ice dam on the Shaksgam River draining the north slope of the Karakoram.

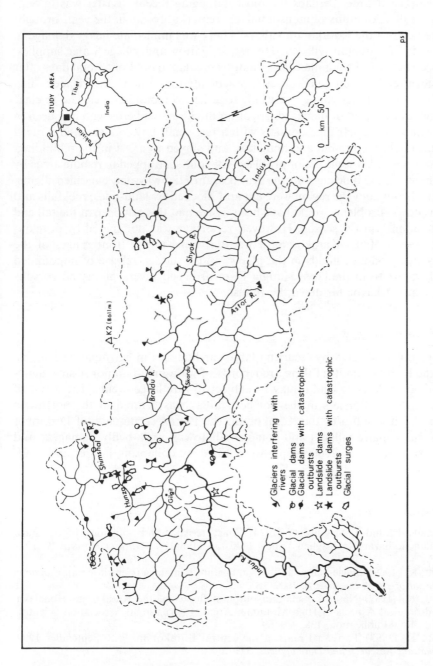

Fig. 9.4 Locations of some of the large natural processes widely occurring in the upper Indus Basin that have been associated with disasters (after Hewitt, 1985)

259

Severe storms

Destructive rainstorms, triggering local landslides, sometimes very large debris flows, damaging crops and blocking roads, have occurred almost every summer for the past decade. Perhaps the most damaging recent disaster was in September 1992. An influx of monsoonal air, remarkably late in the year, brought severe rains to the Karakoram valleys. There was immediate heavy damage in hundreds of mountain villages. Homes, irrigation and village water supplies, terraced fields and communication routes were damaged. For several days after the storm ended, you could observe practically everyone in the village – men and women, old and young – working together to rebuild homes, and to clear and rebuild clogged or damaged irrigation ditches. Shepherds who had been in the high pastures left their animals with a few small boys on watch and raced down to help the injured and homeless, and stayed to assist in the rebuilding.

There were also some major problems for tourists trapped in remote areas by the innumerable blockages to the Karakoram Highway and cancelled flights. Roads throughout the region were cut and blocked by landslides, rockfalls and debris flows. Problems would have been worse had this not been at the tail end of the main visitors' season. However, reportage was dominated by developments south of the Karakoram. Torrential rains in the front ranges of the Himalaya produced floods in the plains, and a sudden release of impounded flood waters from the large Mangla Dam drowned several thousand persons and wrought havoc far down the Indus.

Earthquake disasters

The region is affected by recurring large earthquakes in surrounding regions. For the Karakoram itself there are records of widespread, minor seismic activity in recent decades, and it may be subject to rare large events. The worst of recent natural disasters in the northern areas was centred in the northwest Himalaya just south of the Karakoram: the Pattan earthquake of December 1974. In it many villages and some small towns were badly damaged, and perhaps as many as 7000 people were killed by collapsing houses and landslides.

Suggested reading

de Scally, F. and J. S. Gardner (1994) Characteristics and mitigation of the snow avalanche hazard in Kaghan Valley, Pakistan Himalaya. *Natural Hazards*, **9**, 197–213.

Hewitt, K. (1969) Glacier surges in the Karakoram Himalaya (Central Asia). *Canadian Journal of Earth Sciences*, **6**, 1009–18.

Hewitt, K. (1982) Natural dams and outburst floods of the Karakoram Himalaya, Hydrological Aspects of High Mountain Areas. International Association Scientific Hydrology Publication, **138**, 259–69.

Hewitt, K. (1993) Torrential rains in the central Karakoram, 9–10 September 1992. *Mountain Research and Development*, 371–5.

Kreutzmann, H. (1994) Habitat conditions and settlement processes in the Hindukush–
Karakoram. *Petermanns Geographische Mitteilungen*, **138** (6), 337–56.
Shroder, J. F. Jr. (1989) Hazards of the Himalaya, *American Scientist*, **77**, 564–72.

Contexts and interpretation of risks

Natural disasters were a serious part of life in the last century, perhaps more than now, and are reported in oral histories from earlier times. Excepting wars, they continue to be the main occasions of destructive and lethal events. What is less clear, but certainly important, is the extent to which social upheavals and economic change have made and continue to make people more vulnerable and less able to respond creatively to natural extremes. In the last century that was mainly a question of political strife and wars with economic decline for many of the inhabitants. Today it is related to the social upheavals, environmental and technological changes associated with rapid economic development.

A great transformation in the conditions and forms of danger was marked by the building of an all-weather highway in the 1970s. Linking the plains of Pakistan to China, it is known as the 'Karakoram Highway', or KKH, in this region. The highway and a host of secondary roads throughout the valleys have brought great changes in the life and conditions of work for the villagers. Nearly all women remain in more traditional roles but are affected by the changes just the same. Many have an ever greater responsibility for the traditional subsistence economy, especially food, domestic matters and village life, as husbands look for paid work. Children and the elderly tend to play greater roles in the domestic and subsistence economy. Men spend a growing amount of time in, or looking for, wage work as porters and guides, in the army or construction work, in the towns, down country or even overseas earning 'hard currency'.

There is an increased strengthening, and quite rapid links to the cities of Pakistan and, through them, to the economic and recreational interests of distant, industrial nations. Travellers from London, Tokyo or New York can, if they really want to, be in the Karakoram within 48 hours of leaving home – weather and local arrangements permitting. They can be on the slopes of some of the 8000 m peaks within a few more days, although it is necessary to take more time to acclimatise to higher altitudes.

Suggested reading

Dani, A. H. (1989) *History of northern areas of Pakistan*. National Institute of Historical and Cultural Research, Islamabad.
Keay, J. (1979) *The Gilgit game: explorers of the western Himalayas, 1865–95*. John Murray, London.
Kreutzmann, H. (1991) The Karakoram Highway: the impact of road construction on mountain societies. *Modern Asian Studies*, **13** (1), 19–39.

Contexts of danger

These remarks give no more than a brief background sketch to a huge, complex region and its dangers. Hopefully, they do counteract the stereotypes and popular notions of *remote* mountain fastnesses, *inaccessibility*, 'unspoiled peoples' and even, of 'empty' or poorly utilised if 'fragile' lands. These imaginings tell us how outsiders see, or romanticise, the region. They reveal little about the life worlds of residents, or the sources and forms of risks on the ground. In that respect, at least five situations may be recognised in which the hazards outlined become dangers, or disasters occur and responses must be made:

● risks that affect indigenous peoples within their former settlement areas and traditional activities;
● dangers from recent developments, shifting patterns of land use and social violence, for residents and in their traditional activities;
● hazards affecting modernised developments or modern technologies, such as the highways, air links, and expanding urban areas;
● dangers in the places and activities where indigenous people and visitors or government agents are in close contact, whether in economic development or construction projects or in mountain treks and expeditions; and
● risks for outsiders, especially tourists and recreational or mountaineering groups.

Many damaging events, especially disasters, involve more than one of these contexts. Most involve natural hazards, but social violence and technological dangers seem to be growing. The dangers from them are often interwoven with and exaggerated by natural risks.

Suggested reading

Hewitt, K. (1976) Earthquake hazards in the mountains. *Natural History*, **85**, 30–7.
Hewitt, K. (1989) The altitudinal organisation of Karakoram geomorphic processes and environments. *Zeitschrift f. Geomorphologie N.F.*, Supplement-Bd, **76**, 9–32.
Miller, K. J. (ed.) (1984) *The International Karakoram Project* vol. 2, 'Housing and natural hazards', 245–358. Cambridge University Press, Cambridge.
Kreutzmann, H. (1994) Habitat conditions and settlement processes in the Hindukush–Karakoram. *Petermanns Geographische Mitteilungen*, **138** (6), 337–56.

Mountain hazards in a rapidly urbanising area: lower mainland, British Columbia

The region around Vancouver in western Canada is a mountainous area subject to multiple natural hazards (Figs 9.5 and 9.6). The threat that they pose has been magnified by recent rapid urbanisation. The growth has been what some call 'post-industrial', with a mobile, technologically sophisticated population, and a high demand for modern services and facilities. It involves considerable suburban, satellite centre and ex-urban development, as well as an

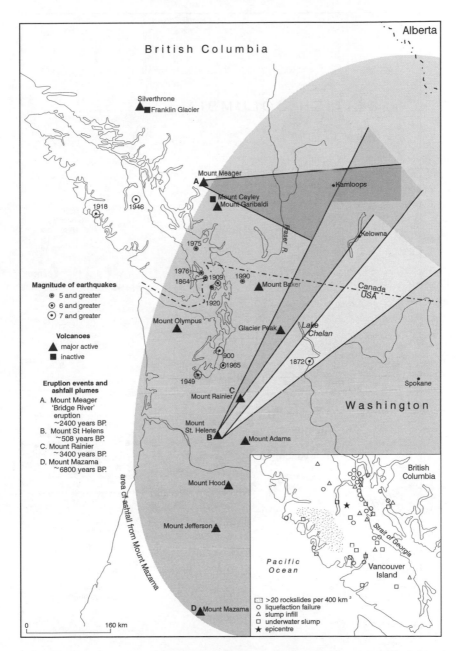

Fig. 9.5 Large earthquakes and volcanic eruptions affecting southwest British Columbia. The shaded areas show ash-falls (= tephras) from reconstructed past
eruptions that affected southern British Columbia. Large earthquakes are
either measured, reported in historical records or deduced from surface
effects. Inset shows distribution of landslides triggered by the 1946 Vancouver
Island earthquake (after Monger (ed.), 1994)

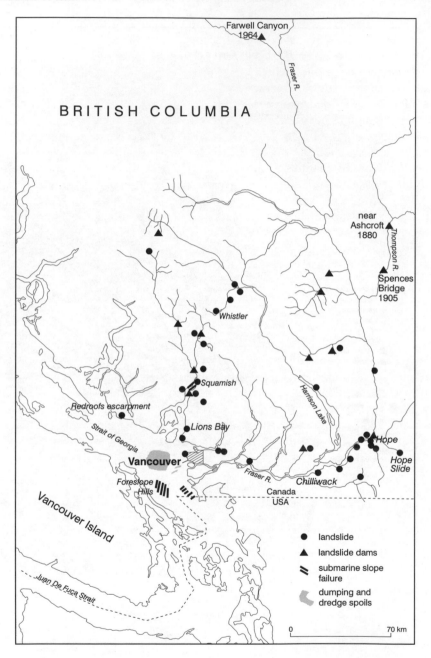

Fig. 9.6 Landslides and related events in southwest British Columbia (after Evans, 1986; Monger (ed.), 1994)

intensification of building, infrastructure and functions in the main centre, Vancouver.

The whole region is seismically active, with many moderate earthquakes and occasional very large ones. Much of the recent development has been on seismically vulnerable sites, involving encroachment on steep slopes, former tidal flats, marshes and flood plain areas of the Fraser River delta, and a string of coastal communities hemmed in between the seashore and the steep walls of the coast mountains. The Fraser and Squamish River deltas, and many occupied areas where steeply falling streams reach the coast, are subject to major floods, debris torrents and channel changes. There is a history of large and countless small landslides on the adjacent mountain slopes and access corridors. They are made more likely by recurring atmospheric hazards of heavy rains and torrential spring snowmelt, and by earthquakes. In the surrounding mountains is evidence of repeated natural dam formation from landslides and glaciers, and rare but devastating outburst floods. A number of active volcanoes exist whose past ash-falls have been identified in the Vancouver area. These dangers are added to by humanly induced erosion and large-scale dumping, and contamination of ground, river and coastal waters. There are actual and potential social problems due to congestion, housing shortages and high costs, and sharp economic swings.

The hazards seem to outstrip the safety measures of even this affluent society. This is despite the presence of a remarkable number of earth scientists and other professionals, evidently with the knowledge and skills to deal with most of these problems. Yet they, among others, believe that, for example, a large earthquake is likely at any time that could cause damage in the tens of billions of dollars. Insurance companies are said to be looking hard at reducing their exposure to property losses, if not pulling out, for fear of this eventuality.

Suggested reading

Rogers, G. C. (1980) A documentation of slope failure during the British Columbia earthquake of 23 June, 1946. *Canadian Geotechnical Journal*, **17**, 122–7.

Evans, S. G. (1986) Landslide damming in the Cordillera of Western Canada. In Schuster, R. L. (ed.) 93–130.

Monger, J. W. H. (ed.) (1994) *Geology and geological hazards of the Vancouver region, southwestern British Columbia*. Geological Survey of Canada, Bulletin 481, Ministry of Energy, Mines and Resources, Ottawa.

CHAPTER 10

Risk in the city

Historically, many cities developed on sites prone to floods, earthquakes, landslides, droughts or hurricanes; but two factors limited the impacts on human
populations. First, the most dangerous sites within the locality were avoided,
second, cities were relatively small and less densely populated . . . Today, there
are many large cities in areas prone to natural disasters and many of the most
dangerous sites have been occupied.

Hardoy and Satterthwaite (1989, 203)

Urban problems

The significance of urban risk arises mainly in the actual and potential scales of
disaster in cities. There are, however, special problems of vulnerability and
response in urban communities and for their members. They involve the
scope and complexities of the technological and socioeconomic context, and
relations to the urban habitat. Increasingly, in this century, the dangers are
being magnified by rapid growth resulting in an ever greater concentration of
people and property in urban centres (Table 10.1).

Danger in a rapidly urbanising world

If present trends continue, about half of humanity will live in cities by the end of
the century. In the more fully industrialised countries, the proportions have
been higher than that for many decades. In the 1990s, the largest, often the
most rapid, rates of city growth have been in countries with predominantly
agricultural and rural populations. The more or less enforced flight of the latter
to urban areas lies behind some of the most intractable urban problems, and
unusually vulnerable urban communities. Equally striking, however, is the
enormous spreading out and intensification of built-up areas throughout the

266

Table 10.1 The degree and rates of urbanisation. Estimates of the percentage of selected world regional populations in urban centres 1950, 1970, 1990 and projected to the year 2000. Selected national figures indicate the considerable variation within different regions, but also the general picture of rapid catch-up of all countries in the last half of the century

Region (country)	1950	1970	1990	2000
WORLD	29	36	45	51
EUROPE	56	68	75	80
Belgium	92	94	97	98
United Kingdom	84	89	93	94
Germany	72	78	84	86
France	56	71	74	76
Spain	52	66	78	83
Norway	32	65	74	77
NORTH AMERICA	64	74	74	78
United States	64	74	74	75
Canada	61	76	76	79
LATIN AMERICA	41	54	69	77
Chile	58	75	86	89
Mexico	43	59	73	77
Brazil	36	56	77	83
Peru	36	57	70	75
AFRICA	15	22	30	44
South Africa	43	48	59	65
Egypt	32	42	39	55
Nigeria	10	20	35	43
Uganda	3	8	10	14
SOUTH ASIA	15	18	25	39
Pakistan	18	25	32	38
India	17	20	28	34
Nepal	2	4	10	15
CIS (former USSR)	39	57	68	71
Russia	–	–	74	–
Ukraine	–	–	68	–
Kazakhstan	–	–	57	–
E. and SE. ASIA	15	22	30	38
Japan	50	71	77	78
Indonesia	12	17	28	37
China	11	20	26	31
AUSTRALASIA and OCEANIA	60	65	72	75
Australia	75	85	86	86
New Zealand	73	81	84	85
French Polynesia	28	56	65	70
Papua New Guinea	1	10	16	20

267

industrialised world. There has been a proliferation of tall buildings for business and housing. Suburban and ex-urban growth have greatly expanded the built-up areas and the scope of urbanised activity in wealthier countries or cities.

The emerging 'megacities', those exceeding 10 million in population such as Mexico City, São Paulo and Shanghai, do raise exceptional problems (Table 10.2). They have extraordinary numbers of densely concentrated and impoverished residents, who may be uniquely vulnerable to most hazards. However, the largest numbers of new urbanites, in total, have congregated in the many smaller, but rapidly growing, regional cities. They appear more often in the disaster lists.

All of these developments are associated with changing risks. Usually, if not necessarily, there is increasing vulnerability to natural and technological hazards, epidemic disease and social strife, at least for some. Of course, few disasters are either only rural or only urban, even when most damage is concentrated in one or the other. This follows from the roles and material connections of any central place with surrounding areas and smaller communities. Urbanisation tends to mean greater centralisation, so that urban disasters have greater impacts upon increasingly dependent hinterlands, while cities play an ever larger role in response to rural disasters. This should not be obscured by the tendency of professional studies and administrations to think in terms of either urban, or rural, conditions. Nevertheless, the built environment, and modern institutions and practices do create distinctive conditions of risk in cities.

Suggested reading

Drakakis-Smith, D. (1987) *The Third World city*. Routledge, London.
Mitchell, J. K. (1990a) Natural hazard prediction and response in very large cities. *Proceedings: IDNDR Workshop on Prediction and Perception of Hazards*. WARREDOC Centre, Perugia, Italy.
Solway, L. (1994) Urban developments and megacities: vulnerability to natural disasters. *Disaster Management*, **6** (3), 160–9.

Table 10.2 Emergence of the megacities: cities with more than 10 million residents in 1950, 1980 and projected for 2000 (after Garrett, 1994, 314)

1950 (2)	New York, London
1980 (10)	*Americas:* Buenos Aires, Rio de Janeiro, São Paulo, Mexico City, Los Angeles, New York
	East Asia: Beijing, Shanghai, Tokyo
	Europe: London
Projected 2000 (24)	*Africa:* Cairo
	Americas: Buenos Aires, Rio de Janeiro, São Paulo, Mexico City, Los Angeles, New York
	East Asia: Beijing, Shanghai, Osaka–Kobe, Tokyo–Yokohama
	South Asia: Dacca, Bombay, Calcutta, Delhi, Madras, Jakarta, Baghdad, Teheran, Karachi, Bangkok, Manila, Istanbul
	Europe: Paris (London no longer on the list)

Hazards and urban disasters

The hazardousness of cities

Cities cover a tiny fraction of the area of most countries, and much less than 1 per cent of the Earth's total land area. It might seem, therefore, that urbanisation, especially with substantial migration from rural areas, does not bring an equivalent increase in risk. The concentration of people and property should be offset by lower exposure to extreme natural events and technological accidents that originate outside, or in other, cities.

However, smaller spatial exposure may be counteracted when cities are established in relatively more dangerous locations. They may favour flood- and storm-prone lake shores and coastlines; flood plain areas and river junctions; droughty and arid coastlines; or mountain foot sites vulnerable to earthquake and landslide. Meanwhile, there is often a concentration of more dangerous transportation, industrial and construction technologies. Where more and more vulnerable people settle on dangerous sites or take dangerous occupations, the thresholds at which hazardous agents may do damage are lowered. As we have seen, conditions in cities may promote civil strife, or offer attractive targets to violent groups. Looking at the diversity of risks, it seems that urbanisation in much of the world has increased overall disaster risks. Given that cities are places of greater longevity now, and most are growing by internal increase, the toll of chronic risks, notably fatal diseases, may be proportionately less than in rural populations.

The range and incidence of recent urban disasters help to define the problem (Figs 10.1, 10.2). A survey of urban disasters reported in news media for Latin America and the Caribbean (1974–1994) indicated that nearly two-thirds of the deaths and damage originated from natural disasters, and rather more than a third from technological or social disasters (Table 10.3).

A more detailed inventory of short-term damaging events in southern Ontario provides a view of dangers in a highly urbanised and industrial region (Table 10.4). Mapping the events or principal locations of damage shows that the greater incidence occurs in urban areas. Over one-third of the events reported occurred in the major centre, Greater Toronto (Fig. 10.3).

Canada is often identified as having an exceptionally low ratio of population to land, with well-promoted images of vast wilderness areas and of 'northernness'. Yet 80 per cent or so of Canadians reside in or near a string of urbanised areas in the south of the country. Here, despite its huge area, Canada is closely packed, if not 'overcrowded', for many of its citizens and activities. Moreover, a generally high standard of living and mobility, with high levels of energy use and waste production, enormously magnifies the impact of given numbers of people on each other and the urban and surrounding environment. The hazardousness of its cities and the damaging events identified in southern Ontario result from, or are aggravated by, these conditions.

269

Suggested reading

Hewitt, K. and I. Burton (1971) *The Hazardousness of a place: a regional ecology of damaging events*. Department of Geography, Research Publication 6, University of Toronto Press, Toronto.
Sylves, R.T. and W. L. Waugh (eds) (1990) *Cities and Disasters*.

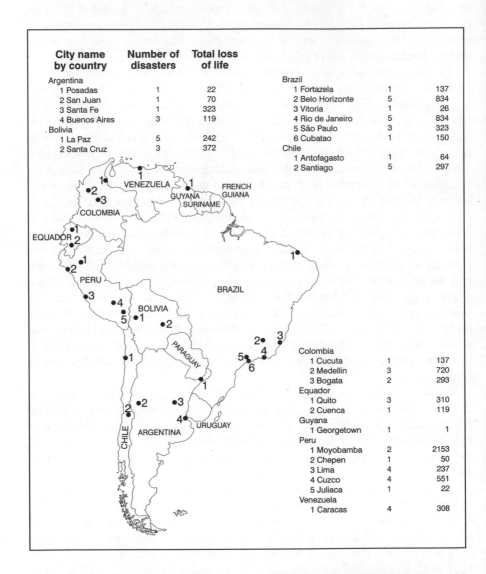

City name by country	Number of disasters	Total loss of life
Argentina		
1 Posadas	1	22
2 San Juan	1	70
3 Santa Fe	1	323
4 Buenos Aires	3	119
. Bolivia		
1 La Paz	5	242
2 Santa Cruz	3	372

Brazil		
1 Fortazela	1	137
2 Belo Horizonte	5	834
3 Vitoria	1	26
4 Rio de Janeiro	5	834
5 São Paulo	3	323
6 Cubatao	1	150
Chile		
1 Antofagasto	1	64
2 Santiago	5	297

Colombia		
1 Cucuta	1	137
2 Medellin	3	720
3 Bogata	2	293
Equador		
1 Quito	3	310
2 Cuenca	1	119
Guyana		
1 Georgetown	1	1
Peru		
1 Moyobamba	2	2153
2 Chepen	1	50
3 Lima	4	237
4 Cuzco	4	551
5 Juliaca	1	22
Venezuela		
1 Caracas	4	308

Fig. 10.1 Disasters affecting urban centres in South America, 1974–1994, reported in mass media. The information gives number of events per city affected and aggregate death tolls in them (from a survey of the *New York Times*, *The Encyclopedia Americana Annual*, *Collier's Encyclopedia* and US OFDA, 1988)

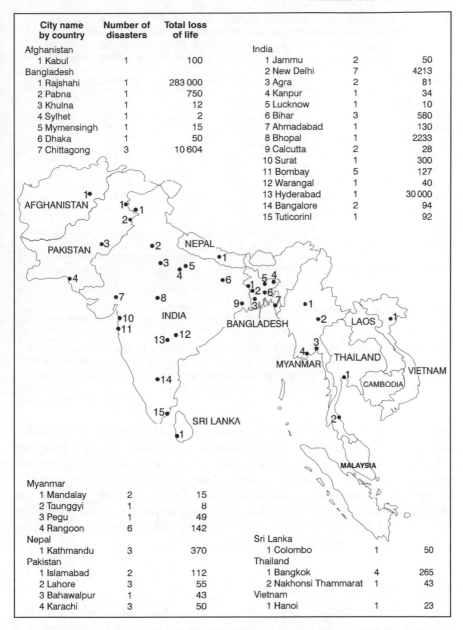

City name by country	Number of disasters	Total loss of life
Afghanistan		
1 Kabul	1	100
Bangladesh		
1 Rajshahi	1	283 000
2 Pabna	1	750
3 Khulna	1	12
4 Sylhet	1	2
5 Mymensingh	1	15
6 Dhaka	1	50
7 Chittagong	3	10 604
India		
1 Jammu	2	50
2 New Delhi	7	4213
3 Agra	2	81
4 Kanpur	1	34
5 Lucknow	1	10
6 Bihar	3	580
7 Ahmadabad	1	130
8 Bhopal	1	2233
9 Calcutta	2	28
10 Surat	1	300
11 Bombay	5	127
12 Warangal	1	40
13 Hyderabad	1	30 000
14 Bangalore	2	94
15 Tuticorinl	1	92
Myanmar		
1 Mandalay	2	15
2 Taunggyi	1	8
3 Pegu	1	49
4 Rangoon	6	142
Nepal		
1 Kathmandu	3	370
Pakistan		
1 Islamabad	2	112
2 Lahore	3	55
3 Bahawalpur	1	43
4 Karachi	3	50
Sri Lanka		
1 Colombo	1	50
Thailand		
1 Bangkok	4	265
2 Nakhonsi Thammarat	1	43
Vietnam		
1 Hanoi	1	23

Fig. 10.2 Disasters affecting urban centres in South Asia, 1974–1994, reported by mass media (from a survey of the *New York Times, The Encyclopedia Americana Annual, Collier's Encyclopedia* and US OFDA, 1988)

Table 10.3   Disasters involving major damage in cities of Latin American and Caribbean countries, 1974–1994, ranked by death tolls and numbers of events (from a survey of the *New York Times*, *The Encyclopedia Americana Annual*, *Collier's Encyclopedia* and US OFDA, 1988)

Hazard agent	Deaths caused	Events (frequency rank)
Earthquakes	35 268	30 (3)
Civil strife	32 000	3 (9)
Vulcanism	21 814	3 (9)
Enforced displacement	20 000	4 (8)
Hurricanes	10 501	18 (5)
'Accidents'*	4247	42 (1)
Floods	3682	41 (2)
Landslides	2246	21 (4)
Other storms	361	9 (6)
Fires	139	5 (7)
Totals	130 258	176

*Industrial, energy systems, transportation, etc.

Vulnerability in the city

> The work of urbanisation goes on but we are unwilling to face its implications and, less and less does the traditional view that the city appears and grows in response to favourable geographic conditions, describe the realities of modern urban growth.
>
> D. W. Brogan (1963, 161).

As human artefacts, cities must raise questions regarding human responsibility and the social conditions of risk. The balance of attention in risk assessments shifts from physical dangers or hazards to the conditions of living or their transformations. This favours a vulnerability perspective. Vulnerability is usually very varied among the persons or activities of a city, but we can recognise certain common features of cities that, working together, create distinctive urban risks. If not taken into account in public safety measures, they set the stage for concentrated and complex disasters.

The built environment

In built-up areas, building safety often lies at the heart of vulnerability. This applies to property risks, and it creates dangers for people who spend most of their time in and around built structures. Densely populated neighbourhoods and large, heavily occupied structures are concentrated in cities. Often, destruction in one district, or the collapse of just one or two big buildings, may bring a calamity. At Skopje, 1963, or Guatemala City, 1987, earthquake destruction was highly concentrated in certain districts and for certain classes of people. 'Worst-case' disasters, as in the fires at San Francisco in 1906 and Tokyo in

Table 10.4 Short-term damaging events affecting the general public in southern
Ontario, 1973–1984, reported in the *Globe and Mail* (Toronto) newspaper

Agent	Number of events	Events in urban areas (%)
Natural	44	34 (77.3%)
flooding	20	
snowstorms	10	
other atmosph.	9	
high winds	5	
Other (rural only)	7	0 (0.0%)
icing damage	3	
forest fires	2	
drought	1	
landslides	1	
Epidemic disease	2	2 (100.0%)
polio threat	1	
Lassa fever scare	1	
Technological	255	214 (83.9%)
industrial	55	
transport	92	
serious fires	89	
explosions	13	
Social	6	6 (100.0%)
bombings	4	
bomb threats	2	
Totals	314	256 (81.5%)

1923, or at Nagasaki in 1945, show the devastation overtaking almost all
buildings exposed in a large city. Then again, in the core area of the 1985
Mexico City earthquake, only (!) 4 per cent of structures were actually
destroyed (Chapter 8), but still creating the worst natural catastrophe there.

Dangers of dependency

These arise partly from what is called the 'metabolism' of a city: the way it
must rely upon a constant inflow of resources, information and goods; and an
outflow of products, information and waste. People dependent upon built
infrastructure for life support and economic productivity are at special risk if
it is destroyed, but the city is also a mosaic of personal, societal and economic
dependencies. A majority of urban people, without productive land or other
private resources, depend upon wage employment, barter or trafficking to live.
Dependency in that sense also singles out the most vulnerable members of
communities and families who, from personal misfortune or lack of rights
and opportunities, have to depend upon others. These and many others have

273

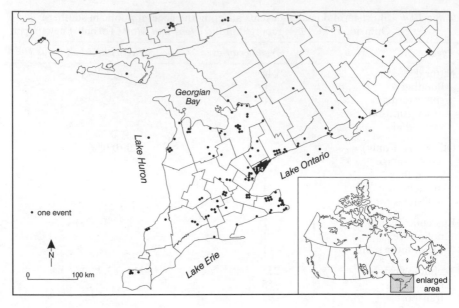

Fig. 10.3 Location of short-term damaging events affecting the general public in southern Ontario, 1973–1984, listed in Table 10.4

to rely upon social security measures, mutual aid, charity or emergency relief to survive personal accident or collective disasters.

Lethal forces

Modern industries and infrastructure have matched the growth of cities with increasing scales and concentrations of energy or physical forces. Ever more forms and quantities of dangerous substances are being manufactured in urban areas, stored in warehouses and even in people's kitchen cupboards, or passing through on the highways and railroads. More, and more varied, techniques of violent environmental modification – high explosives, biocides, heavy earth-moving and tunnelling equipment – are used in the city. The range of such forces, or their proximity to densely occupied areas, magnifies the scope of dangers and likelihood of technological disaster.

Dangers of congestion

The numbers and densities of persons, or the frequency of exchanges among them, increase the potential for dangerous encounters. Vulnerable persons are more likely to be exposed to adverse conditions. The stresses of crowded living can make persons who would otherwise be fairly robust, more vulnerable, most obviously in traffic accidents or the spread of infectious diseases. Congestion reduces options and constrains flexibility.

Dangers of complexity

These arise from the diversity of persons, artefacts and functions, and the range of interactions among them. The city's complexity lies in the many different kinds of thing that happen within relatively small areas and over short time frames. Think of a large railway station or harbour front, or the density and mixture of persons and exchanges at a major city market or a great religious or sporting event. Above all, an awareness of how and for whom vulnerability arises requires a sense of the social differentiation of the city. Cities, especially for newcomers hurried into them, can 'socialise' some into more dangerous occupations or living sites, reduce the strengths and options of others, perhaps children or the elderly, often wives and women workers.

Dangers of the 'mass order'

All of the above add up to what has been called, rather unhelpfully, *mass society*. Perhaps this term is a way to describe communities too numerous for any one person or family to have face-to-face and personalised contact with a majority of those whose worlds they share. Many important encounters are with 'strangers'. Crucial decisions or careless acts can affect unknown thousands, even millions.

Such are the social environments and geographies of civil society, but they can be, and often are, doubly endangered by institutions and attitudes, manipulations and plans, that treat people *only* 'in the mass'. Everyone in 'mass society' is still a person to themselves, a member of a family, school, team, group or other more intimate and shared unit. Advantage or misfortune, happiness, fear, or pain are only felt by and shared among individuals. More importantly, for collective security, only the person and the group are fully aware of these predicaments and can say whether they are adequately addressed by the larger society. Otherwise – or for those who think otherwise – what happens (at least to others) does not matter. It all becomes 'process' and statistics.

Increasingly documented, for example, is how the official, mass, 'objective' approach has ignored the very different predicaments and vulnerabilities of women. Little regard has been paid to how legal, economic, community and family status, in so many societies and classes, places them at higher risk, with less protection against misfortune. The special vulnerability of female 'sex workers', and especially of women spouses, in the AIDS pandemic provides harrowing and growing testimony to women's lack of rights and protection almost worldwide.

One of the gravest dangers of a 'scientific' and technical 'mass' modernity lies in purely bureaucratic and detached notions of order and administrative practice. This becomes especially calamitous when coupled with elite, religious, political and racial or other biological or geographical theories of superiority. These have allowed and excused urban developments that deliberately ignore the safety needs of some or most citizens. They are made most dangerous, and most often lead to calamities, through:

Urban violence and cities of misrule

Urban communities are most vulnerable when political, economic and military strength are used not to protect them but to exploit or terrorise them. When state powers are used against ordinary citizens, especially politically weak and minority groups, against women or children, singular calamities take place, since the victims are often concentrated and readily targeted (see Chapter 12).

It is under such conditions that we realise why safety in the city and for the majority of folk must be underpinned by 'civility' – tolerance, if not enjoyment, of the differences and diversity of persons in the city; protocols and rights to enjoy the freedoms of the city, backed by just and evenly applied laws. This is what defines the behaviours and practices of a civilised, collective life. They provide the basis of an urban security. Urbanites of different backgrounds are socialised or contracted to participate in the goals and values of the larger society, with a sense that it will improve their safety and prospects.

Of course, bombs placed in a crowded subway in Paris, a pub in the English Midlands or a discotheque in Germany threaten not only 'the authorities'. Horrible and indiscriminate casualties among ordinary citizens in public places destroy the security of everyone else.

Suggested reading

Zwi, A. and A. Ugalde (1991) Political violence in the Third World: a public health issue. *Health Policy and Planning*, **6** (3), 203–17.
Scarpacci, J. L. and L. J. Frazier (1994) State terror: ideology, protest and the gendering of landscapes. *Progress in Human Geography*, **17** (1), 1–21.

Cities besieged

In war, cities may be targeted, besieged and annihilated. In the last few decades, the cities of Bosnia and Croatia, Cyprus, Algeria, Lebanon, Iraq, Iran, Afghanistan, and Korea, to name a few, have been scenes of appalling destruction in conflicts. Most often the bombs, rockets and sniper fire have indiscriminately killed and maimed resident civilians, and wrecked their homes and support systems. Women, children and the elderly are still present in wartime cities. They are most vulnerable both to armed assault and because disruption of their other cares and needs threatens health or survival. The urban prospect became much bleaker when air power and weapons of mass destruction were turned against cities. Contemporary cites are totally vulnerable if the weapons now available are turned against them. We will return to that overarching problem at the end of the chapter.

Suggested reading

Ashworth, G. J. (1991) *War and the city*. Routledge, London.
Cuny F. C. (1994) Cities under siege: problems, priorities and programs. *Disasters*, **18** (2), 152–9.

Identifying the most vulnerable

> . . . the sites are cheap because they are dangerous.
> Hardoy and Satterthwaite (1989, 159)

In cities everywhere, but led by the industrialised countries, wealthier and more influential classes have left or are leaving the inner city. They move to the suburbs, small, favoured towns or ex-urban living. The job opportunities and social services are still, often, concentrated in downtown areas. They draw the poor to live within reach of them, and demand expensive communications infrastructure for the better-off commuters. But these trends are also usually associated with impoverishment, and declining safety and services in inner city neighbourhoods. A declining tax base in the core cities is often the reason or excuse for doing little about this, but abandonment of the urban problem by more influential groups is also a large factor.

Newer, poorer residents and other disadvantaged groups are most likely to be neglected or mistreated, making them more vulnerable. As highlighted in Chapter 6, the most vulnerable tend to be those who have multiple sources of weakness, lack of protections and options. This applies, most obviously, to the millions of what Hardoy and Satterthwaite (1989) described as 'squatter citizens' – a not unintentional contradiction in terms:

> . . . large clusters of illegal (sic) housing often develop on the steep hillsides, flood plains or desert land. Or they develop on the most unhealthy and polluted sites – for instance, around solid waste dumps, beside open drains and sewers or around industrial areas with high levels of air pollution. Poor groups do not live here in ignorance of the dangers; they choose such sites because they meet more immediate and pressing needs. Such sites are often the only places where they can build their own house or rent accommodation. The sites are cheap because they are dangerous. [but] Polluted sites next to industries are close to jobs.
> Hardoy and Satterthwaite (*ibid.* 159)

An almost identical picture was painted by Maskrey (1989, 12) of the risks faced by the most vulnerable people in Lima, Peru, suggesting that:

> Low income families in Lima only have freedom to choose between different kinds of disaster.

And as was emphasised earlier, if a class, a district, or a social group is unusually vulnerable, the chances are high that some of its members will be the most vulnerable of all. Commonly this will apply to women, perhaps to girl children or widows, sometimes to the elderly or disabled. Others in the group may, in fact, be unusually robust. We must resist the temptation to reduce everything to mass statistics or stereotypes. However, mitigating risk and the chances of disaster most often means identifying, helping and empowering those who are most vulnerable. Whatever else happens in the city, disasters will continue and probably grow – unless the social geography and injustices that force some into more vulnerable living are addressed.

Suggested reading

Hardoy, J. E. and D. Satterthwaite (1989) *Squatter citizen: life in the urban Third World.* Earthscan, London.
Gertel, J. (1993) Food security within metropolitan Cairo under conditions of 'Structural Adjustment'. In Bohle, H.-G (ed.) 101–30.

Active perspectives and terms of reference

> There is, however, a danger in labelling density, per se, as the culprit.
>
> Newman (1973, 195)

Something needs saying here to balance our emphasis upon risks and disasters. Most of the features identified above as sources of urban vulnerability can be equally *sources of strength*. Each also involves or arises from the means to make urban society more safe, or to offset dangers. Nearly every city, taken as a whole, is unusually well-endowed in wealth, political and technical sophistication, and the capacity for organised commitment. If an adequate portion of the latter is devoted to public safety, then we may find unusual successes in limiting dangers and responding to crises. Cities generally contain most, if not all, of the modern instruments for risk reduction and disaster response. Here are most of the medical, engineering, fire-fighting and other professionals; the best or best-endowed hospitals and research laboratories; well-entrenched public and private organisations; the seats of government, head offices and branches of businesses; and the purse-strings to provide resources for emergency action and public assistance. In cities is the authority to deploy the military and other official units used to respond to crises, to request and permit, or to refuse, international assistance. The mass media mostly ply their trade from the major centres. They can make the tragedies and, if they choose, official action or inaction, publicly and internationally known.

There is no inevitable relation between density and danger over a broad range of options available in most cities. One city deals well and more safely with much higher concentrations of people and activities than another. Only rarely is congestion itself the problem, even for the poorest and most vulnerable. Given adequate facilities and services, better and more secure economic opportunities, or more public attention to their basic needs, they can live more safely and prosper. Historically, people have often preferred to congregate very closely together, finding satisfaction in gregarious and convivial living, and a greater sense of collective security. In other words, dependency, congestion, complexity and mass society often are, but *are not necessarily*, dangerous. Nor are they the underlying reasons why people are endangered. Only if other standards and protections are missing do they lead to vulnerability.

In many disasters, urban dwellers can fare better or have more options for survival. Cities, even in the less wealthy nations, are places to which rural and small town folk go to seek refuge and assistance. Displaced or impoverished by famine, civil war, modernisation of agriculture or environmental degradation, that is where they find the only possibilities, if seemingly precarious, of a new life.

The twentieth century experience includes many cities, or parts of cities, with remarkable and effective commitments to public safety. Some cities pioneered methods that drastically reduced or eliminated some great dangers of the past. Public health measures and disease control, fire-prevention and building codes stand out – albeit in response to dangers brought about largely by urbanisation and overcrowding in cities. Hundreds of cities have managed periods of good or adequate administration and civility, avoiding unusual social conflict. It is essential to recognise these 'successes', even though our focus is upon the times when they fail and the places lacking them.

As in most of our concerns, therefore, the dangers and the benefits of urbanism are far from separate questions. Disaster-proneness, and those most likely to be victims, goes hand in hand with economic and political transformations that greatly increase the wealth and influence of cities. Those who control, or are well-entrenched in, urban institutions generally benefit economically from urbanisation. They often can and do gain in material security. Meanwhile, greater commercial use, or central control, of agricultural and other rural resource lands is a major process behind urban growth. Wealth from mines, forests, fishing grounds or wheat belts ends up, disproportionately, in central business districts and urban-based enterprises. This applies as much to Canada or Australia as to Brazil or Pakistan. The influx of cheap labour into cities, and increased numbers and more concentrated, materially dependent consumers – even though, and sometimes because, many are poor – improve the overall commercial prospects and productivity of cities. As a result, however, the greater cause for concern is not urbanisation itself, or even concentrations of vulnerable persons or high-risk technologies in cities. It is the widespread absence of strong commitments to improving public safety measures. This is a problem in North America and other industrial nations. Newly emerging, large cities and conurbations provide even more calamitous results. However, *the untapped potential for safer living* in cities sets the stage for the dangers and disasters that are our focus.

Suggested reading

Seley, J. E. and J. Wolpert (1982) Negotiating urban risk. In Herbert D. T. and Johnston R. J. (eds) vol. 5, chapter 8, 279–306.

'Natural' hazards

The smaller fraction of all natural disasters occur in cities, but these are generally the more lethal and costly ones. Some natural hazards are especially dangerous for built environments. Costly and extensive windstorm damages can occur in, and may be aggravated by, tall or close-packed urban buildings. Fog and snowstorms have been transformed, primarily, into threats to motorised mobility in and between cities (see Chapter 3). The worst tornado disasters tend to be where the funnel passes through a densely built-up area. Examples include three major disasters in the United States in 1953: at Waco, Texas, Worcester, Massachusetts, and Vicksburg, Mississippi (see Table 3.3). These were the subject of pioneering studies of disaster (Wallace 1956; Perry *et al.* 1956).

Flood disasters threaten unusual damage to the built environment, or paralysis of transportation. Alexander (1993, 340) has drawn attention to the way flood risks in cities are 'exceptional because the urban fabric is generally the most impervious category of terrain.' Water from rain or melting snow, and from river and shoreline floods may be ponded in paved-over and low-lying areas, or in basements or backed-up drains.

Many large and influential centres are coastal and port cities. Otherwise favourable locations and harbours may be unusually at risk from tsunamis, storm surges and wave action in the highest tides. The larger part of the Japanese population, for example, lives in coastal cities facing the Pacific Ocean. It is the windward shore for nearly all typhoons that reach the islands. Immediately offshore are active submarine faults identified with large, tsunami-generating earthquakes. The record of typhoon and tsunami disasters is almost 2000 years long, but the emergence of the country as a major industrial, commercial and, for a time, military power brought enormous growth in the coastal cities. That growth brought more pressure to utilise shoreline, delta and estuarine areas, and to increase coastal land by reclamation or artificial fill. Such developments in Osaka, Japan's second largest city, are associated with a succession of typhoon disasters and damage due to inundation (Fig. 10.4).

Among geological hazards, the more lethal and costly earthquake disasters have been those striking cities. Damage in the 1995 Kobe earthquake is recent, disastrous testimony to the way waterfront developments in Japan, just described, also create exceptional seismic dangers (see Chapter 8). Large land-

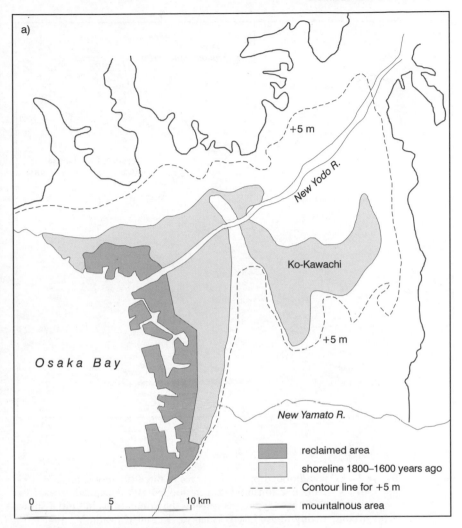

Fig. 10.4 Typhoons and the city of Osaka, Japan.
a) Topography of the urban area of Osaka, showing old shoreline and area reclaimed from the sea.

slides and avalanches are relatively rare in urbanised areas, but they threaten great devastation and large numbers of casualties if they do occur. They are especially a threat in earthquakes.

Some important cities and many local centres lie on or close to active volcanoes. When lava flows, hot ash-falls, lahars (volcanic mudflows) and other volcanic mass movements sweep into built-up areas, the devastation can be total. Few urban risks have fed the 'imagination of disaster' more than the

281

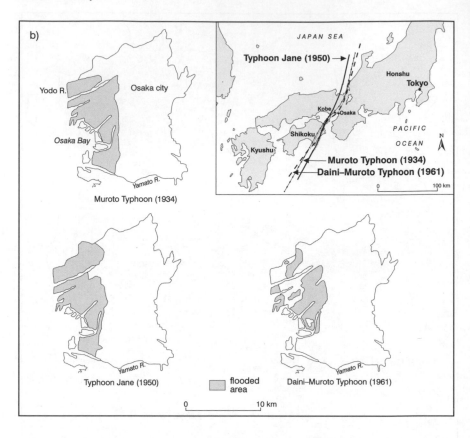

Fig. 10.4 b) Areas of Osaka flooded by storm surges in three typhoons, in 1934, 1950 and 1961. Note the relationship between flooded and reclaimed areas. Inset shows paths of the three typhoons causing the inundations in Osaka shown in the accompanying figure (after Tsuchiya, Y. and Kawata, Y., 1988)

complete annihilation of Pompei and Herculaneum in the eruption of Mount Vesuvius, 24 August, AD 79. The former was smothered in hot ash, the latter buried and burned in a high-temperature lahar. There is no reason to suppose that today's heavily urbanised Bay of Naples is not at risk from a repetition in the future. In the 1902 eruption of Mont Pelée, Martinique, a rapidly moving avalanche of hot ash, or 'nuées ardentes' (glowing clouds), engulfed and destroyed the port of Saint Pierre, killing 30 000 people. In the Nevado del Ruiz, Colombia, eruption of 1985, three successive lahars overwhelmed the unprepared town of Armero (pop. 38 000), destroying 80 per cent of its buildings and killing 22 000 people.

Suggested reading

Sheets, P. D. and D. K. Grayston (eds) (1979) *Volcanic activity and human ecology.* Academic Press, New York.
Torry, W.I. (1980) Urban earthquake hazard in developing countries: squatter settlements and the outlook for Turkey. *Urban Ecology*, **4**, 317–27.
Alexander, D. E. (1989) Urban landslides. *Progress in Physical Geography*, **13**, 157–91.
Jiminez Dias, V. (1992) Landslides and the squatter settlements of Caracas. *Environment and Urbanisation*, **4** (2), 80–9.

Epidemics

Congested populations are unusually at risk of harbouring or spreading contagious diseases. Most ordinary communicable diseases are readily passed on in schools, barracks, day-care centres, dormitories, hospitals and markets. Prisons, shelters for the homeless, and cheap hostels or hotels are also concentrated in cities. They can involve conditions or behaviours that more readily promote exchange of diseases. In epidemics, the dangers tend to be worse for urban populations, the disease more readily passed among them or having more concentrated impacts.

Since cities increasingly interact and depend upon each other, with a continuous flow of people and goods between them, the potential for rapid spread of contagious diseases also increases. A globalised economy links together all cities in countless fast exchanges involving persons and material contacts, and these have been growing, the great increase in electronic communications notwithstanding.

Rapid urbanisation can overwhelm sanitation systems and medical care. Only rarely is it accompanied by improvements in these and in living conditions, in health education or encouragement of popular participation in health care. Outbreaks of mosquito-transmitted dengue fever and malaria are associated with newly urbanising areas in the tropics, and aggravated by occupation or the creation of poorly drained areas. Congestion and overloading of health care increases risk of epidemics that can originate at medical facilities. Inadequate or aging equipment, unreliable funding and over-worked, under-informed or careless medical staff can exaggerate this problem. Recent outbreaks at, or spread from, medical facilities include deadly Ebola (Ebola virus), hepatitis B, major outbreaks of Lassa fever in West Africa, the legionnaire's disease (legionella) bacterium in the United States and, probably, swine flu. Diseases spread by respiration are singular risks in crowded hospitals and, if long-lived in the air, through centralised ventilation systems.

Where there is in-migration of rural folk, they encounter and may be more vulnerable to urban diseases and crowded, polluted environments. They may also bring rural diseases that become 'urbanised'. Chagas disease appears to have been transmitted by organisms endemic to rural Central and South America. Recently, however, the protozoan concerned has been turning up in contaminated blood supplies in cities such as Brasilia, Buenos Aires, and Santa

Cruz in Bolivia. Apparently the blood was taken from rural folk visiting or moved to the cities.

Suggested reading

Walmsley, D. F. (1988) *Urban living: the individual in the city*. Longman, London.

HIV/AIDS as an urban pandemic

> Only in the big cities, where the ghettos spill over into the entertainment, banking, and commercial districts, is there any significant contact between the majority and the infected minority.
>
> Perrow and Guillen (1990, 12)

Three out of four persons infected with HIV live in cities. Most of the initial infection of rural people seems to be by individuals returning from urban areas (Table 10.5). Dangerous contacts are more possible and more likely in cities. Sex work and drug trafficking, long-distance travel and migrant work forces, with the possibility of anonymous liaisons, are essential to the chain of infection, and the key role of cities. In North America, thus far, the epidemics have been highly concentrated in certain large cities, and among distinct populations there (Table 10.6). About one-fifth of AIDS victims have been infected through drug taking. They are also concentrated in cities, especially of the eastern and western United States seaboard, and they comprise up to half the cases in some parts of New York and New Jersey.

Drug taking is identified with a third of Latin America's victims and more than half in southern European countries. High rates of HIV-infected hard drug users, recently reported in Thailand, Burma and parts of India and China, are mainly in urban areas.

The devastation of rural areas in wars has driven huge numbers of more or less destitute people to towns and cities, where they may become involved in activities that will expose them, and make them more vulnerable, to AIDS. Military bases act as magnets for the unemployed, and for sex- and drug-driven activities. All but a tiny fraction of sex workers are women; most of the remainder are children (Mann *et al.*, 1992, 376). Few of them have the option to decide how safe the encounters they have will be. Research suggests that AIDS in southern Africa has spread along the routes of military movement and around the towns and bases where military actions have taken place.

This is not to underestimate the present scale and future potential of AIDS in rural communities, and some peculiarly tragic and intractable aspects there. Most often the path of infection is a man returning from the city to his spouse or women friends and children. Few societies will then take the woman's part but, rather, blame her. In Uganda, where a tenth of the people are HIV-

Table 10.5 HIV/AIDS and urbanisation. Urban-to-rural and gender ratios of persons infected with HIV by major world regions, and rates of growth of urban populations. Estimates as of 1 January, 1992 (after Mann et al. 1992, 20, 30)

	Sub-Saharan Africa	North America	Latin America	Western Europe	SE. Asia	Caribbean	NE. Asia	SE. Mediterranean	Eastern Europe
Total HIV	8 772 500	1 183 000	1 035 500	725 000	699 000	326 000	41 750	36 000	27 200
urban/rural	3.6 : 1	3.2 : 1	2.3 : 1	5 : 1	6 : 1	3.6 : 1	5 : 1	12 : 1	3.2 : 1
urban growth*	5.3	1.0	2.5	0.6	4 : 1	2.5	4.0	3.9	1.4
male/female	1 : 1	8 : 1	4 : 1	5 : 1	2 : 1	1.5 : 1	2 : 1	5 : 1	10 : 1

*Projected annual rate of population increase (%), 1990 to 2000, and in most cases represents urban in-migration.

Table 10.6 The HIV/AIDS epidemic in the United States: some basic statistics to 1989 (after Perrow and Guillén 1990, 56)

Persons with HIV (1989)	1–1.5 million
AIDS cases	cumulative, as of November 1989: 115 158
Transmission categories	60% homosexual or bisexual male 21% intravenous drug user 7% both homosexual or bisexual male and intravenous drug user 5% heterosexual contact 2% recipient of blood transfusion 1% pediatric 1% haemophiliac 3% other
In New York and San Francisco	Over 50% of gay men have AIDS or HIV
New cases	About 686 new cases were reported every week during 1989, as opposed to 618/week during 1988
Deaths	As of November 1989: 68 441, or 60% of all cases
Persons with HIV in New York City	200 000
Cases of AIDS (N.Y.C.)	Cumulative as of November 1989: 23 066; active: 10 263
Deaths (N.Y.C.)	As of November 1989: 12 803, or 56% of all cases
Living situations for persons with AIDS (N.Y.C.)	In hospitals: 1531 In shelters for the homeless: 1000–2000 In subsidised apartments: 777 Municipal beds: 600 Other and unaccounted for: 1479–2479

positive and 100 000 die of AIDS each year, almost half are women. But, as one writer was told in Uganda:

> When a man dies of AIDS, a huge stigma attaches to the widow. They are often driven out of their homes by the in-laws. Sometimes they are accused of causing the disease by witchcraft – even though the husband probably caught it from one of his girlfriends. I could introduce you to hundreds of women in that situation.
>
> Taylor (1994, 119)

Uganda is unusual in that three out of four with AIDS are in the country-side. Overall, however, the highest proportions of infected people are in cities, and AIDS is more prevalent among urban women.

Suggested reading

Cohen, I. and A. Elder (1989) Major cities and disease crises: a comparative perspective. *Social Science History,* **13** (1), 28–63.

Perrow, C. and M. F. Guillén (1990) *The AIDS disaster: the failure of organisations in New York and the nation.* Yale University Press, New Haven, Conn.

Mann, J. M. *et al.* (eds) (1992) *AIDS in the world.* Harvard University Press, Cambridge, Mass.

Technological disasters

> Every tendency in modern society points to accelerated mobility . . . Everybody is going places, but what is happening to us . . . along the way?
>
> Cox (1966, 43)

Most technological hazards are in urban centres. Those derived from large-scale transportation of people, goods and fuels converge upon, and are more concentrated in cities. An inventory of the 112 worst civilian or peacetime military air crashes, from 1945 to 1990, shows 91 were in or close to urban centres, most often their airports (Davis, 1993, 3–5). Concentrated residential neighbourhoods and commercial districts, and the industries and energy supplies of cities, involve large amounts of inflammable materials. These can create exceptional risks of mass fires and have been associated with many of the most disastrous fires. Urban mass fires in peacetime are rare these days, yet hardly a week goes by without a deadly fire in a residential, industrial or public building somewhere in a city. Industrial spills and accidents in the transporting of dangerous substances most often occur in and near urban centres (Table 10.7).

Bhopal, India, 3 December, 1984[1]

The worst peacetime chemical disaster to date involved Union Carbide India's Bhopal plant, in production since 1969. Certain highly toxic and volatile chemicals occurred in the intermediate processes of pesticide manufacture, notably methyl isocyanate (MIC). After initially lucrative performance, demand for the pesticides declined through the late 1970s and early 1980s. There were major cost-cutting measures at the plant. People were laid off. The plant was run by fewer and less qualified workers. Safety equipment, inspections and emergency arrangements that are commonplace elsewhere, where such lethal materials are stored, were lacking.

Meanwhile, when the plant was founded Bhopal was a modest city of 300 000. By 1984 there were about 900 000 inhabitants. Large numbers of

[1]Background research by Dr Jayati Ghosh, in a (currently) unpublished comparative research project on industrial disasters.

Table 10.7 Examples of large evacuations from urban areas in the United States as a result of chemical accidents (after Cutter, 1993, 108)

Date	Place	No. evacuated	Chemical
1986	Miamisburg, Ohio	40 000	white phosphorus
1987	Salt Lake City, Utah	30 000	ammonia
1981	San Francisco, Calif.	30 000	PCB-contaminated lubricating oil
1969	Tallahatchie, Mass.	30 000	vinyl chloride
1980	Newark, NJ	26 000	ethylene oxide
1988	Los Angeles, Calif.	23 500	chlorine
1980	Somerville, Mass.	23 000	phosphorus trichloride
1918	South Amboy, NJ	20 000	TNT
1985	West Chester, Pa.	20 000	pentaerythritol
1988	Springfield, Mass.	20 000	chlorine
1982	Taft, La.	18 500	acrolein
1987	Nanticoke, Pa.	17 000	sulphuric acid
1988	Henderson, Nev.	17 000	perchlorinated ammonia
1987	Pittsburgh, Pa.	16 000	phosphorus oxychloride
1981	Blythe, Calif.	15 000	nitric acid

squatters had occupied land close to the chemical plant where, in other countries producing these dangerous chemicals, there is a wide, unoccupied safety perimeter. Indeed, in May 1984, prior to an election, the position of these thousands of squatters was regularised, apparently by incumbents hoping to win their votes.

These indirectly related developments set the stage for the disaster of the night and early morning of 3/4 December, 1984. There was a major leak of MIC. Being heavier than air, the gas cloud moved downslope, spreading through the low-lying squatter camps and on over a radius of some 25 km. According to a recent summary:

> Approximately 200 000 persons were exposed to methyl isocyanate in the Bhopal gas leak . . . 4037 deaths have resulted . . . and about 50 000 persons are estimated to be suffering from long-term health effects.
>
> Ramana Dhara and Kriebel (1993, 281)

At least half the dead and injured were women. Many have suffered extraordinary and continuing health problems. Perhaps half the survivors are permanently weakened or disabled and face early permanent blindness. Sterility, increased still-births and birth deformities, higher cancer rates and lowered resistance to other diseases threaten them. As at Hiroshima and Chernobyl, 'the disaster' continues for years and decades after 'the event'.

Suggested reading

Shrivastava, P. (1989) Bhopal: anatomy of a crisis. Balliger, Cambridge, Mass.
Tachakra, S. S. (1989) The Bhopal disaster: lessons for developing countries. Environmental Conservation, 16 (1), 65–6.

Cities under fire

> But this much must be conceded. As soon as war had become one of the reasons
> for the city's existence, the city's own wealth and power made it a natural target.
>
> Lewis Mumford (1961, 43)

Unusual opportunities for, and dangers of, social violence arise in cities. They
may be seen in high incidence of violent crime and terrorist bombings. Urban
terrorism, by dissident groups and foreign agents, and by state forces, has
reached epidemic levels in some regions and periods of this century. Ashworth
(1991, 11) pointed out that 'it is European cities that have borne the brunt of
the urban terrorism offensive since 1970, with 33 per cent of total terrorist
incidents.' Such events terrorise and can undermine the security of a city.
The bombings of the World Trade Centre, New York, 1991 and the Federal
Building in Oklahoma City, 1994 struck at those serving in focal institutions,
and also at those depending on them. However, the worst and most widespread
of these disasters of violence in cities are identified with state forces and occupying troops. In the last couple of decades they include the 'death squads' and
'disappearances' in various Latin American countries; and extra-judicial killings, acts of terror and crimes of war in Bosnia, East Timor, Tibet, Kashmir
and Kampuchea. Civilian casualties are everywhere the highest. Rape and
torture of women on a large scale, and by 'regular forces', are documented
for most of these conflicts, although not only for urban women.

In general, the links between modern wars and cities are strongly drawn,
whether as instruments of war-making or objectives of armed assault. Chapter
11 looks at urban and civil consequences of armed assaults upon industrial
cities in the Second World War. However, there is an impression that most
conflicts since then have been 'small wars', fought in rural and remote areas. A
few cities, like Beirut, Kabul or Sarajevo tell a different story and have widespread parallels. The Second Indo-China ('Vietnam') War, 1961–1975, is also
often seen as a war of the countryside. And, indeed, the fighting, casualties and
environmental devastation were concentrated overwhelmingly in rural South
Vietnam. Nevertheless, the war had massive urban consequences that illustrate
a variety of these predicaments for cities.

Tragic contrasts: Hanoi and Saigon 1955–1975

> The American military build-up and the way we fought the war basically restruc-
> tured [South] Vietnamese society from predominantly rural to predominantly
> urban.
>
> G. C. Hickey, quoted in Thayer (1985, 222)

There were huge, enforced displacements of civilians out of or into cities in
North and South Vietnam. The dramatic changes in population in the two

capitals, Hanoi and Saigon (now Ho Chi Minh City) provide an indicator of these effects (Fig. 10.5). They also reveal sharply contrasted results for civil life. Greater Saigon's population grew much more rapidly and continuously than Hanoi's, reflecting the effects of very different conditions and goals in the war.

In the North, much stronger, direct planning and control of Hanoi was exercised by the regime, including a policy to decentralise industrial capacity. There were large evacuations in the mid-1960s from the threat of bombing and, in 1972–73, from actual raids. When bombing was first resumed in 1972, up to 750 000 people were moved out, including 75 per cent of inner city residents. In December, raids killed almost 3000 persons and destroyed as much as a quarter of Hanoi's housing. Further evacuations took place. When bombing stopped again, people returned, slowly, and from 1974 there was again slow but sustained growth.

In the South, most of the wealth and jobs created by the war and huge American presence were in the cities. The North had no capacity to bomb the cities or exercise control there. Meanwhile, in the countryside, terror, military devastation and draconian 'pacification' or resettlement programmes drove millions to the cities.

At least two-thirds of the entire population was displaced in the course of the war (Table 10.8). The huge growth of Greater Saigon was part of that. From a survey of South Vietnamese civilians in 1972, Thayer (1985, 223) reports:

> When asked if they were native to the area 75 percent of the urban respondents said no. Only 25 percent were native to the city they were living in.

According to Kolko (1985, 202):

> The South's urban growth was the product of physical necessity, security from combat being the most important reason urban dwellers . . . gave for being there.

City life and population were affected by huge, unorganised flights of war refugees. In that sense, they appear unplanned. Nevertheless, forcing people into the cities was policy in the South. Huntington (1968) called 'forced draft urbanisation' a major politico-military strategy. Wherever rural peasants were not under South Vietnamese government control, they were 'encouraged' or forced to move to the cities that it did control. Once there, most lived in congested refugee camps, squatter settlements and overcrowded quarters. A high proportion were the elderly and children with women-headed households. Older folk were locked into a life of boredom, unemployment and acute worry about their families. Women and children, desperate to survive and care for dependants, became thieves, beggars and 'sex workers'. Drug trafficking and the black market provided strategies for survival, or war profiteering. Some put the numbers of prostitutes in the South by the war's end at 600 000, plus the number of drug addicts at 500 000. Of course, the end of the war did not mean the end of upheavals, displacements and, for some, violence. More than

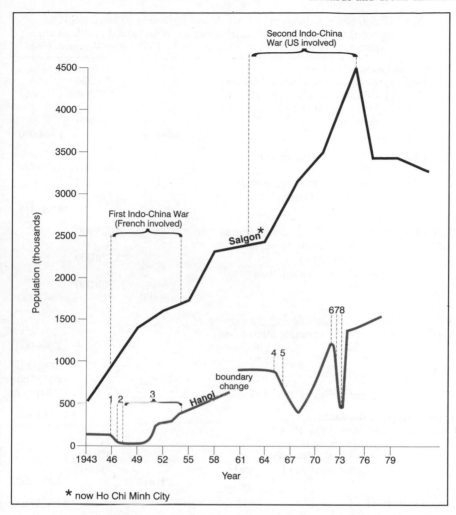

Fig. 10.5 War and cities. Changes in the populations of the capitals of former North and South Vietnam, Hanoi and Saigon (now Ho Chi Minh City), between 1943 and 1979. The changes reveal the large impacts of war (after Thrift and Forbes, 1986, 147 and 154)

Notes: 1. August, 1945, Declaration of Independence, founding the Democratic Republic (DRV).

2. Fighting between DRV forces and returned French colonial troops, causing major destruction of Hanoi.

3. French occupation of Hanoi.

4. First order to evacuate Hanoi and beginning of bombing campaign, Operation 'Rolling Thunder'.

5. April, 1966, order to evacuate all but essential personnel from Hanoi.

6. 17 April 1972, US bombing resumed.

7. December 1972, Operation 'Linebacker II' causes serious damage to Hanoi.

8. December 1973, operations of war cease in the North.

291

Table 10.8 Geographical calamities of war: violent uprooting as the context of enforced urbanisation and deurbanisation. A statistical profile of the civilian war victims in South Vietnam, 1954–1975 (after Wiesner, 1988)

Displaced and in-place victims:			
1954–1962	1) Refugees from North Vietnam	928 000	
	2) Official relocation programmes	230 000	
	3) Montagnard (Highland) people resettled	90 000	
	subtotal		1 248 000
1961–1968	1) Relocatees, evacuees and internal refugees	3 848 700	
	2) 1968 'Tet' Offensive victims[1]	892 454	
	3) 1968 May Offensive	179 164	
	subtotal		4 920 318
1969–1974	1) Internal refugees and evacuees	4 323 700	
	2) In-place victims of warfare (homeless, etc.)	1 320 000	
	3) Repatriates from Cambodia (1970)	200 000	
	subtotal		5 843 700
1975 (to April 30)	Internal refugees fleeing the fighting (some 140 000 became international refugees)		2 000 000[2]
Other[3]			700 000
	Total		14 012 500
(less 'double-counted' estimate[4])			(3 000 000)
Total individuals displaced			12 512 500
Civilian war casualties, 1965–1974			
	1) Deaths (estimated)	247 600	
	2) Admitted to hospital	570 600	
	3) Lightly wounded (estimate)	342 600	
	Total		1 160 600
GRAND TOTAL			13 673 100
Population of South Vietnam *c.* 1970		18–20 million	
Total direct civilian war victims:		68–75% of population	

[1] Composed of 863 158 registered refugees plus 12 737 civilians killed and 21 253 wounded. Also, 108 115 houses more than 50% destroyed and 102 308 with damage less than 50% (Weisner, 1985, 161).
[2] Hardly more than a 'guesstimate' given the conditions.
[3] Estimate displaced under Communist control, and those not registered from 1972 offensives.
[4] i.e. uprooted twice or more.

700 000 fled the country in the first decade after the war, bringing a new term into the vocabulary of human disasters, the 'Boat People'.

Suggested reading

Thrift, N. and D. Forbes (1986) *The price of war: urbanisation in Vietnam 1954–1985.* Allen and Unwin, London.

Safe cities or defenceless spaces?

> The paradox here is that by the time ninety per cent of the population are urban, the city has really ceased to have any meaning in itself.
>
> Boulding (1963, 143)

The conflicting images of progress and disaster in modernity, discussed in the introductory chapter, seem to apply especially to cities. They are 'engines' of growth and innovation, sociopolitical instruments to organise society and make it more productive, sometimes more civilised and sophisticated. In the process, however, large populations have also become uniquely vulnerable to destruction by natural, technological or armed forces. This was so in the nineteenth century for cities that later became remarkably prosperous and relatively secure. It is so, today, in the more rapidly urbanising countries and has returned to haunt some cities once identified with affluence and civic pride.

Any realisation of the goals of disaster reduction currently on the global agenda would seem to require, above all, investment in the safety of those majorities in, or soon to move into, cities. This seems unlikely, however, without a drastic revision of attitudes towards urban life and its future, at least among some dominant actors on the world scene. And a revised sense of priorities for cities must cope with certain fashionable and negative visions of the city in the West.

These are notions of the city variously discussed as the conditions of 'post-industrial' or 'post-civilisation' societies and, more generally, post-modernity. Boulding's comment, above, exemplifies the mood of rejection or pessimism. The perspective was elaborated upon and its bases identified by Martindale (1958, 61):

> The destruction of the city no longer represents the extinction of the institutions of social life. Modern government, business, and religion are more interlocal ([i.e. 'of national and international rather than civic scope'] with every passing year. At any time they choose, the decisively militarily competent social formations of the modern world, the national states, can crush the city . . . The age of the city seems to be at an end.

These visions follow partly, perhaps mainly, from the way cities have become virtually 'defenceless space' in relation to modern weapons. Many have felt

293

that the prospect of urban progress through urban solutions was annihilated at Hiroshima. As Martindale suggests, to those for whom political power, backed up by the threat of armed force, is the decisive factor in and of *security*, the city is 'dead meat'. Many in the West have come to some such conclusion, but they have also accepted various developments, from traffic problems to inner city crime, as also signalling the demise of the city. They do not think cities, at least according the ideas of recent centuries, are viable frameworks for living, or at least safe living. If true, or if believed by influential groups, this undermines all the terms upon which urban risk and disaster reduction are based – excepting deurbanisation, which seems even less likely to happen in the foreseeable future, or Orwellian police state methods, which seem more likely. Any methods and hopes for making cities safer have to be defended against, or changed in relation to, these interpretations of the city.

I suggest, however, that behind all of this is the dominant Western assumption that development or dangers ultimately reflect, or must be addressed in terms of, large, impersonal forces – in fact, a 'hazards paradigm'. It may be overpopulation and 'overcrowding' as an inevitable consequence, or some unavoidable result of 'global warming', technological innovation or capitalism, or man's propensity for violence. How far does pessimism about the future of the city, or ways to make it more safe, reflect material realities or excluded possibilities? Could it be more an expression of attitudes and preoccupations in the wealthy countries and enclaves, of a prevailing intellectual climate, or what suits corporate and elite agendas?

Again and again, in every city, we find people who, given half a chance, will work and invest to improve the safety and conditions of their neighbourhood. Yet as long as the dominant political, economic and social institutions are distanced from these realities, as long as the protection of cities or civil life is not a priority, cities must become ever more vulnerable. Improvement seems likely only if civil government, hence civil society, again becomes the focus and basis for policy and public safety. This requires a much greater commitment to policies and practices that give those whose lives are primarily dependent upon the safety and quality of life in the cities much higher priority and powers. Most obviously these are persons whose lives turn strongly upon domestic and civil space – that is, mainly families and communities in inner city or other major residential areas: women, children, the elderly and those who serve their health and well-being, their cultural and spiritual life, directly.

Again, only with a strengthening and adoption of alternatives to military power and conflict as the basis of political and social security can cities be protected. Is that so far-fetched in the post-Cold War world?

For the moment, the problem is at least partially the absence of a continuing and systematic diversion of a sufficient or fair portion of revenues or taxes, the profits and skills *generated in cities*, to equitable risk reduction in urban life. Less and less of the wealth and control generated by further urbanisation is invested in making cities safe, let alone safer, for a majority of their residents. This is especially so in inner city neighbourhoods, the congested areas of impoverished communities and those disadvantaged by age, gender, health or prejudice within such communities. Yet it is here that we find most actual damage and casualties occur in urban disasters. Indeed, here is often the only reason there is a disaster, rather than a manageable crisis, when an earthquake, storm or epidemic occurs.

Unfortunately, in cities or neighbourhoods that are conspicuously safer, although not necessarily conspicuously wealthy or influential, it has usually taken a calamity, a 'body count' and social outrage to mobilise the administrative and humanitarian resources for improvement. At least that is evidence against the view that risks and disasters in cities arise mainly from impersonal forces, or are inevitable. But the prevailing pessimistic or anti-urban rhetoric threatens to undermine even these possibilities.

CHAPTER 11

Place annihilation: air war and the vulnerability of cities

I saw many dreadful scenes after that – but that experience, looking down and finding nothing left of Hiroshima – was so shocking that I simply can't express what I felt . . . Hiroshima didn't exist.

quoted in Lifton (1957, 29)

[What a] weird experience to drive through a city you'd once known intimately and in order to get your bearings – to recognise the section of the town you were in – have to base your calculations on the indestructible river, or the tower of the gutted cathedral.

James Stern (1947, 2 in Frankfurt-am-Main)

There is a subtle psychological tie-up between the citizens, however apathetic, and the centre, the heart of their city.

Mass Observation, UK 1941 Rept. #538 (6/1/41) 5)

The disasters of air war

Aerial bombardment has been used to attack populous cities in more than 100 international conflicts in the twentieth century. It has caused massive death and destruction for civilians even when aimed at other targets in cities. The air weapon has been employed in countless smaller actions by colonial and state air forces within their own territories, or in undeclared wars. Countless small settlements have suffered bomb destruction and civilian casualties throughout Eurasia and Africa, and in much of Latin America and the Pacific islands. Most 'Old World' cities have undergone some bombing. In these ways air war has played a large role in urban destruction and civilian casualties of warfare, and their greater proportions in recent wars (see Chapter 5). Civilian air raid deaths and injuries now number in the millions. Those bombed out, or otherwise forced to leave their homes due to air attack, number in the tens of millions. Hundreds of millions have been indirectly traumatised as the families and neighbours of these victims, or from measures against and fear of raids.

In bombing campaigns against cities in the Second World War, civilians were by far the largest numbers of victims, whether in terms of casualties or loss of property and livelihood. In this case study, we will examine the raids against cities of Britain, Germany and Japan as demonstrating the risks for cities and a continuing, global hazard for settlements and civilians (Table 11.1).

The chapter is based largely on Hewitt, 1983, 1987, 1992 and 1994.

Air war: the view from below

A huge literature is devoted to air war, but it deals largely with warfare, air forces, weapons and air raid stories, or how leadership, science, and intelligence gathering and other conditions of war relate to them. The concerns of air war are certainly important, yet they present a partial and misleading picture of the dangers of raiding. Then, domestic and communal conditions, with which most civilians remain concerned and that define their vulnerability to air attack, disappear or are made to seem trivial. And, in fact, the majority of those whose bodies, families, homes and neighbourhoods have been destroyed in urban air raids are ignored or treated only in military terms.

To understand the risks of air attack for civilians requires a view from 'under the bombs'. In wartime, civilians must make do with reduced equipment, a greater burden of manual work in the domestic and civic tasks left to them, and getting about on foot or bicycle. They know and respond to air raids mainly through evidence of their own senses and local services. We are concerned mainly with actions confined within the small compass of a home area. The remarkable developments in air weapons and war technology have no equiva-

Table 11.1   The urban impact of war: summary estimates of the urban and civilian losses from raids on cities in Britain, Germany and Japan during the Second World War (after Hewitt, 1983)

	Britain	Germany	Japan
Civilian casualties			
Deaths	60 595	600 000	> 450 000
Severe injuries	> 86 000	c. 800 000	> 1 500 000
Built area destruction			
Number of cities	c. 45	70	62
Area destroyed (sq. km)	c. 15	333	425
Proportion of built-up-area	3%	39%	c. 50%
Housing losses:			
housing units	tens of thousands	2 164 800	2 500 000
Persons made homeless	c. 500 000	7 500 000	8 300 000

lent for civil life in wartime and have generally outstripped methods of civil defence.

Suggested reading

Iklé, F. C. (1958) *The social impact of bomb destruction*, 1st edn. University of Oklahoma Press, Norman.

Sherry, M. S. (1987) *The rise of American air power: the creation of Armageddon*. Yale University Press, New Haven, Conn.

Raid hazards and vulnerabilities: the urban risks of air war

An urban and civilian perspective draws attention to the experience of, and material damage in, cities. In industrial cities raided in the Second World War, we can identify an escalating series of raid risks and damaging events for civil life:

Air raid alerts without bombing

Every town had many 'false alarms': raid alerts not followed by bombing. In Europe, the larger cities, and many small centres on the bombers' routes to preferred targets, had between 800 and 2000 air raid alerts during the war. In the heavily and frequently raided places, a couple of hundred alerts were followed by bombs falling. No more than a handful, perhaps just one or two, presaged attacks that affected a large area of the city and caused large losses, but even the 'false alarms' brought fear, hardship, loss of sleep and disruption of everyday life.

In terms of civilian vulnerability an obvious fact, easily passed over in descriptions of warfare, needs emphasis here: the demands of everyday life, such as sleep, care of children and other dependents, do not stop because of war. They are hardly less urgent when a raid fails to materialise. This was made clear in the recollections of survivors:

> "This everlasting running to the cellar; pulling the child out of bed – a child doesn't like to have his sleep disturbed, and it isn't good for him."

> "The household suffered terribly for my wife had to spend much time in the cellar."

> "Towards the end my [little] daughter wouldn't even leave the bunker to eat – I had to take her her food."

> "The milk soured in the bunker and my child was often sick."
>
> (quoted in Hewitt, 1994a)

Aimless bombings

Most actual raids or incidents, in nearly all cities, involved bombs falling more or less at random, and with relatively inconsequential, widely dispersed damage. Wartime propaganda spoke of air attacks against specific targets, especially when these had been hit. It suggested that the bombs could be dropped, as one phrase put it, 'into a pickle barrel'. However, the weaponry gave no possibility of such accuracy, least of all in the predominantly massed, night-time urban attacks.

Raids could be severely disrupted and dispersed by bad weather, navigational error or aerial defences. 'Creep back' and 'spill over' of bombs from designated targets spread them over hundreds of square kilometres in every large attack. Some fell upon hundreds of places not on the target lists. There were countless small, 'nuisance' or 'decoy' attacks, and 'dumping' by individual aircraft that failed to find their main target.

The larger fraction of all bombs and raids caused this sort of 'aimless' impact, but civilians never knew in advance which it was going to be. As one woman from near Köln expressed it, 'bombs can fall anywhere. You had to be ready for anything' (*ibid.*). Air raid protection personnel were hardly better able to anticipate the difference between these highly inaccurate bombings and the next two categories. This sense of an unpredictable, if not purely 'chance', process added to the difficulties of dealing with the hazard. Even though it was planned destruction, for its victims it appeared more unpredictable than many natural hazards.*

'Collateral' damage

This refers to civilians and their property hit in attacks aimed at factories and other war-supporting installations. Even when called 'precision' attacks, raids were normally quite inaccurate. Attempts to overcome inaccuracy included simply drenching the target area with bombs. Instrument or 'blind bombing' through clouds made for an indiscriminate fall of bombs. Civil damage would tend to be greatest for communities around industrial areas, military camps, railway stations and airfields, and it could be very great in closely built-up areas.

However, in terms of scale and concentration of civilian and urban harm the most significant raids were those aimed at city-wrecking and terrorising civilians. Among them, we can distinguish two crucial developments:

Progressive demolition

Most urban raids that found and dropped a tenth or more of their bombs on the chosen city could cause large losses. They brought concentrated harm in

*Half a century and billions of dollars in weapons development later, the citizens of Baghdad, Beirut, Grozny or Kabul will tell the same story.

particular districts, and 'disaster incidents' in which hundreds of civilian casualties occurred in one small area. Large cities underwent several, sometimes dozens, of these attacks. In many ways progressive demolition in such attacks describes the impact of the raiding on cities like London, Berlin, Köln and Essen: a kind of 'war of attrition' against cities.

Disaster raids

In some cases, and those that concern us most, single large attacks achieved massive concentrations of bombs in densely populated and vulnerable urban areas. In a few hours they might destroy whole districts, most of the inner city and, sometimes, nearly all of the built-up area. These were calamities for the city as a whole, with thousands, sometimes tens of thousands, of civilian casualties. They would place huge stress on medical services, water supplies, sewage treatment, etc., which could cause another round of problems for the civilians. Such attacks dominated the entire profile and distribution of damage for urban life, and they are set apart as unique calamities in the memory of each city where they occurred.

These categories identify a scale of *increasing* severity and share of urban destruction, civilian hardship and loss of life, but a *declining* frequency of raid event. Aimless raiding, the most frequent, caused the least damage; disaster raids the most. The classes also identify a geography of raid experience *within* cities. Where buildings were most dispersed, mainly in the suburbs, wealthier enclaves or surrounding villages and small towns, damage was generally scattered and minor, the casualties few. When calamitous destruction occurred in one or two disaster raids, it was concentrated in inner city and the most congested residential districts.

Many have compared the experience of air attack to natural and other disasters. Less often recognised is the overwhelming role, in urban and civilian losses, of the most destructive attacks. The devastating raid had a significance out of all proportion to its share of all raids, wartime, or a person's life time:

" . . . in 10 minutes during the terror raid I had nothing but rubble."

(*ibid.*)

Suggested reading

Harrison, T. (1976) *Living through the Blitz.* Schocken Books, New York.

Urban armageddon

In strategic counter-city bombings, especially against Britain, Germany, Italy, China and Japan in the Second World War, civilians comprised by far the largest numbers of victims. This applied to casualties and forced evacuations, and loss of homes, jobs and possessions.

The attacks of main concern are referred to as '(urban) area attacks' or 'town raids', sometimes as 'morale' or 'terror' bombing. Their distinguishing features were attacks by massed bombers, arriving in waves and intended to lay down as dense a carpet of bombs as possible. The 'aiming point' was usually the city centre, or the most populous and densely built-up districts. The most destructive attacks were those that set great mass fires, usually by dropping large quantities of incendiary bombs. Fire was the greatest cause of urban destruction in each country.

From the 'Big Blitz' on Britain in 1940–41, through to the great fire raids on Germany and Japan, and the A-bomb attacks of 1945, certain raids are singled out not just by exceptional destruction but by where and to whom they occurred (Fig. 11.1).

The raid on Coventry, England, in November 1940 was, and still is, cited as a yardstick of the severity of urban air raiding in the Second World War. Few will doubt it was a calamity for the city. After several hours of bombing on one night, 568 civilians were dead, some 20 000 people were bombed out, and a square kilometre of the inner city lay in ruins, including its medieval core and cathedral. Among those killed, 52 per cent were women and children. Almost half the adults killed were classed as 'housewives' and 'widows'. The destruction was overwhelmingly to homes, mainly terraced houses in the denser, working class districts of the inner city. Most other damage was to buildings serving the urban community, including three out of four shops in the central area, and many municipal and public service buildings. Thus it was an event whose scale, diversity and intensity of destruction had all the attributes of a catastrophe for the people of Coventry. It overwhelmed civil defence measures and other community institutions and safety systems, as well as causing an unusual amount of damage and disruption to industries (Longmate, 1976; Hewitt, 1994).

The Battle of London 1940–41, more commonly known as 'the Blitz', was a relentless attack night after night and appears as a single crisis. Yet in any one part of the city and even as a whole, death and destruction were dominated by a few raids. The most severe ones came in April and May 1941. Four of them caused almost a quarter of the total death toll. The most lethal, on 10/11 May, killed 1436 civilians. The most destructive raid was on 29/30 December, 1940. Although relatively small it set some 1500 fires. Strong winds fanned those within the City of London into uncontrollable mass fires. Warehouses full of combustible materials, including millions of books in publishers' premises, old churches, apartment blocks and other close-packed buildings, fed the fires. It has been called the 'Second Great Fire of London' (Johnson, 1988), recalling that which had razed much of the medieval city in the seventeenth century.

There were 14 raiding episodes on Britain that killed more than 500 civilians (Table 11.2). They accounted for a quarter of total air raid deaths, but only 3 per cent of aircraft dispatched, about 6 per cent of those reported over cities, and a much tinier fraction of bombs dropped. Our concern is not with what

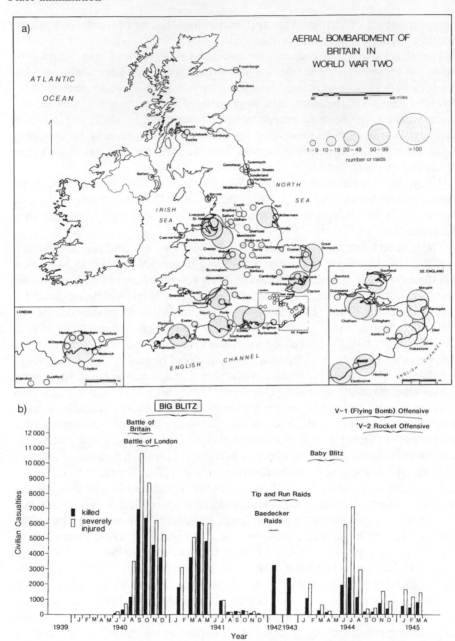

Fig. 11.1   The bombing of British cities, 1940–1945. a) Numbers of air raids on main cities and towns. b) Chronology of civilian air raid casualties, including the V-weapon attacks in 1944–45, illustrating the overwhelming role of 'the Blitz' 1940–41, and extremely concentrated, episodic nature of the losses (after UK Ministry of Health, 1947; Titmuss, 1950; Collier, 1957)

Table 11.2   The most lethal raids against British cities in the Second World War: those that killed more than 500 civilians (after Front Line, 1943; Collier, 1957; Hewitt, 1990 and 1994)

City	Number of raids (large)	Total civilians killed	Worst attack deaths	date
London (Metro)	360 (71)	29 890	1436	10/11 Apr., 1941
Liverpool	68 (12)	2568	1800[1]	7/8 Apr., 1941
Birmingham	77 (8)	2241	615	19/20 Nov., 1940
Glasgow–Clydeside	? (5)	1828	1235	13/15 March, 1941
Bristol	56 (9)	1238	257	16/17 March, 1941
Plymouth	71 (9)	1172	336	20–22 March, 1941
Coventry	(3)	1160	568	14/15 Nov., 1940
Hull	76 (10)	1156	450	7–9 May, 1941
Belfast	(2)	946	750	15/16 Apr., 1941
Portsmouth	72 (9)	855	?	9–10 Apr., 1941
Sheffield	(2)	668	668	12/13 and 15/16 Dec., 1940
Southampton	67 (5)	630	137	30/31 Nov., 1940
Manchester	(3)	556	376	22–24 Dec., 1940

[1]Includes Merseyside (total Liverpool–Merseyside deaths, 4100 +)

that says about the efficiency of the bombing. It is with the sort of risks that are implied for civilians on the ground.

German cities suffered at least 117 attacks with civilian casualties equal to or greater than Coventry or greater urban devastation – in most cases both (Fig. 11.2). These attacks caused over two-thirds (approx. 320 000) of civilian air raid deaths, and a still greater proportion of housing destruction and bombed out people (Table 11.3).

Berlin's worst fatalities occurred in three consecutive RAF raids between 22 and 27 November, 1943, killing 3758 civilians and bombing out almost half a million, the majority in the first raid. The most destructive attack on the capital was on 3 February 1945 by the USAAF. Some attributed 20 000 deaths to it, but the most comprehensive source gives 2541.

The overriding impact of calamitous fire raids on congested inner districts can be seen at Hamburg. Out of 213 recognised attacks on the city, 104 caused civilian casualties. However, over 90 per cent of these and homes destroyed came from three raids of the July 1943 'catastrophe' (Brunswig, 1987). And 95 per cent of those occurred in the firestorm of 27/28 July, the most destructive attack of the war in the European theatre (see Table 11.3). It determined the overall wartime balance of buildings lost, yet a much greater weight of bombs would fall on Hamburg after August 1943 than to that time!

Damage in Japan was of similar or greater magnitude, but much more compressed in time (Fig. 11.3). Most Japanese cities were razed by one, or a small number of, such attacks (Table 11.4). The great conflagration set by the

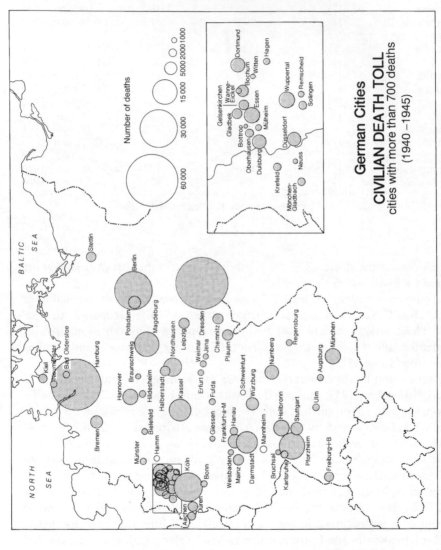

Fig. 11.2 The bombing of German cities: urban centres with civilian death tolls in excess of 700 from air raids in the Second World War. The largest fraction of these casualties occurred in one or two 'disaster raids' on the city concerned (after Hewitt, 1993)

Table 11.3   Summary of civil losses in the twelve most lethal attacks on German cities (after Hewitt, 1994)

City	Date	Civilians killed	Bombed out (approx.)	Urban area razed (km²)	Comments
Hamburg	27/28.07.43	c. 40 000	800 000	13.2	firestorm
Dresden	14/15.02.45	c. 40 000	250 000	6.8	firestorm
Pforzheim	23/24.02.45	17 600	50 000	4.5	firestorm
Magdeburg	16/17/01.45	c. 16 000	190 000	3.1	firestorm
Darmstadt	11/12.09.44	10 550	49 000	2.0	firestorm
Nordhausen	03/04.04.45	8800	20 000	1.8	great fire
Kassel	22/23.10.43	7000	53 800	7.5	firestorm
Heilbronn	04/05.12.44	c. 7000	50 000	1.4	firestorm
Potsdam	14/15.03.45	c. 5000	40 000	–	great fire
Würzburg	16/17.03.45	4500	56 000	1.7	firestorm
Köln	03/04.05.43	4377	230 000	–	great fire
Wuppertal-Barmen	29/30.05.43	3 400	130 000	2.6	firestorm

9/10 March 1945 'Big Fire' raid on Tokyo was the most destructive in the history of warfare. It swept through the inner, most congested districts, burning out an incredible 41 km². Under the smoke lay more than 130 000 civilian dead (Daniels, 1975). A huge exodus followed, depopulating large parts of the capital's inner wards (Fig. 11.4).

Suggested reading

Havens, T. R. H. (1978) *Valley of darkness: the Japanese people and World War Two*, 1st edn. Norton, New York.

The vulnerability of urban areas and civil life

The social geography of bomb destruction

The primary destructive impacts in all three countries fell upon residential districts and domestic life. Less often realised is that during the worst bombings, cities in these countries generally contained 80 per cent or more of their prewar resident populations. More than 90 per cent of air raid deaths were of civilian residents. Much greater numbers were bombed out of their homes. Housing destruction, by numbers of buildings, areal extent or replacement costs, comprised the greatest property losses. However, the gross damage statistics do not show *where* losses were concentrated.

In Britain's capital city, over the war, a greater weight of bombs fell on outer, suburban boroughs, but their dispersed buildings suffered much less. Some lost barely a dozen homes and their civilian casualties were measured in single figures. By contrast, 26 inner boroughs had over 400 civilian air raid

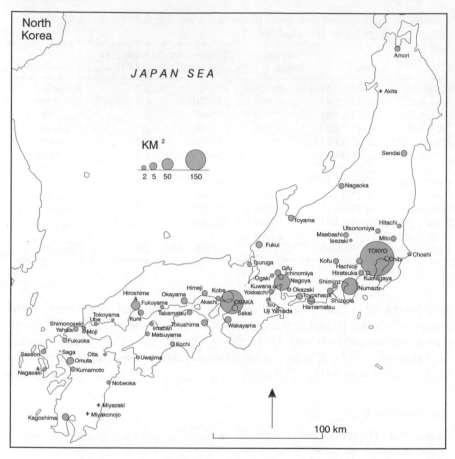

Fig. 11.3   The bombing of Japanese cities. In this case the impact is represented by the urban area razed in the attacks. Again it largely represents devastation in the main, central built-up areas, and by a small number, often just single, disaster attacks on each city. Destruction was mainly due to fire, even in the case of the A-bomb attack on Hiroshima. Except for the A-bombs, nearly all of this devastation occurred between early March and the middle of June, 1945. It represents the most concentrated annihilation of urban areas and their civilian residents in world history (after Hewitt, 1983)

deaths (Fig. 11.5). The tolls in Lambeth (1470 killed) and Wandsworth (1253) were exceeded only by the cities of Liverpool, Birmingham and Glasgow. The geography of property damage in London, especially housing, broadly followed that of casualties (Hewitt, 1995). Housing losses exceeded 80 per cent of all aircraft-caused losses in most boroughs, and over 95 per cent in six inner city ones.

In the industrial city of Birmingham in the English Midlands, 2241 civilians were killed and almost 10 000 were injured. Some 5000 homes were demolished,

Table 11.4 Hamburg and the disaster raids: distribution of civilian casualties in the seven most lethal attacks, having at least 300 civilian deaths (after Hewitt, 1993)

Raid rank	Date	Civilians killed				Weight and type of attack				Target designation
		total	women. (%)	children (%)	W & C (%)	RAF/ USAAF	planes dispatched	incend- iaries %	day/ night	
1	27/28.07.43	40 000	60%	20%	80%	RAF	787	53%	N	area (firestorm)
2	24/25.07.43	1500	50%	12%	62%	RAF	791	43%	N	area
3	29/30.07.43	1300	–	–	–	RAF	777	53%	N	area
4	25.10.44	779	47%	18%	65%	USAAF	532	23%	D	port, town, industry
5	18.06.44	679	33%	7%	40%	USAAF	633	8%	D	port, shipping
6	20.06.44	354	12%	3.7%	15.7%	USAAF	467	–	D	industry, oil
7	26/27.07.42	337	48.7%	11%	59.7%	RAF	403	46%	N	area

307

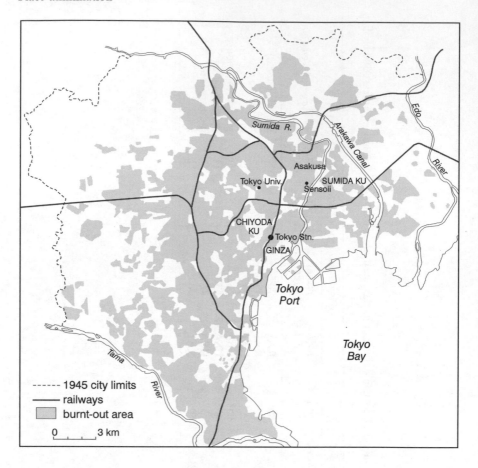

Fig. 11.4 The most destroyed capital city: the area of Tokyo burned out in the 1945 fire raids (after Hewitt, 1987)

more than 20 000 made uninhabitable, and 140 000 damaged to some extent. Two-thirds of these occurred in the 'central ring' of older, mostly poor residential neighbourhoods, which contained only one-ninth of the urban area and less than half the homes.

The picture in the larger cities was, in many ways, intensified in smaller ones. The port of Plymouth in the southwest of England illustrates this. Of 169 000 who had been living there in March 1941, just 67 000 remained by May; 1172 civilians were killed, 1092 seriously injured and 2177 slightly injured. A report on raids in April tells us:

The civil and domestic devastation exceeds anything seen elsewhere, both as regards concentration throughout the heart of the town, and as regards the ran-

Table 11.5 Summary of damage in the most destructive air attacks on Japanese cities in the Second World War (after Hewitt, 1987)

Bombs dropped (tons); USSBS (1947b)	
all of Japan	160 000
urban area	104 000
all Tokyo attacks	16 500 (12 500 incendiaries)
Civilian casualties	
deaths: all raids	459 000
fire raids	270 000
A-bombs	180 000 (immediate) + 160 000 (to 1950)[a]
severely injured	> 150 000 (Hewitt, 1983)
Fire burns as cause of death	56–84% (five cities) (USSBS, 1947c)
total casualties	~8 million (Kosaka, 1972)
Built-up area destruction (62 cities):	
complete destruction:	
total	425 sq. km (USSBS, 1946)
proportion	~50% (USSBS, 1947b; Craven and Cate, 1953)
buildings destroyed	~2.2 million (USSBS, 1946)
Greater Tokyo	0.86 million (USSBS, 1947b)
Hiroshima	70 000[a]
Nagasaki	19 587[a]
by structure evacuation	614 000[a]
Specific damage for civil life	
housing units destroyed	2.5 million (Havens, 1978)
persons made homeless (raids)	8.3 million
structure evacuation	3.5 million (Havens, 1978)
persons evacuated (to October 1944)	2.1 million (USSBS, 1947a)
(to August 1945)	8.3 million (USSBS, 1947a)
Tokyo (to August 1945)	4.1 million (57%) (USSBS, 1947a)
hospitals destroyed (fire raids and A-bomb)	969 (USSBS, 1947c p.10)
hospital beds lost (fire raids)	519 235 (USSBS, 1947c, p.10)
pharmaceutical factories destroyed	200 (32%) (USSBS, 1947c, p.10)
food stores destroyed	221 891 tonnes (5%) (USSBS, 1947b)
service professionals killed in Hiroshima[a] (%)	
physicians	90
pharmacists	80
dentists	86
nurses	93

[a]Source: Committee for Compilation of Materials on Damage Caused by the Atomic Bombs at Hiroshima and Nagasaki, 1981.

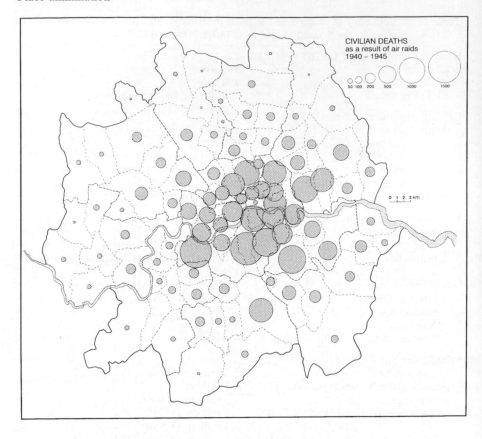

Fig. 11.5 Civilian deaths from air attacks on London shown by borough (after Hewitt, 1994)

dom shattering of houses all over. The *dislocation of everyday life* also exceeds anything seen elsewhere and an enormous burden is being placed on the spirits of the people.

Mass Observation, File Report 903, 6/10/41 (my italics)

Turning to German civilian losses, in Berlin they were also concentrated in inner, more congested areas. Civilian casualties were heaviest in Kreuzberg (1697 killed), Charlottenburg (1621) and Mitte (1463). Housing demolition and numbers bombed out were even more concentrated in inner districts (Fig. 11.6). Mitte, the smallest district, had the highest proportion of losses: almost 54 per cent of its housing and population, closely followed by Tiergarten and Friedrichshain. Mitte and Tiergarten, with many apartment blocks, lost proportionately more individual homes per building destroyed. Not all densely populated areas were equally affected. High-density Prenzlauer Berg fared better than the less dense Steglitz and Wilmersdorf due to the way in

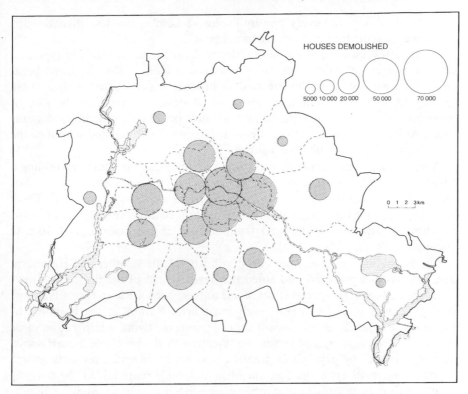

Fig. 11.6 Houses demolished by air attacks on Berlin during the Second World War, by district (after Hewitt, 1994)

which large attacks approached the city from the southwest or northwest. Large, low-density districts like Spandau, Reinickendorf and Zehlendorf suffered least, although containing the important war industries, and being targets of many industry raids.

Severe raids were primarily calamities in one or two districts. For example, in the 'disaster raid' of 23/24 August 1943, half the 899 deaths, 57 per cent of the roughly 104 000 persons bombed out, and two-thirds of the building damage occurred in just one district, Steglitz. While damage occurred in 17 other districts and civilian deaths in 12, this was, above all, the Steglitz 'catastrophe'.

In the well-known 30/31 May 1942 'thousand-bomber raid' on Köln, bombing was extremely scattered, but destruction was concentrated in the inner city and upon housing. Aerial photographs after the attack showed 61 per cent of visible damage in the old city and congested surrounding residential areas. Most of the 411 civilians killed, 5027 injured and over 45 000 bombed out and 70 per cent of housing destroyed, were in the inner city. An ominous indication of the future was the role of fire. In the inner city, fires destroyed

311

about 18 buildings for every one by explosive bombs, but less than two for every one in industrial and suburban areas.

Before 1943, damage in German cities was similar to Britain's in type and scale. However, once the large forces deployed at Köln in the 'thousand-bomber raid' were linked to the fire raid techniques experimented with in 1942 against Lübeck and Rostok, a new scale of devastation sealed the fate of most inner city areas. Uncontrollable mass fires became the rule in the disaster raids. At least 12 German cities suffered firestorms. This included six out of the eight largest disaster raids (see Table 11.3).

A firestorm is a meteorological as well as an incendiary process, resembling a tornado with a fire at its centre. Air rushes in to replace the fiercely rising hot air at the fire centre. The winds generated can exceed 200 km hr^{-1}, fierce enough to uproot large trees and pick people up bodily, flinging them toward the fires. Temperatures, deduced from melted objects, can exceed 1000°C (Brunswig, *op. cit.*; Ebert, 1993).

The possibility of a firestorm can depend upon the weather. At Hamburg, hot, dry weather preceded and aided the development of the firestorm. However, unlike Japan, no special effort seems to have been made to take advantage of dry weather in plans to start such great fires. Rather, in the bombing of Germany, dry spells usually meant more favourable flying weather, and clear nights better navigation and target identification by the bombers. Nevertheless, although helped by dry, windy weather and set by incendiaries, a firestorm depends above all upon the amount of inflammable materials in the burning neighbourhoods. It was this that made congested, inner city areas more vulnerable and the targets of choice in fire raids. With that came an intensification of the harm to domestic and inner city, civil space.

In Hamburg, the social distribution of harm reflected the overwhelming role of the fire, and the firestorm, in congested districts. Of all buildings destroyed, 89.9 per cent were residential. Moreover, for every single family home damaged, there were 7.5 homes in multi-home buildings. This ratio exceeded 8 to 1 for destroyed homes. A higher fraction of multi-home dwellings exposed to bombing was actually hit – 53 per cent damaged compared with 31 per cent of single family homes. A higher proportion of the former were destroyed – 53 per cent of those damaged, compared with 46 per cent of single homes (Hansestadt Hamburg, 1951).

Enforced evacuation of inner districts of Hamburg further demonstrates this social geography of destruction (Fig. 11.7). After the firestorm, large parts of the central residential area became abandoned 'dead zones'. Most outer districts grew in population as people were billeted in largely untouched houses there. Within two months of the July–August raids, occupancy of remaining housing moved from 0.73 persons per room to 0.97, and to 1.03 by February 1944 (Iklé, 1950). A similar picture can be found in all the cities subjected to raids that set uncontrollable mass fires in central urban areas.

Darmstadt, Heilbronn, Würzburg and Pforzheim were old towns of great historic and cultural but little military significance. People in each of these cities

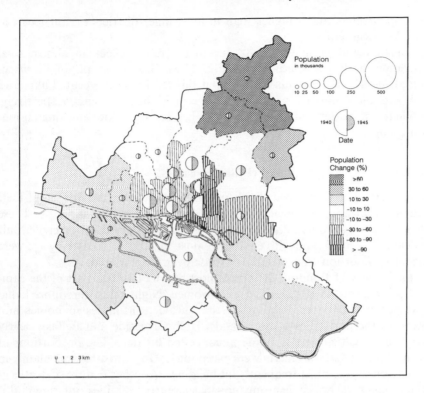

Fig. 11.7 Population change in Hamburg due to the disaster raids, especially the fire-storms in July–August, 1943. The map shows changes by district, 15 July 1943 to 10 October 1943, in number and percentages

said they had not expected to be raided. They had few or no aerial defences and quite inadequate civil defence. This helped the bombers to attack and lay down their bombs with minimal interference and small losses. Meanwhile, the compact cores, often of medieval origin and with many half-timbered buildings, fed the flames and generated firestorms very quickly.

Such consequences of fire raids apply even more completely to Japan. In Tokyo, density of occupancy was again the great determinant of who lived and who died. An incredible 116 km^2 of the most built-up area was burnt out, more than 90 per cent of that in three fire raids of March and April 1945, but almost 95 per cent of civilian deaths were in the 9 March raid and conflagration. The waterfront districts, crowded around the Sumida River, comprised barely 7 per cent of the city area but held over a quarter of its population. They were estimated to have an average of 40 000 persons per square kilometre, rising to 55 000 in some parts. Nine out of ten civilian deaths occurred here. The built-over area in Asakusa and Nihombashi wards was 75 per cent and 68 per

313

cent, respectively – much higher than in most inner districts of European or North American cities.

Evacuation and population loss due to fire raids reflect an almost total annihilation of living space there (Fig. 11.8). The most populous ward, Honjo, had 241 000 residents in 1944, but this fell by 95 per cent. Fukugawa and Asakusa lost 93 per cent and 89.5 per cent of their residents, respectively. The outermost, newer and less dense wards had few casualties and small population change.

Lethal space: the composition of civilian casualties

The disaster raids determined the distribution of civilian casualties, and their unique concentration by gender, age and (usually low) social status. These raids explain why women comprised almost half (49 per cent) of civilian air raid deaths in Britain, and with children almost 56 per cent. Losses by age were highest in proportion to exposure for those over 55 years old.

In the County of London, the inner, most heavily bombed part of the capital, 51 per cent of the 17 811 killed were women. Night raids were more lethal for women and children, relating to their heavier casualties in homes and shelters. In the port cities of Merseyside, Hull, Clydeside and Belfast, nearly all civilian casualties were at home in congested districts. The proportions of women, children and the elderly were exceptional. Home deaths were mainly in terraced houses, the most frequently hit buildings. Another feature of deaths in working class areas was a disproportionate share from families suffering multiple deaths.

At Clydebank in Scotland, three-quarters of civilian deaths occurred on 13/14 March 1941, and all but two of the remainder on the following night. Females comprised 51 per cent of 448 killed; women and children, 68 per cent. Adding men older than 50 years, these 'definitive' non-combatants formed over 81 per cent. In 58 households with three or more members killed, *almost half were children and 82 per cent women and children together*! Nothing showed more clearly and tragically that 'ordinary civilians' and domestic life were most vulnerable and that, whether by default or design, they were the social 'ground zero' of these attacks. This was an extermination of 'non-essential' persons, not soldiers or even war workers. It was largely an annihilation of domestic and civil space out of the war-making capacity of the enemy. This is why we call these 'disasters of war' and humanity.

In most German cities, more adult women were killed than men. Almost 14 per cent of resident civilian deaths were under 15 years, and nearly one-fifth were over 65 years old. Disaster raids largely determined the overall balance, with more women dying in most of them, and more women and children in all (cf. Table 11.3). In this too, Hamburg shows the extreme role of the great fire raids. In the 'catastrophe' of July 1943, of the 30 482 deaths actually registered, 95 per cent were in the firestorm, and *60 per cent were women*, which ensured the predominance of women and high casualties among children for the whole

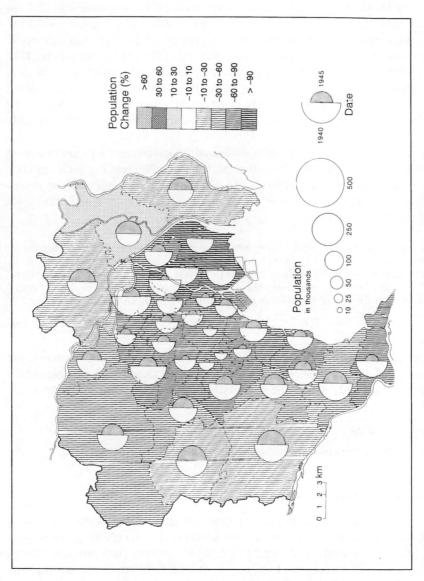

Fig. 11.8 Population change in Tokyo between 1940 and 1945 by city wards. Almost all the changes shown actually occurred between February and May 1945 (after Hewitt, 1994)

war. It accounted for about three-quarters of civilian men killed there too. The saddest indication of the social geography of danger was the much higher proportions of death among women in districts with the highest overall mortality. Some 70 per cent of all deaths occurred in the central districts, such as Hamburg Mitte, where deaths per thousand were 59.6, compared with 22.1 for the whole city. However, the latter involved 15 900 women killed, or 70 per thousand, compared with 10 065 or 48.2 per thousand men. In Hammerbrook, the worst-hit part of Hamburg, 7571 women died (328 per thousand) and 4410 men (203.5). Ortsteil #123, with the most concentrated mortality in the firestorm, lost almost one in two women residents, or 435 per thousand, compared with 282 per thousand for men (Hansestadt Hamburg, *op. cit.*). Again, the more lethal an attack, the more it affected the domestic scene and especially women.

Various sources refer to women and children casualties as predominant in Japanese cities. Considering the nature of the attacks, largely fire raids, and late stage of the war, this seems likely. In Hiroshima and Nagasaki the demographic and social distribution of wartime A-bomb casualties closely resembles those in the worst fire raids:

> explosion-time injuries were heavier among women than among men . . . among men in the cities, the effects were heaviest in the forty-to-fifty age bracket . . . A-bomb deaths (were) high among spouses [= wives] and heirs [= sons], especially in the central bombed area [since] Many people were in their homes or somewhere close by, on the day of the bombing . . . [and] . . . among children, girls suffered more than boys.
>
> (Committee for the Compilation . . . , 1981, 376)

Men accounted for 51.9 per cent of 'bomb-time' fatalities at Hiroshima, but 53.7 per cent of identified casualties were women. Death rate was higher for wives than husbands within the inner city, but the reverse outside that.

Suggested reading

Edoin, H. (1987) *The night Tokyo burned.* St Martin's Press, New York.

'Slum raids' and 'low city' disasters

In all three countries the impact of these disasters differed sharply among classes and parts of the cities. Those affecting large industrial cities were described as 'slum raids' and seen as directed at working class districts. Official reports on housing damage in London's East End, Southampton, Birmingham, Liverpool, Bootle, Hull and Belfast, among others, found the greatest losses in 'crowded terraces and courts', 'mean, narrow streets', 'back-to-back' and 'congested 19th century housing', or as 'house property of a slum or semi-slum character and shoddily built'.

After the 'thousand-bomber raid' on Köln, a woman there observed:

"It was mostly directed towards the centre and the Altstadt . . . Unfortunately, the poorer class of people and children had to suffer most."

(quoted in Hewitt, 1994)

A Japanese journalist drew a similar picture of the 9/10 March raid on Tokyo:

"One night, one of the first raids, 100 000 people were killed when they hit a slum area."

(*ibid.*)

He added, what was also widely thought, that:

"If the raid had come in a 'better' section of town, the victims would have been able to spread more concern."

(*ibid.*)

As Sherry (1987, 286) says, these attacks 'hit what was physically the most vulnerable but socially the least effective component of the city.' They give additional and pointed meaning to William Bunge's (1971) description of the densely occupied central core as 'the city of death'.

This is not to say that all the wealthier city dwellers were spared. The oft-quoted words of novelist Virginia Woolf (1989), for example, describing the loss of fine churches and squares in London, attest to that. So did the writer and peace activist, Vera Brittain, whose apartment in London's West End was destroyed by bombs. In the 'Baedeker raids' on historic cities of Britain, and the attacks upon historic towns like Darmstadt, Freiburg and Würzburg in Germany, many better-off persons and officials were severely affected. In the worst raids there could be a terrible 'levelling', socially as well as physically. A middle class woman after the Darmstadt firestorm:

" . . . this night we have all become beggars and homeless."

However, deprivation, as always, was relative. People coming from wealthier, higher status groups could generally obtain more and speedier relief, and, in effect, call upon more 'entitlements'. The congested neighbourhoods were the most devastated and emptied of people. War measures increased the presence of people normally associated with greater social disadvantage in peacetime. There were many more women alone, and *de facto* 'single-parent' or female-headed families. Often there were more elderly, youths, children and others needing care. In the inner city, especially, there were relatively more people with lower education, unskilled and without wage earnings. Many were families of ordinary soldiers at the fronts. However, as the war continued, or the raids got worse, civilians of all classes had less and less influence.

Place annihilation

Inner city disasters

All of the disaster raids are distinguished by devastation in, often annihilation of, inner city areas (Fig. 11.9). In addition to the effects in congested residential districts, this put an end to the most distinctive civil, cosmopolitan, commercial and historic features of city life. In most cases these were concentrated in inner districts. Retail and commercial activity was focused there, including the larger department stores and specialty shops, the headquarters or main branches of banks, and other financial and insurance businesses.

In the downtown areas were most, and usually the finest, hospitals, hotels, colleges, theatres, museums, art galleries, and religious and civic buildings. There were historic streets, squares and monuments of all kinds. They were razed indiscriminately in these attacks. As one historian of Japan observed: 'no one could ever count the books, documents, paintings and other treasures that went up in the flames' (Ienaga, 1978, 2). This too was an assault upon civil life and its shared heritage, which, if not confined to the inner city, was concentrated there.

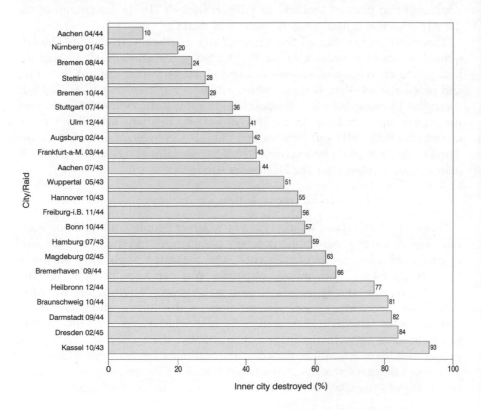

Fig. 11.9 Inner city devastation in Germany. Proportion of inner city built-up area razed in air attacks 1943–1945. The data are derived from air photo reconnaissance following the raids (after Hewitt, 1993, 38)

Place annihilation: an interpretation

The civil ecology of violence

War histories often convey the impression that all but 'essential' workers were evacuated from cities. Instead, we have found that the cities still contained other, civil concerns, a majority of ordinary civilians, consumer economies and people's space. These remained much more prevalent and substantial than appears in the target geography of air war. It is widely assumed that the essential purpose of the raids was to destroy or undermine the war-supporting functions of cities, perhaps to damage the morale of munitions workers, but 'non-essential' civilians, and their space and support systems, actually loom much larger in the toll of damages. This was despite sometimes draconian war measures intended to make cities as productive as possible, and to minimise 'unnecessary' activities and persons.

A profile of the principal raid damages is a kind of reverse inventory of the requirements for survival of people in cities: a negative civil ecology. They were:

- fatalities concentrated largely among resident civilians;
- a predominance of casualties among 'definitive civilians'; i.e. women, children, the elderly and the infirm, who were not just non-combatants, but were mainly involved in continuance of, and dependence on, civil life;
- an attack on the domestic foundations of the city, through physical damage predominantly to homes, with death and injury mainly occurring in homes and shelters;
- the second focus of destruction: commercial and public services, schools, hospitals, places of entertainment and worship, that is, mainly civil support systems and urban communities;
- destruction of buildings of historic and artistic significance, ancient landmarks, symbols of identity and continuity of urban settlements;
- enforced uprooting of resident populations – directly through the bombed-out and indirectly in the evacuation and separation of families and neighbours – the dismemberment of their shared worlds;
- indiscriminate destruction of the inner city and reflecting its social geography, rather than war functions or political power;
- a landscape of violence, whose rubble and the dead buried under it converted the living city into a necropolis.

This profile of destruction reflects the demise of that particular form of human ecology we can call civil. It also defines the major consequences for urban geography.

This evidence of the hazards and disasters of war against cities leads to my view of them as 'geographic calamities'. They attack and destroy human settlements. This raiding of cities was literally a destruction of settlement geography and, as human experience and history, *place annihilation*.

The raiding affected and destroyed 'places' in the two senses in which the word is used in social geography. First, 'place' is a shorthand way of identifying the contexts, positions, situation and lives defined by particular geographical names. This is to emphasise the particularities, and local and specific features, of life and landscape. They are what distinguishes a place and differentiates it, or the condition of those living there, from others. These social, material and active places, or every person's place in them, were annihilated in the disaster raids.

Second, there is the notion of place as something embracing all, if not more than, the sum of these particulars. It belongs with the kind of 'geography' introduced in Chapter 3, and in Dardel's (1952) notion of geography that matters. Place, in this sense, considers how intelligent (i.e. language-using and sharing), gregarious creatures construct habitats, and share and socialise living space.

As a minimum such a place refers to a context in which one moves around – knowingly, purposefully and most of the time without even reflecting upon the fact – between many indoor and outdoor spaces that are not intervisible, *without needing a printed map*. Here, one can converse and make arrangements with a more or less large group of other persons, in relation to a full range of the sites of social and cultural life, without doubting that they know where and what you are talking about, again without needing to refer to a map. Socially, and usually geographically, a place also defines the difference between 'insiders' or, at least, residents – those for whom this is their home place – and 'outsiders' and 'strangers'. In addition, a 'place' is a setting one can remember and describe in intimate detail, even long after it has ceased to exist, whether in itself or as part of one's life.

The core of the idea is that places are *shaped* and reproduced by the personal and shared work, the purposeful exchanges and the intelligent participation of resident communities. Our places are literally composed and handed down, providing clues by which strangers know they are somewhere, and somewhere distinctive. Urban places are the definitive constructs of civilised existence. To destroy the residential, historic and civic areas of cities, to kill and uproot their long-time inhabitants, is to deprive them, and those who come after, of their places.

In terms of human geography the urban air attacks, and especially those identified as disaster raids, caused a catastrophic unmaking of civil places. The intentional razing of cities by means of large bombing raids, targeting whole districts with weapons of mass devastation, was a process of place annihilation. Whatever their war-fighting objectives, the attacks fell mainly on those whose primary role was the maintenance of domestic and communal life – the 'place-makers' of civil society. The destruction of the inner city and residential districts annihilated living space, buildings and streets dense with meaning and habitation. With that expired many of the characteristic places of civil society and Old World urbanism.

CHAPTER 12

The Holocaust: genocide and geographical calamity

From 800 to 2000 people died every day. We estimated that if you were Norwegian, Dutch, Danish, you might survive. If you were Belgian or French, your chances were slightly poorer. If you were Czechoslovakian or Hungarian, they were even poorer. As a Polish prisoner, you had a life expectancy of three weeks. The Jews, of course, were a totally separate category, brought to Buchenwald for the express purpose of being exterminated.

Reidar Dittmann, Holocaust witness. Lewin (ed.) (1990, 17)

Introduction

The Holocaust refers to the destruction under National Socialism ('Nazism') of the Jews of Europe. The term comes from an old word meaning to sacrifice by consuming in fire – in Hebrew the *Sho'ah*. Its defining atrocities were systematic, mass killing in death camps using poison gas, and incineration of the victims' bodies in factory-like crematoria. Even greater numbers were killed in organised massacres and random killings; by privation and disease due to deliberate starvation and exposure; by 'destructive labour' (heavy work without adequate food or medical care); in 'death marches', and the infamous rail 'transports'. The full count of those who committed suicide is not known, but thousands took that route. Tens of thousands were killed trying to escape or hide from their persecutors.

A great many peoples and groups were caught up in the plans and destructive processes of the Holocaust, but the most determined and concentrated killing was of Jews. It is now generally accepted that no less than 5.2 million were murdered, possibly more than 6 million (Fig. 12.1). Thousands of communities, in which they had lived for centuries throughout mainland Europe, were annihilated. The Romany and Sinti ('Gypsies') were also targeted for

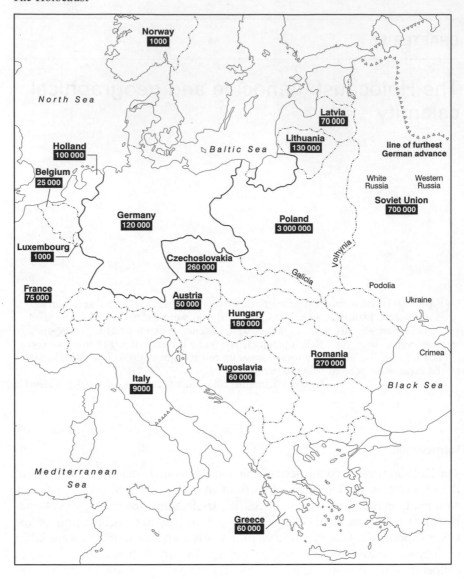

Fig. 12.1 Estimates of Jews from different national areas of Europe killed in the Holocaust (after Hilberg, 1985)

extermination; some 220 000 were murdered (Fig. 12.2). Anyone found to be homosexual was marked for death, and their death rate in the camps was especially high. Thousands with congenital impairment, crippling diseases or feebleness in old age were killed in euthanasia centres. Meanwhile, related to and intertwined with the focused killing of particular 'racial enemies', was an

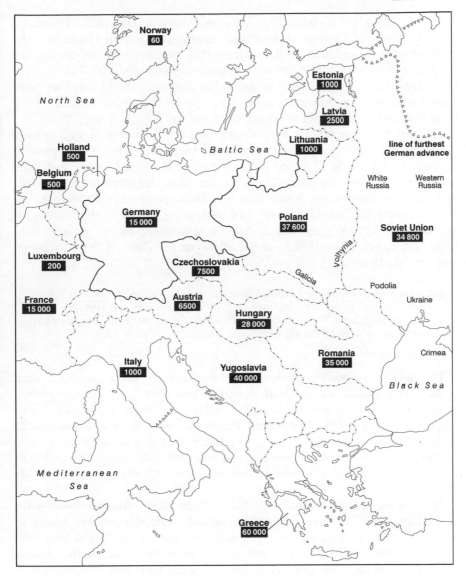

Fig. 12.2 Estimates of Romany and Sinti in different national areas killed by the Nazis (after Kenrick and Paxon, 1972, 183–4)

enormous death toll of political prisoners and civilian nationals of territories taken by the German army, especially in Poland, Ukraine and Byelorussia.

These events comprise one of the greatest calamities of this or any other century. The statistics of lives destroyed by no means exhaust and do not reveal

323

the full nature of the disaster. Even if we consider one small part – France or Italy, the Jews of Croatia or Rumania, or any one city in Germany – the evidence of how and why, and to whom this was done, speaks of singular calamities. Whole communities were brutalised and murdered. People were killed regardless of age, gender or any other attribute, except that they were classified as Jews, or from certain other groups deemed to be 'enemies' of the Nazi State, or a 'blood threat' to the German *Volk*.

The victims of the killings were almost invariably defenceless, and already suffering from severe privations and brutal treatment. They were unarmed civilians at the mercy of armed police guards and 'special forces' dedicated to killing. Children were often among the first to be murdered in the pogroms and extermination camps. More than a million Jewish children were killed, infants having the lowest survival rate of any age group in the Holocaust. Out of some 130 000 women and children sent to the major women's concentration camp, Ravensbrück, about 90 000 were killed. At the Auschwitz-Birkenau death camp,

> throughout the twenty-eight months of selections . . . the procedure was rule of thumb. Children under fourteen, men over fifty, and women over forty-five went to the gas chambers. To save the SS difficulties, all mothers who accompanied young children went to the gas chambers, irrespective of their age.
>
> Reitlinger (1961, 118)

The elderly, and anyone conspicuously weakened by privation and in need of care, were singled out for death at once. Then again, a former war hero, a scientist or a valuable munitions worker, if Jewish or Romany, or a known or suspected member of the Communist Party, was more likely to be summarily murdered than a convicted felon, especially in the concentration camps. Nothing upset committed Nazis like the idea that they somehow needed or might owe 'these people', might somehow be 'tricked' into pitying or saving them.

There is no complete consensus, but it seems likely that the decision for extermination of Jews was not made, or not made the primary object of Nazi policy, until 1941, that is, two years into the Second World War, and eight after Hitler's accession to power. We must be aware of the shifting geography and the time frame of events. The early treatment of Jews in Germany, horrible as it was, remained well short of what was to happen after 1939. The same may be said of early moves in the conquered territories of Western Europe in 1940–41. However, in Eastern Europe, from the very beginning in 1939, practices were ruthless and uncompromising. Even there, however, a drastic change in the scale and forms of killing occurred in 1942. Browning (1992, xv) points out that:

> in mid-March of 1942 some 75 to 80 percent of all the victims of the Holocaust were still alive, while 20 to 25 per cent had perished. A mere eleven months later, in mid-February 1943, the percentages were exactly the reverse.

That is before the most infamous killing devices – the main, factory-like gas chambers with attached crematoria – were in operation at Auschwitz-Birkenau. And there were two and a half years of killing still to come.

Any attempt to present an overview of disaster in our time can hardly do so while ignoring the Holocaust. There are good reasons for feeling inadequate for the task. As someone who was not there, one can never adequately reconstruct the events. Not having been subjected to anything remotely like the abuses that were commonplace throughout this whole episode, one cannot 'know' the experience. Even Holocaust survivors admit they are unable to re-imagine or represent it adequately. After lifetime study, scholars remain baffled by so many aspects, as were virtually all the victims at the time. 'The feeling of fracture, of being subjected to things incomprehensible and unutterable, was felt by the witnesses themselves as they recounted their suffering' (Trunk, 1979, 72).

Nevertheless, it seems worse to keep silent or, in the present context, to ignore these events. They record 'unnecessary' human suffering and 'untimely' death on a scale and in a form that outstrips most of the other risks we are considering. Something else that singles out this calamity and should draw our attention is the extraordinary literature of the Holocaust.

The literature of the Holocaust

The first aspect of this literature that I would single out is a determination, even in many studies of 'the system' and the statistics, not to lose sight of the faces – both of killers and victims. Although it was a disaster of 'populations', it is felt as the brutalisation and pain of one person at a time. The 'geography', also, is as much about the landscapes of violation and places of ignominy as the maps of uprooting and the demographics of race-murder.

Many of the best-known and most influential works are personal memoirs. The considerable published testimony of victims is an essential resource, and still vaster archives await the researcher. However, the scholarly work is distinctive too, so much of it being written by survivors, members of their families, or those who have lived among them since 1945. This contrasts with most of the literature of disaster, or that which research tends to emphasise. And it challenges a common view that 'involvement' and 'subjectivity' are prescriptions for bad scholarship. Rather, it *is* involvement, not detachment, that gives this literature its power and credibility. Moral and spiritual commitment are the cornerstones of the unrelenting search for solid evidence, authentic description and sound interpretation. There have been disagreements, even bitter ones, about what is important, which remembrance does and does not serve the memory of the victims well, and even about whether to speak at all. But we are reminded constantly that it is the victims rather than 'science', especially those who did not survive, who require truthfulness and humility in scholarship. In that, the Holocaust literature is a guide and inspiration, if a sobering one, for thinking about approaching the whole field of modern calamities of social violence.

The Holocaust

Geographical calamities

Another, crucial, aspect of the Holocaust is the extent to which it was a disaster *of*, and *by*, 'geography'. It annihilated ways of life and worlds. It destroyed 'geography that matters', or that mattered enormously to its victims (see Chapter 2). The first violations were to people in their home places: the undermining and destruction of communities, of urban and rural living space. Usually, the worst was still to come, but afterwards they were transformed into 'another people'. They lived and died in another, alien, geography – uprooted, dispersed, imprisoned; as refugees and exiles; as victims and survivors.

It was a process of destruction *by* geography. First, Jews were ostracised in the places where they had been settled, and denied the ordinary norms and securities of home and civil life. Then they were uprooted from places where they had, to a greater or lesser extent, 'belonged' for centuries. The lucky ones were driven into exile. The remainder were deported and concentrated in places of misrule, slavery and extermination. What we usually mean by human geography is revealed in opposite and negative images; landscapes of violence; streets and journeys of terror; a 'space economy' of extermination and places of death.

Suggested reading

Gilbert, M. (1993) *Atlas of the Holocaust*, revised edn. Lester Publishing, Toronto.

When the state kills

Little by little the grand plan of the Nazis came to light – the extermination of the Jews.

P. Girard (in Dimsdale (ed.) 1980, 76)

The Holocaust can also be described in terms of 'hazards' and damage, the vulnerability of victims, and what they or others did and did not do to counter the threats and harm. As always, there is a tendency to focus on the worst happenings and see the problem as essentially one of explaining them. Certainly, such places and moments of the Holocaust – the overcrowding and great death from starvation and disease in the Warsaw ghetto; the huge massacres by 'death squads' following the invasion of the Soviet Union; the Auschwitz-Birkenau extermination camp; the 'rebellions' at Treblinka and Sobibor death camps – have a significance out of all proportion to their share of the whole disaster. The statistics of mass killing, the images of brutal treatment and terror, of helplessness and extraordinary courage in death, speak directly to the experience of the victims. They remind us of what Nazism brought to the streets and countryside of Europe, yet they can also hide developments and crimes that were fundamental in other ways, especially in making the Holocaust possible.

Throughout, comprehensive, exhaustive murder required the replacement of the mob with a bureaucracy; the replacement of shared rage with obedience to authority.

Sabine and Silver (in Dimsdale (ed.) 1980, 329)

In the terms applied to other dangers, the hazard here – the societal equivalent of a hurricane or a drought – was nothing less than the German state under Nazism. This disaster occurred, and could only occur, when the instruments and agencies of government were deployed *against* its subjects. The Holocaust was not just a massive 'return to barbarism'. It could be carried out only by using all the instruments of the modern state. Its scope and thoroughness reflect the full use of legislative powers and the state's monopoly of armed force. Once in power, the Nazi leadership, and committed party members whom it had placed in all levels of administration, acted to entrench and expedite the policies of racial discrimination and, in due time, of race-murder.

The fascist principle demanding absolute obedience to 'der Führer' (the Leader) did enable Hitler to impose policies and actions by decree with little regard to whether or how they might violate existing laws and practices. What is sometimes called the SS State, or secret police 'state within the state', acquired extraordinary powers or pre-empted those of the regular administrative structures. This was where Nazism was clearly essential to, and fully responsible for, the priority given to racial policies and uncompromising pursuit of genocide.

Yet the fully committed Nazis, and their institutions, were not adequate to cause the Holocaust. They had to involve all ministries of the German government and the public infrastructure. There is a functional aspect of the Nazi project to eliminate Jews that resembles a national crash programme to deal with toxic waste or drug trafficking – a problem for project planning and implementation. Not least important, vicious treatment and killing were 'routinised' within administrative practices. Behind the direct brutality lay a paper chase of reports, orders and quotas passed through every level of government.

To appreciate the scope and scale of what was done, and how it could be done, one can recognise a number of processes or phases of the calamity. Each involved successive escalations in the scale of destruction or severity and life-threatening hazards for the victims (Table 12.1). Although not following neatly one from the other, each provides essential insights into the overall calamity.

Suggested reading

Dimsdale J. E. (ed.) (1980) *Survivors, victims and perpetrators: essays on the Nazi Holocaust*. Hemisphere Publishing, New York.
Hilberg, R. (1985) *The destruction of the European Jews*, 3 vols. Holmes and Meier, New York. (I am especially indebted to this massive and compelling study of the system and processes that are here identified with the 'hazards' of the Holocaust.)
Furet, F. (ed.) (1989) *Unanswered questions: Nazi Germany and the genocide of the Jews*. Schocken Books, New York.

Table 12.1 Processes of the Holocaust: the series of measures taken against the victims of the Holocaust, roughly in order of implementation by the Nazi state and increasing roles in the destruction of lives

---

Prelude to annihilation: targeting, segregation and deprivation

definition, discrimination, selection	e.g. pass laws, 'Jewish Star'
disenfranchisement, denial of rights, 'special' laws	Nürnberg Laws
segregation, ostracism, confinement	special houses, anti-mixing laws
arbitrary harassment and denied protection	intimidation, arrests, torture, searches, violence to persons and property
denial or destruction of livelihoods	dismissals, exclusion from employment, closure of businesses, 'special' (i.e. low) rations
'economic' despoliation	expropriations, liquidations, fines, seizures of property
exploitation	forced labour, quotas, unfair trading, sexual abuse, payment of costs by victims

---

Constructing the space of annihilation: the 'concentrationary' process

'transporation'	round-ups, expulsions and deportations, forced relocation
ghettoisation	urban 'incarceration'
the concentration camp system	'political' and forced labour camps, extermination camps

---

Processes of extermination: demographic violence and genocide

forced sterilisation	
deliberate deprivation, neglect, overcrowding	exposure, starvation, disease, suicides of despair
arbitrary brutality	unrestrained beatings, punishments, torture and killings
destructive labour	intentionally working people to death, inadequate food, etc.
lethal experiments	internees as 'guinea pigs'
euthanasia	'medical killing'
mass executions	pogroms, massacres, 'ethnic cleansings', mobile gas units
'death marches'	
death camps	

The hazards: processes of the Holocaust

Precursors of disaster: definition, targeting and segregation

... in consecutive steps, the Jewish community was isolated socially, crowded into special houses, restricted in its movements, and exposed by a system of identification.

Hilberg (1985 vol. 1, 180)

The Holocaust was largely a calamity for people 'chosen' by official policy and discriminating laws. The groundwork was laid by violations of basic civil rights, the *official* removal of civil protections, and substitution of punitive and demeaning regulations – matters rarely encountered in our work on natural or most technological disasters.

The 'Nürnberg Laws', promulgated in September 1935, denied Jews citizenship and a range of basic rights within the German state. Various 'Supplementary Decrees' barred Jews from almost all areas of social life and intercourse with non-Jews, demanding that they carry special identity cards and use officially approved names. Later on, to make their exclusion from society fully visible, they had to wear an insignia on their persons – the 'Jewish star' – and place identifying signs on the buildings they occupied.

These acts of discrimination on racial grounds created, or greatly increased, the space of separation for Jews in German society. They did so by making 'special' use of the governmental devices more commonly used to control and administer such matters as taxation, health care and professional certification. Here they were made tools of a police state. In fact, these hazards on the road to genocide reflect why the strenuous defence of civil liberties is regarded as an essential safeguard in civil society. This includes strong restrictions on the ability of governments or other organisations to hold, demand or use personal information.

This first phase of discrimination in civil and economic affairs reconstituted social geography under German control. Jews were 'uprooted' and deprived of living space *before* being physically expelled. The racial laws, decrees and rules *took away rights of property and domicile*. Jews were rendered *non-citizens and 'non-persons' by state decree*. They were made *'stateless' in the country and places of long-term residence: 'displaced persons' in their own homes*. Nevertheless, this was also basic to setting them apart physically.

Once 'defined' and 'identified', Jews could be targeted for a range of measures intended to separate them from surrounding inhabitants. The 'anti-mixing laws' forbade not merely intermarriage and any other sexual liaisons, but even friendliness between 'Aryans' and Jews. Jews were denied access to schools, hospitals, hotels, resorts, beaches, movies and libraries, and had restricted access to public transport and shopping. They were expelled or barred from the professions and from holding public office. In time, they were forbidden to use telephones or have a driver's licence.

Once these steps were complete, Jews were disempowered, subjected to constant surveillance and to easy and arbitrary police actions. They no longer enjoyed elementary forms of public safety and protection. New 'standards of behaviour' left them open to physical abuse and exploitation by official and unofficial thugs. Businesses were expropriated or liquidated. There were confiscations and theft of property. The first form of enforced concentration in urban areas was begun, or 'ghettoisation' – a word that is, at least, less ugly than the reality it conveys. Nazi officials attempted to gather them in 'Jews-only' dwellings.

Early on, emigration was encouraged or not prevented. About half of Germany's Jews, some 250 000, fled to all parts of the world, and about 85 000 from Austria. This episode in itself was an enormous mass flight full of hardship and fear, of journeys into an unknown exile. However, it brought a backlash from reception countries, where anti-Semitism was common too. Nazi Germany was forced to restrict and soon banned the flight of Jews. From then on, they were entirely at the mercy of Nazi policies, where the threat of arbitrary and severe punishments conditioned many to be submissive and turn inwards, to accept visible segregation and police control. This made them much more vulnerable to later, even harsher measures.

When the Second World War began, a night-time curfew was applied to Jews. They were excluded from the wartime community's 'togetherness' by being identified with the enemy. Their radios were confiscated and listening to them forbidden. Once Allied bombing began, Jews could not use communal air-raid shelters and had to provide their own protection against air attacks.

These abuses were accompanied by a constant barrage of accusations of activity against Germany, of claims that they posed a threat to 'public order' and the health of the population. In this way, as D. A. Hamburg (1980, xiv) expressed it, conditions were created under which 'the destruction of the powerless was made palatable to the powerful.'

'Captive populations': the constructing of a carceral geography

> The only adequate solution to problems posited by the racist worldview is total and uncompromising isolation of the pathogenic and infectious race [sic] . . . through its *complete spatial separation or physical destruction.*
>
> Bauman (1989, 76, my italics)

Hitler and his minions, like other fascist or totalitarian and militaristic governments, were deeply enamoured of techniques to imprison 'elements' they disapproved of so as to exercise total physical and mental control. This was used to discipline and punish political and racial 'enemies' and, eventually, for their controlled extermination. For such purposes they created a new and ever-expanding *carceral geography*. Eventually, it involved a vast system of mass confinement, especially ghettos and concentration camps to deal with 'captive populations'. This term conveys the enormous numbers involved and that they were, indeed, persons of all ages and backgrounds.

The history of, and places in, this carceral geography help to identify the hazards of such captivity. Following the early phase of segregation, it became a *concentrationary* system of mass imprisonment. However, to create that, vast numbers of people had to be transported across all of Europe.

The transports

> We were forced into railroad cars, 100 to 120 in one car, like sardines, without food, without water, without sanitary facility. The cars were sealed and we stood there for maybe half a day before even moving . . . There were children in our car, and old people. People got sick, died, and some went insane . . . I really don't know how many days and nights we were in that living hell on wheels.
>
> Fred Baron, in Lewin (ed.) (1990, 9)

An essential 'hazard', in the geographic scope of the Holocaust, was the European railway system. It facilitated the destruction of the previous geography of settlement and the creation of a new one of mass imprisonment. Only in this way could whole populations be moved across a continent. Railway time-tabling and rolling stock became an organised system of punishment and 'attrition' as well as deportation. The routes bringing people to the death camp at Auschwitz show the geographic scope of this 'enterprise' and its nightmare journeys (Fig. 12.3). At least 1000 persons were usually assembled per train and over 8000 in one terrible journey. There were endless hold-ups and delays, 'inefficiencies' borne by the victims as added tortures. Out of 4000 on an average train, between ten and twenty would have died before the end. However, they were always travelling to places where the chances of survival were much slimmer, often zero.

The Holocaust was, overwhelmingly, about killing in Poland and other territories of Eastern Europe. However, it affected all German-occupied territories. Most of the Jews from Western Europe and the Mediterranean lands who died were killed by transportation to, or slave labour in, ghettos and death camps in Poland.

Certain major changes in the context of risk after 1939 also became critical. Germany was now at war. Rule by decree and excuses for draconian measures increased enormously, as did opportunities for secrecy and control of information. The geographical focus of anti-Jewish measures shifted eastwards, especially to Poland, bringing millions more Jews under German control. But Nazi racism extended to Poles and other Slavic people, treating them as lesser humans. Actions were much less constrained by any inconvenience or affront they might cause to populations among whom the Jews lived.

Meanwhile, the people in the other conquered territories were themselves in the midst of an enormous calamity; anti-Semitism was well-entrenched in Poland too. The chances of help for the Jews were greatly reduced. There

Fig. 12.3 The European railway system as hazard: routes for the transport of Jews to the death camp at Auschwitz-Birkenau

was little willingness to assist Jews, and some readiness to turn them over to the occupying forces.

Britain and, quite soon, the United States, went to war, ostensibly to save 'the little nations' gobbled up by Germany. But now they could offer little or no protection for Jews, could not intercede diplomatically and were rarely interested in any other forms of help. The Jews remaining in continental Europe were totally isolated and at the mercy of Nazi designs. These turned to a more complete and life-threatening form of 'ghettoisation' from the middle of 1940.

Incarceration in the city: the second phase of ghettos

> The ghetto was a captive city-state in which territorial confinement was combined with absolute subjugation to German authority.
>
> Hilberg (1985 vol. 1, 234)

Most of the newly conquered areas of Poland were to be 'Jew-free', the land and property being earmarked for German-speaking peoples. Jews here were deported to the so-called *General Government*, comprising central and southern parts of former Poland. Jews already there were not to live in small communities or towns but in a few larger cities; and in these they were to be confined in specially designated districts: the ghettos. In some cases they were in or around existing, predominantly Jewish quarters. Others were newly assigned, usually in undesirable and unsanitary districts. All were to be walled in and, officially at least, cut off from the surrounding city. Internal order and administration were given over to Jewish councils and police.

The largest imprisoned city-within-a-city was the Warsaw ghetto. The Polish capital housed the largest concentration of Jews in prewar Europe, about 380 000 in 1939. In October 1940, a part of Warsaw where 280 000 Jews were already living was made a closed ghetto of some 307 hectares. More than 50 000 Jews from other areas of Warsaw were forcibly taken there. Early in 1941, 70 000 people from the region surrounding Warsaw were added and, subsequently, large numbers from France and other parts of Europe. In the most congested period, as many as 450 000 people were confined inside. Food and amenities provided by the Nazis were not adequate for half that number. Jews managed to smuggle in extra food and raw materials, but never enough for all. Hunger, exposure, accumulating human waste and insanitary living all helped to cause severe epidemics. Between October 1939 and the middle of 1942, almost 100 000 people died, mainly from starvation and disease. In the harshest winter month, January 1942, 5560 persons died. The death toll, from similar causes, in the next largest 'captive city-state' at Lódz, was 45 000. The Nazi authorities planned and expected even greater losses.

As the genocide moved into its final phase of deliberate mass killing, ghetto populations were deported to concentration camps. Between 22 July and 21 September, 1942, 300 000 survivors of the Warsaw ghetto were shipped to the Treblinka death camp and murdered there.

Suggested reading

Sakowska, R. *et al.* (1988) *Warszawskie getto (Warsaw ghetto) 1943–1988*. Wydawinctwo Interpress, Warszawa.
Hilberg R. *et al.* (eds) (1989) *The Warsaw diary of Adam Czerniakow*. Stein and Day, New York.

The concentration camps

> The history of the concentration camps graphically demonstrates that political persecution, economic coercion and ideologically based extermination were firmly linked in the National-Socialist system.
>
> Falk Pingel (1984, 17)

The Nazis founded their first concentration camps in 1933, mainly to imprison and degrade political enemies. These were separate from other prisons and, although not in law until 1939, were under the control of the SS. Cruel and inhuman punishments and arbitrary executions were commonplace. The early camps played their part in the Holocaust by later expansion or conversion, or as training and experimental stations for the much larger system it would require. Ultimately, in scale and organisation, in isolation from the rest of the world and in extreme conditions, the camp system became a 'concentrationary universe' (Rousset, 1947). It extended from the Pyrennees to the Black Sea and from Estonia to the edges of the Sahara Desert (Fig. 12.4).

There were different types of camp, or of functions in different parts or phases of the same camp. Confinement and punishing of political opponents remained an important function throughout. A second development, in preparation for war and during the conflict, were camps or parts of camps for slave labour. Most were run on the principle of destructive labour. Inmates were expected to work until they dropped or, when deemed unfit for work, were selected for extermination. However, as labour and economic problems grew more severe, the SS found it advantageous to hire prisoners out to private companies. Sometimes they were hired out for agricultural labour. This became a major source of funds for the SS, but prisoners might receive better treatment, and could find ways to improve their lot.

Following the invasions in Eastern Europe, camps were built for political enemies and resistance elements from the civilian populations. Some were set up first to accommodate and then to murder Soviet prisoners of war by starvation or mass execution. They included Auschwitz, where the practice and techniques of mass extermination using the poison gas Zyklon B were first tried out on them.

The camps or parts of camps unique to the Holocaust were those to exterminate Jews, Romanies and other 'racial enemies'. Some were purely 'killing centres'. Notable were Chelmno, and the 'Operation Reinhard' camps at Belzec, Sobibor and Treblinka. About 2 million Jews and 52 000 Romanies were gassed in them. Some of the most infamous death camps, such as Auschwitz-Birkenau, combined extermination with the other functions described above (Table 12.2).

At their height in 1943–44, the camps housed over 1 million persons. Between 3 and 4 million Jews were killed in the major death camps in Poland. Several hundred thousand were worked to death or died of other privations in

Fig. 12.4 Carceral geography: the larger camps and main types of camp in the 'Holocaust system' of Germany and the occupied countries during the Second
World War (after Gutman and Saf (eds), 1985). Hundreds of other, smaller
camps are not shown here

a host of small camps. Rarely are numbers known with any precision, but the
proportion of inmates killed in some camps is reasonably well-established:

> in Mauthausen-Gusen [a concentration camp in Austria], out of the total popula-
> tion of prisoners, 37% were killed; in Plaszow [a concentration camp near Cracow
> in Poland], 53%; in Majdanek, 72%; in Auschwitz-Birkenau, 91% [both death
> camps in Poland].

> Madajcyzk (1985, 53)

Table 12.2 The death camps: high and low estimates of the numbers killed (after
Gutman and Saf (eds), 1985)

Chelmo	310 000	152 000
Belzec	600 000	600 000
Sobibor	250 000	250 000
Treblinka II	900 000	900 000 (700 000)
Majdanek	200 000	120 000
Auschwitz-Birkenau	2 500 000	over 1 200 000 (1 500 000)
	4 760 000	3 222 000

As the battlefronts advanced towards the concentration camps, near the
war's end, many were hastily disbanded, their records destroyed, their remaining prisoners killed or driven out. Surviving prisoners were forced to walk for
days to other destinations in what proved to be 'death marches'. Thousands,
often the majority in each column, died or were executed when they fell down
exhausted or tried to break away. When Allied troops finally reached the
camps, appalling images of what had gone on there reached the world news
services – thousands of emaciated, starving prisoners; pits full of the dead.
Many who had survived until then succumbed to hunger, disease and exposure
before adequate help could be provided.

Suggested reading

Gutman, Y. and A. Saf (eds) (1985) *The Nazi concentration camps.* Proceedings of the
Fourth Yad Vashem International Historical Conference, Yad Vashem, Jerusalem.
Rousset, D. (1947) *L'Univers concentrationnaire (the other kingdom).* Boyer and Hitch-
cock, New York.

Vulnerability and genocide

. . . anyone who opposed or did not fit into the Nazi worldview was vulnerable.
Carol Rittner (1990, xii)

The Holocaust involved a special array of vulnerabilities. They stemmed from
relations between a modern state and its subjects, specifically minorities, when
the government in power not merely fails to protect, but uses its powers to
oppress them. Pushed to the extreme in Nazi Germany and some other criminal regimes, the exposed population can seem almost totally vulnerable. All
courses of action appear desperate if not doomed to failure unless other powerful groups intercede or the oppressor is less than thorough.

The vulnerability of 'subjects'

> So there we were, ninety-six weak, emaciated women, marching down the high-
> way with all these guards with rifles. Then the German guards told us to run into
> the woods. The snow was so deep, up to our knees, and most of us were barefoot,
> frozen, our feet were blistered . . . Then the guards began to shoot.
>
> quoted in Lewin (ed.) (1990, 54)

When government organisations and armed violence are turned against
unarmed civilian populations, there is a basic rift between how we can describe
the vulnerability of victims, and the hazards they face. The Nazis mobilised a
modern state and some of the latest technology. They had a relative abundance
of resources, in particular weapons. Their business was carried out by specially
trained, well-equipped and fit personnel: predominantly male police, para-
military and military units.

By contrast, those at risk were vulnerable as families and civilian communities facing disaster. They were attacked first in their own homes and neighbourhoods: men and women, spouses, children and the elderly, together. They
were people from all walks of life still preoccupied with their everyday business;
encumbered with property and dependents; sharing the hopes and values of a
lifetime, and the habits of peaceful existence.

From 1933, practically every Jew living within reach of the Nazis felt the
whisper of fear enter their homes and everyday lives. Overtly racialist attitudes
and policies spilled from radios and newspapers. Indignities, vile accusations,
and violence to their persons and property reached into neighbourhood streets
and residences. Parents lay in bed at night terrified for their children as official
thugs roamed the streets knocking on doors, summarily removing a husband
or a brother, or secretly raping a sister or a wife.

A survivor recalled the day he, his wife and small infant were rounded up for
deportation:

> You want to do something and you know you are in a corner. You can't do
> *anything*. And when someone asks me now 'Why didn't you fight?' I ask them
> 'How would *you* fight in such a situation?' My wife holds a child, a child stretches
> out their arms to me.
>
> quoted in Langer (1991, 97)

His wife and child, and also his own mother and grandmother, were
shipped to Auschwitz and gassed. We will return later to this relation between
vulnerability and the inability, or extreme perils of trying, to resist the
dangers.

Any comfort and shared activity towards survival for victims came from
bonds of family, religion and shared misery. More rarely, there was support
from people in the surrounding community. Later, of course, victims were
separated, often in all-women or all-male groups, or as uprooted and isolated

individuals from different parts of Europe, in the life-threatening anonymity of the camps.

Geographically, the perpetrators created an elaborate, continent-wide system, but each victim had to respond within the small space of his or her immediate environment. They were progressively cut off from and ignored by the larger world. Once the full disaster of uprooting and killing was in progress, there was no help from the international community. Connections with other Jewish communities were broken, and their own were increasingly fragmented. Over time, survivors lost touch with their closest family and friends. Their familiar worlds were destroyed, to be replaced by ghetto walls and blows, by barbed wire and punitive barracks discipline.

The Holocaust was *planned* annihilation. For the perpetrators, it was about administration and order, but the world they created for their victims was one of 'chaos': chaos in everyday life and the most basic means of survival. The intention was to undermine the victims' adaptive and survival capacities, and to cause as many as possible to die from privation. Where their adversaries drove, rode and flew, the victims had to walk carrying their belongings. In a modern industrial state, their work was hard manual labour, their means of survival reduced to the most primitive. Their energies were absorbed in endless marches and waiting, in line-ups for supplies that did not arrive or were totally inadequate when they did. They were made to suffer constant, debilitating overcrowding in homes, transports, ghettos and camps.

Their vulnerability as social and intelligent beings was made the more extreme because nothing made sense any longer. The 'policemen' committed the worst crimes. There was sudden, brutal treatment over trivial 'transgressions'. Countless men were shot or beaten to death for stepping aside to relieve themselves. People were murdered rather than given simple medical care. Women were summarily killed *because* they had small children or were pregnant.

Many Jews and their organisations were convinced that the best hope of not being deported, or of surviving the ghettos and camps, lay in hard work and making themselves useful to the Germans. Yet even in the desperate and overstretched German economy of wartime, destructive labour was the rule. Skilled slave workers, even a whole factory workforce, might be killed by the SS to demonstrate that they were not needed.

At one level, vulnerability did seem to depend upon the crudest of Darwinian criteria – 'the survival of the fittest'. The most vulnerable were usually those more severely weakened by hunger or illness, too young or old to walk far or to stand hard labour, or not able to fake their age. A woman survivor from Lithuania conveyed what was true, at least for a time, in many of these captive populations: 'as long as we were strong and useful, we would survive' (quoted in Lewin (ed.) 1990, 52). She described the day when fit adults returned from work to find that all of the children and elderly in the ghetto had been taken away and shipped to Auschwitz – 'useless mouths sent to the gas.' To be dependent, in need of care or not able to work was to be absolutely vulnerable.

As we have seen, however, the plan for the 'strong and useful' was to work them to death. Leading Nazis expressed the view that Jews who could survive all these assaults would *have to be killed*. By their crude Darwinian criteria, these had proved themselves the most resourceful and strong, therefore the most dangerous 'racial enemies'.

In the short run, gender, age, physical strength or a particular professional or artisan skill might make a huge difference. The fittest of those arriving at all except the purely killing centres were commonly kept alive for construction projects, or work in the quarries, forests and mines. Sometimes, however, carpenters, medical staff or chemists were needed, and they might survive where others who were fitter did not. The largest fraction of Holocaust survivors were those selected for such tasks, at least in the later stages, when they had some chance of not being worked to death before the Allied troops arrived. Nevertheless, far greater numbers, assigned to forced labour, still died of the extreme conditions or were later selected for the gas.

The question of differences in vulnerability, or the criteria on which they might be based, is a vexed one throughout. Common sources of strength, or useful abilities in one context, could prove lethal in others. In the camps, as the quotation at the head of the chapter indicates, vulnerability came down to probabilities of length of survival based on the Nazi's racial hierarchy. Life and death for many in the ghettos, on the marches and at the 'selections' can seem to have resembled a lottery. In part this is because the SS directed its main assaults against potential sources of strength or resistance, or in order to deceive and co-opt them.

Thus, rational assessments of vulnerability and response break down. Except in certain cases such as pregnant women or Jewish communists, neither the victims nor later analysts could say who was more vulnerable among Jews, or which responses might aid survival. Theories of social adaptation and disadvantage, or 'adjustments' to hazards, are of little help. The worlds portrayed by Franz Kafka, or George Orwell's *1984* and *Animal Farm*, speak more convincingly of the condition of the victims.

Then there is the question of the defencelessness of Holocaust victims. Vulnerability is usually offset by protections and rights within society and is not just about the physical exposure and frailty of those at risk. But Jews were not only deprived of adequate food, shelter, medical care and safe employment. They had no recourse if officials or anyone else stole from them, beat them up, or raped them. They could be imprisoned, tortured, worked to death or killed – but *no* institution or witness would intercede, or even had the capacity to. These victims were deprived of the right to exist.

Gender and vulnerability

Lack of protection and unusual vulnerability are especially distressing and ugly features of the Holocaust in relation to women, children and the elderly. The testimony of women survivors and such studies as have been done suggest that

339

their vulnerability and survival chances were rarely identical to and often worse than those of men. As noted, in the camps, children and the elderly, and most women with small children or pregnant, were selected for death at once. In this and other ways the crude biology of Nazi racist notions placed women in a specially dangerous or ambiguous position.

Maternity, the sexuality and fertility of Jewish and Romany women, or of any with some kind of disability or visible difference, tended to single them out for ruthless treatment. The infamous sterilisation experiments in Block 10 at Auschwitz targeted Jewish women, as did forced sterilisations in the women's camp at Ravensbrück. Castration was widely practised on Jewish men. Although sexual relations between 'Aryans' and Jews were proscribed, in the ghettos and camps, guards would single out 'good- or Aryan-looking' and still healthy Jewish women as temporary sexual partners or camp prostitutes. Moreover, even these 'options' were welcomed by some women as the only or best chance to survive, at least for a time.

Sometimes the Nazis were looking for women as cooks, seamstresses or laundresses in the camps, and these might be less dangerous jobs with a better chance to survive. A few women also came from relatively emancipated or independent backgrounds. Occasionally they could take advantage of that, like the women inmates' orchestra at Birkenau, recalled in Fania Fénelon's *Playing For Time* (1977). But most women were from more 'traditional' backgrounds. Initially, at least, they were captured with their families, and caught up in the duties and emotions of caring for them. There were countless instances of daring and successful hiding and saving of their loved ones, at least in the initial round-ups, in the ghettos and transports. Women from the same family or community might manage to stay together and quietly help each other survive.

In the more fanatical and extermination camps, however, and under the extreme conditions near the war's end, race hatred inspired murdering and working women to death. A terrible aspect of the women's predicament was the inability, in all but a very few cases, to save their loved ones. This is added to the disproportionate death toll of women themselves and the pain of survivors whose mothers were killed.

Suggested reading

Heinemann, M. E. (1986) *Gender and destiny: women writers and the Holocaust*. Greenwood Press, New York.

Lifton, J. R. (1986) *The Nazi doctors: medical killing and the psychology of genocide*. Basic Books, New York.

In all, then, the Holocaust can seem a disaster brought about and explained wholly by the Nazi system for genocide: a rather complete case of the hazard paradigm in action. The victims can appear absolutely vulnerable. Few were able to save themselves or members of their families. They were never given adequate, and usually, no protection by surrounding, unharmed communities, or relief through organised, international assistance. The scale of the calamity itself is revealed by the fact that the Nazis came so close to their goal of exterminating every Jew who remained in Europe. But is that all there is to say about the condition of the victims?

Alternative adjustments to genocide

These and innumerable others, died not despite their valour but because of it.

Primo Levi (1988, 63)

Table 12.3 summarises the kinds of action that were attempted in response to the Nazi assault on Jews, some examples of each, and a tentative assessment of how widespread or substantial they were. It seems fair to say that no actions to stop or substantially frustrate and alter the processes of the Holocaust were successful. World opinion, the enquiries of the International Red Cross, the early boycott of German products by Jewish organisations and diplomatic efforts led to some temporary setbacks for the perpetrators, less often to some relief for the victims. Inside German territory, in the ghettos and camps, there were small acts of resistance going on all the time and clandestine or underground organisations in most places. In one way or another they did things that constituted resistance to oppression. Belonging to such resistance groups probably had its greatest effect in giving members some sense of purpose and an alternative to pure subjugation. There were instances of enormous courage in uprisings, escapes and sabotage. Many forced labourers, realising that their work helped the Nazi war effort and could prolong the war, engaged in courageous sabotage efforts (Dunin-Wasowicz, 1984, 140).

These acts of resistance were of great importance for the self-respect of victims. They remain so in Holocaust remembrance of their descendants, for Jews and for any who champion resistance, however constrained, to fascism and racism. Sometimes, there was direct harm to and temporary difficulties for the oppressors. Rarely, however, did they save lives, at least directly. More often they brought waves of indiscriminate reprisals, a round of executions and accelerated extermination.

That so little was done directly to stop the perpetrators, or that affected the processes of the Holocaust, hardly appeals to the preoccupations of disaster studies or the modern literature of 'social action'. In the terms used here, these look mainly to actions that target and *successfully* stop or modify a source of danger.

Meanwhile, the picture for vulnerability reduction was even grimmer. There was some slowing, cushioning or reducing of the impacts of *ever-increasing*

341

vulnerability. We can find virtually no improvements of strength of protection for the victims, unless very temporary ones.

The only effective method for saving many lives was flight and exile – completely avoiding the danger by leaving the regions where it applied. Hiding, either in rare refuges provided by courageous neighbours, in remote areas or disguised as gentiles, preserved some thousands of lives in total. It was a scattered and very limited option overall. More may have been lost than saved in this way, by being discovered, informed on, or through bad luck.

Indeed, this is a catastrophe, not merely a set of risks, in which other events, and actions directed at other objectives, alone brought real relief and an end to the destruction. The Nazi genocide was stopped and the vulnerability of surviving victims improved only by indirect, although totally effective, means – the defeat of the German armies and the termination of the Hitler era.

But what of the vast majority of those who did not escape and became captive populations under Nazism? After all, most survived months, years or as much as a decade before being killed or rescued. And they did not survive only because the Nazis allowed them to. However, their responses were largely a combination of coping, and finding ways to bear the hurt and loss. Nearly everyone could do *something* in these ways to reduce suffering and save lives, and to find some respite even if, in most cases, it was only temporary. That is what needs to be emphasised here and as a general focus of concern with what oppressed and captive peoples do to survive.

Suggested reading

Bauer, Y. (1989) Jewish resistance and passivity in the face of the Holocaust', chapter 12 in Furet (ed.) (1989) 235–51.

Marrus, M. R. (1989) *The Holocaust in history*. Penguin Books, New York, chapter 7, Jewish Resistance.

Coping *in extremis*

Professional disaster studies have not been very concerned with how ordinary victims cope with disaster. There has been little effort to examine or appreciate the sorts of actions they can take, yet these may be fundamental to survival in extreme situations, and were so almost throughout the Holocaust years.

In the camps, survival might turn on an extra crust of bread, not getting sick, or not showing it when you were. It came down to dealing with lice or blistered feet, winning the favours of a vicious camp orderly; or a mental trick to get you through hours standing in bitter cold for the punitive roll calls. Women had to hide menstruation or overcome added depression after separation from children; and to find ways to avoid sexual assaults by guards and prisoners put in charge, unwanted pregnancy, or stress- and privation-related amenorrhoea.

Table 12.3   Adjustments to the Holocaust. A list of the ways in which Jews or
others could confront the hazards or processes and an estimation of how
important or effective they were in relieving suffering and saving victims'
lives

Hazards reduction (i.e. confronting and mitigating the processes of the Holocaust)	Degree of assistance
External actions	
boycott of German products (1935–1938)	minor
international pressure	minor
exiled Jews joining Allied forces	minor
Red Cross enquiries about camps	minor
Internal actions	
legal, civil, peaceful challenges	'counterproductive'
partisan activity and uprisings	small
sabotage	minor
escapes from captivity	minor
undercover, covert resistance record keeping, etc.	important
Avoidance	
mass exodus	major (life-saving)
hiding, physically or with false identity	modest
Vulnerability reduction	little or none
Coping	
interpersonal support among families, friends, inmates, etc.	major
economic self-defence (trafficking, barter, bribery, theft, begging, scavenging)	major
domestic and community self-defence (initial and in ghettos)	major
cultural and spiritual resistance	major
exploitation and preying on other prisoners	significant
Loss bearing	(always, but adding to vulnerability)
Emergency assistance	
international community	negligible
refugees	significant at times
surrounding populations	variable but minor
Allied military forces (prior to war's end)	little or none

For all, a serious daily crisis attended control of when to relieve yourself, being
ready at times and places allowed, then doing so quickly.

In the transports, survival could depend upon the presence of mind not to
panic; little tricks for staying warm or sleeping on your feet; catching a little
rain water or breaking off icicles to assuage thirst; or working out a system to
deal with excrement. These small acts might save lives or conserve strength for
later ordeals.

For many in the ghettos, each day was an increasingly unequal search for food, shelter and medical care for the sick. The margin of survival for self or family might come from begging, scavenging, theft, prostitution or bribing a guard. It could depend on readiness to do the dirtiest job or deploy and spread family energies effectively. In the ghettos, too, there developed a large-scale 'hidden economy': clandestine trading, barter and manufacturing that added enormously to the resources and survival of the captives. According to one assessment, the illegal trade and trafficking in and out of the Warsaw ghetto reached as much as 40 times that officially allowed. Essential help came from those with the courage and skill to slip over the ghetto wall at night and go dealing, scavenging or thieving in the rest of the city. Hundreds of ghetto children learned these dangerous skills and so brought in means to save many lives. 'The German plan to starve the Jews to death quickly, was foiled' (Hilberg *et al.* (eds) 1982, 13). This was probably the single most elaborate and effective example of organised resistance to Nazi designs, although it lasted only until the annihilation of the ghettos.

In the world of the ghettos and camps, survival depended upon something else we rarely confront. Because the regimen was intended to decimate those who were not killed directly, no one could survive on their exact portion of the total life-support available:

> The conditions of extreme overcrowding, intimidation, hunger etc., were responsible for a situation in which inmates could survive only at the expense of their fellow prisoners.
>
> Madacyzk (1985, 65)

It seems that Jewish committees or agents in charge of ghetto administration could not have distributed food equally to all. For then, at best all would have succumbed quickly to totally insufficient rations. In this, as well as in their role in selecting people for deportation to the camps, some folk were sacrificed to save, or improve the life chances of, others. It was, perhaps, a form of what, in assignments for treatment of battlefield casualties, is called 'triage' – saving those most likely to survive and abandoning those who are not. Some felt this encouraged corruption and favoured those in authority or in the good graces of the committees. More recently, the appalling responsibilities of the Jewish leaders, and a sense that many acted in good faith, if not always realistically, has gained ground. For a time they held the line and saved lives. Many of them protected weaker members, resistance and cultural activities. At worst, they were far less dangerous administrators than the Nazis (Marrus, 1987).

From the camps there are stories of secret killings for a loaf of bread, agonised recollections of those whose hunger drove them to steal from a fellow prisoner who later died. There is what Primo Levi (1988) called 'The Grey Zone'. To survive, considerable numbers of prisoners helped the Nazis by their labour and by carrying out most tasks that kept the camps running. They were rewarded for having disciplined, spied on and even punished fellow prisoners. Cooperative prisoners could obtain privileges, or extra rations, from aiding and doing the dirty work of the perpetrators. Strong ones might prey upon weaker prisoners. Some were made essential 'operatives' in the killing processes – usually to gain just a few days or weeks longer for themselves before being exterminated.

In some ways, the cynical enforcing of prisoner complicity, or what Lifton and Markusen (1990, 235) call 'coerced collusion', was the worst of Nazi crimes. It made the price of survival alienation from the most basic forms of self-respect and identity with their own people. However, there were also those who used their privileges to help others secretly, or to carry out resistance activities. But the regime of the concentration camps was intended to destroy the possibility of 'solidarity' among prisoners: any capacity to intercede for, or cushion the blows, for another prisoner. The SS pursued its goal of removing every vestige of humanity and moral possibility from the camps with great determination. At best one can say that this provided 'negative proof' of the fundamental importance that decency and sharing do have in the survival capacities of any human group. The extreme plight of the Holocaust victims turned upon the progressive destruction of all such possibilities among families, communities and any other support groups.

These things said, we must respect and emphasise the much larger evidence of victims for whom personal survival was not the most important goal. The survival of those they cared for, of their family or group, often came before that of themselves. The woman whose testimony was quoted on page 337 confronts us with the most final form of this basic dilemma. Up to that fateful day when armed guards forced them to run into the snow-filled woods, she, her sister and her mother had managed to stay together. This was, from July 1942 until January 1945, in a ghetto, a series of forced labour camps and deportations. Staying together was surely a major factor in their ability to survive until that morning. Then, however, as they were running through the snow and the shooting began, the mother broke away. She ran back to the guards begging them not to kill her daughters. They shot her. The daughters did get away and were eventually rescued. The witness does not actually say her mother's sacrifice saved their lives, but that she tried, and at such a cost to herself, tells us something about humanity in danger that goes beyond mere survival. And response can only ever be proportional to the means available, and the things that matter most.

This woman was one, we know, of countless victims who did the only thing left to save their loved ones and, indeed, their self-respect – sacrifice themselves. There were also religious Jews, Jehovah's Witnesses and others for whom refusal to compromise on religious observance and principles came before, but sometimes aided in, mere survival. Indications are that caring for, and sharing with, others at the simplest level was more often the basis of personal survival than a ruthlessly selfish approach. However, the Nazis tried to make sure that punishment or death was the most likely reward for anyone who tried to protect or care for the lives of others. They made no

345

objection when Jewish leaders volunteered to go with the deportations to the death camps, although they could have stayed. At the 'selections', mothers and siblings who wished it were often allowed to go to the gas chambers with their loved ones.

As a whole, the chaos that they faced in everyday life created a 'protean' situation for the victims. They had to be unusually, but secretively, alert and cunning: flexible as well as tough. Where they were so, they developed some remarkably creative and ingenious survival strategies. In the long run, perhaps, these saved a very few, only staving off the moment when victims were overwhelmed by the multiplicity of dangers, or chosen for death. And yet those who did survive did so mainly because of others who had not given up as soon as they might have, keeping the Nazis busy that little bit longer than expected. Ultimately, in this process of attrition and annihilation of populations by overwhelming force, the end of the world war was the only real hope. *But any who prolonged their lives against the odds also gave a tiny measure of life to those who did survive at the war's end.* This is something we rarely confront, but to ignore it and the efforts of those who did not survive demeans the most critical survival capabilities *in extremis*. It even pronounces the oppressors victorious, in every sense, over all who did not survive.

For such reasons too, coming through the Holocaust has been a deeply ambivalent and disturbing fact for survivors. Few have felt they were more deserving of life, or that they survived mainly from their own abilities rather than extraordinary luck and some crucial bits of help from others who did not live. There were always so many others whom they cared for, whose qualities they admired or who seemed better equipped for survival, but who died needlessly. At best, they have felt 'chosen' to bear witness to the qualities and sufferings of those who did not survive.

To recognise these predicaments is not to exonerate the perpetrators, exploiters and indifferent bystanders. To speak to the extraordinary, if so often doomed, capacities of the victims to cope with such extreme and degrading distress is not to justify their predicament. We can say only that the initiative and strength were not all on the side of the Nazis, passivity on the other.

And what of the Nazis' 'success' in killing so many, now that they and their vile plans are dust and ashes too? By comparison, the achievements and remembrance of the victims survives to humble and amaze us. They give hope that if such brutality is prevented, or at least mitigated, the human capacity to struggle for the right to live is boundless. The social Darwinism of the architects of genocide was an entirely misguided as well as a criminal stratagem.

Suggested reading

Trunk, I. (1979) *Jewish responses to Nazi persecution: collective and individual behavior in extremis.* Stein and Day, New York.

The lessons versus the uniqueness

Another reason for considering the Holocaust in an overview of risk and disaster in the twentieth century involves the dangers it reveals, and the further possibilities of similar calamity under modern conditions. But this is a complicated and potentially contentious argument. Of course, the Holocaust was, and remains, a pivot of human rights legislation and international declarations in the second half of this century. It continues to inspire the vilification and criminalising of acts elsewhere that are exemplified by the Nazi persecutions and exterminations, from hate literature in North America to genocide in Southeast Asia. But the fear that 'interpretation', in a wider sense, can diminish the remembrance of the victims, blur the sheer evil of the event, even lead to 'exploitation' of the memory and the terror, is justified. These are not trivial issues here if, as I and others believe, this was as much a calamity rooted in moral, intellectual, public and civil hazards as in techniques of violence.

Then again, to see only the uniqueness may endanger the memory and the meaning too, if it does not mislead us as to the basis of the calamity itself. If it was unique, what can we learn from it except pity and blame? May that not play into the hands of the perpetrators, who were very concerned and clever about segregating, distancing, obfuscating and making a 'special project' of these developments and the Final Solution? For them, the standards of conduct and treatment that applied to their own and, indeed, to much of the rest of the world were not applicable here. The Jews were a 'special case' – indeed every racial group was. The standards applying to one were not seen as relevant to others. The Nazis wanted those with doubts to believe that their critics and any they could not simply dispose of were somehow 'different'. Nazi leaders cast themselves and their killing squads in an especially 'heroic' role, doing something to which the usual values and practices of German life had no bearing, except as being opposite.

The Holocaust is widely seen as both a calamity of the Jews of Europe *and* a calamity of, and for, modernity. It showed how the organisation and 'rationality' of modern states and institutions may be used for, and can be singularly effective in, destroying human life. It was a disaster for others to the extent that we are all diminished and threatened when any people are torn from their homes and watch their children violated and killed without legal recourse or proper protection. In extreme form, the Holocaust confronts all modern states with the prospect of what can happen when civil rights, standards of conduct, social justice and rule of law are sacrificed to the desires of the government in power. We cannot even appreciate the uniqueness of the calamity if we think of the Jewish and Romany peoples as only 'different' and 'special', for that too is to play into the hands of the perpetrators. *Either* the protection against the potential violence of the state or other powerful groups applies to all citizens and, hopefully, all of humanity, *or* 'exceptions' can be made for any group that power and prejudice single out as targets. The Holocaust stands apart as a catastrophe of genocide for particular peoples, but it also destroyed the bases of public life in modern society and trust in civil security for all. The Nazis and their collaborators turned a blind eye to or reversed whatever had been termed progressive or enlightened in the dominant culture of our times, that of Europe. And finally, without making odious comparisons in terms of numbers or severity etc., we must recognise that many aspects of the Holocaust speak to some of the gravest dangers and worst calamities around the world to the present day, where states use violence against citizens and target minorities for ruthless treatment.

The heavy silence on this topic almost throughout the literature of modern geography is more than just an abdication of responsibility. It not only risks aligning us with all those who turned a blind eye to violent social engineering and criminal acts between 1933 and 1945. It also ignores or diminishes the meaning of our field. At least, it was intended to show here, in part, that the Holocaust has a special significance for geographers, and it teaches lessons about 'geography that matters' and the nature of violent geographies.

Suggested reading

Marrus, M. R. (1987) *The Holocaust in history*. Penguin, New York.
Bauman, Z. (1989) *Modernity and the Holocaust*. Cornell University Press, Ithaca, NY.
Lifton, R. J. and E. Markusen (1990) *The genocidal mentality: Nazi Holocaust and nuclear threat*. Basic Books, New York.

Concluding remarks: the perspective of ideas

... the complexity of social life cannot be adequately addressed by one perspective (p.53) ... a novel and integrative framework is necessary to capture the full extent of the social experience of risk and to study the dynamic processing of risks by the various participants in a pluralistic society. Such a novel approach cannot and should not replace existing perspectives, but should instead offer a meta-perspective that assigns to each perspective an appropriate place and function.

Ortwin Renn (1992, 78–9)

The case histories and geographies presented above involve a large geographical canvas. The last two may well seem to far exceed the scope or possible competence of a field dedicated to alleviating risks and disaster. In one sense, that is a fair judgement. They are historical events that lie well beyond any ordinary response and, for that reason perhaps, their literatures have taken on a life of their own. Yet if that is so, it should be seen to be so. As well as a remembrance of the victims, there is a need to recall what happens when neither good government nor social understanding can contain the calamitous potential of modern states and enterprises. If the main goal is to find ways to reduce disaster, the implications of failure should also inspire the effort rather than be ignored.

However, even these greatest disasters of the century did not lie in totally different realms from the other dangers that we have looked at. They flag the places where the capacity for destruction was paramount. Such efforts as had been made to create a secure environment for ordinary and defenceless people were fundamentally flawed. Nevertheless, there too we saw how the erosion of common principles and routine safety measures, of public security and the rights of citizens, were like the warning rattle of small stones that precede the great rockslide. There was no lack of people who saw as much at the time, but having been ignored, they were also buried in the avalanche.

349

Moreover, a range of contemporary humanitarian catastrophes, or potential ones, are not so far removed from these 'worst case' examples, in origin or consequences. They are in other places, perhaps. They may not involve genocide or city-wrecking on so great a scale, but the uniquely destructive capacities of modern industrial states have been all too evident in Vietnam and Afghanistan, and for the post-Cold War world, in Kuwait and Iraq in 1991. The genocide in Rwanda, war and famine in Somalia and AIDS in Uganda, present wholesale catastrophes for peoples and worlds. And these are events in which the 'disasters community', or parts of it, have been closely involved.

Then again, it would be surprising if we cannot, at least, learn from the worst disasters of the age. For this it is important to recognise the value of another side of modern disaster studies. Recently, our field has tended to concentrate wholly upon applying a particular expertise, or responding technically and pragmatically, to recurring emergencies – surely an essential part of this work. But there are important insights developed only from a larger historical canvas, or as comprehensive enquiries into calamitous events and the places where they have occurred. This is exemplified, in their different ways, by the work of P. Sorokin, S. J. Prince, Lewis F. Richardson, B. Woodham-Smith, M. Barkun, Susan Sontag, R. J. Lifton, Susan George, D. Craig and M. Egan, Kai Erikson, Elaine Scarry, and Richard J. Evans (see Bibliography). Their approaches require as full and thorough an immersion in the phenomena of misfortune, or conditions and content of events, as the evidence, methods of representing it and mental capacity allow. It is the search for interpretations of events that are delineated historically and geographically, rather than by particular disciplines or techniques and specific managerial responsibilities.

Such work suggests that success or failure in seeking to mitigate great risks is not just a question of 'know-how' or investment of resources. It is also about the adequacy of our reading of events, our ideas about calamity, and the values that guide our efforts. If so, then *the perspective of ideas* needs to be added to those of hazards, vulnerability, context and action. Indeed, it should continually inform and critique the other perspectives. This will be addressed by way of a conclusion here. However, as noted at the beginning of the book, we are in a period of debate, if not conflict, over existing notions and over what the future of the field can or should be. Apart from the difficulties of painful, out-of-control, out-of-place events, disasters prove to be one of the great, intractable problems of modernity. And it is necessary to be aware of how and why the ideas that have prevailed in recent decades are increasingly being contested.

Suggested reading

Lifton, R. J. (1970) *History and human survival*. Vintage Books, New York.
Kirby, A. (ed.) (1990) *Nothing to fear: risks and hazards in American society*. University of Arizona Press, Tuscon.

The dominant view and its limitations

> Focusing on one facet of the historical process, it draws an arbitrary dividing line between norm and abnormality... It diverts attention from the permanence of the alternative, destructive potential of the civilising process, it effectively silences and marginalizes the critics who insist on the double-sidedness of modern social arrangements.
>
> Zygmunt Bauman (1989, 28)

Over the past several decades, a certain approach to risk and disasters has prevailed, at least in the most influential institutions and disciplines contributing to this field. I described this as 'the dominant view' some time ago (Hewitt, 1983, 6). It remains predominant today, although alternatives are having more influence and are guiding the work of increasing numbers of workers and institutions.

The type of work that receives the greatest funding and figures prominently in professional activities rests upon the hazards paradigm. As we saw, that involves research directed at particular, usually extreme, geophysical, technological and social agents, or the conditions they involve. Technical and official responses are made largely in terms of these agents.

In addition, the dominant view is associated with several other general constructs:

- disasters are treated as the result of forces external to, or unplanned failures within, society. The disaster is then seen as an 'accident' or, as it is often called, an 'unscheduled' or unplanned event;
- not merely the experience, but the *explanation*, of disaster is contrasted with 'normal' or everyday conditions. The latter are taken as planned, productive and orderly, so that disaster becomes essentially an out-of-control condition that arises from an extreme and chaotic environment, or unplanned, disorderly social forces. Risk is then reduced to questions of how, when and where the damaging agent or unplanned failure may occur;
- the geography of disaster is also reduced to that of the damaging agents, while the places of crisis or damage are treated as islands or disorder where other, abnormal, rules apply. Hence, the space of disaster appears to arise, as it were, 'vertically' from the immediate association of inhabitants and an extreme event in their environment or places of work. Conversely, the important responses are expected to be 'horizontal', from places where life has supposedly remained normal, or where there is more wealth, knowledge and skills;
- the desired, productive 'normality' is expected to be restored to the disaster zone from outside by an injection of unimpaired order and superior technical expertise;
- disaster reduction is thought to depend upon extraordinary actions directed at 'the hazard' and extreme, emergency conditions, or at people's exposure to them, using the latest scientific, technical and institutional means.

351

Concluding remarks

In this way, the dominant view divides disaster and the risks it represents from on-going or peaceful human geography. And it has another, distinctive geographical consequence – or, perhaps, source.

This view assumes and adopts a vision of risk and disaster from 'the centre', from outside and in effect 'from above'. Risk is reduced to impersonal factors that can be measured in standardised ways, or seen remotely and by (expert and official) strangers. This situates the problem as a governmental or professional task, requiring a technocratic approach. In that regard, technical and official approaches are remarkably similar in governments of all political stripes and geographical areas, and in many other modern institutions. Of course, this tends to be identified, misleadingly I believe, with the requirements of an objective, more scientific view as well.

Suggested reading

Hewitt, K. (ed.) (1983) *Interpretations of calamity: from the viewpoint of human ecology.* Allen and Unwin, London.

Arney, W. R. (1986) *Experts in the age of systems.* University of New Mexico Press, Albuquerque.

Hewitt, K. (1995) Excluded perspectives in the social construction of disaster. *International Journal of Mass Emergencies and Disasters*, **13** (3), 317–39.

Dominating knowledge and its limitations

When the preoccupation of an approach is with impersonal measures, how can it be sensitive to, or aware of, interpersonal knowledge and concerns? If it emphasises modern techniques and formal organisation, how will it deal with the practicalities of people who are unfamiliar with those techniques, not least in the many places that are outside and not part of modern organisations? How can the viewpoint of 'the centre' – whether urban, governmental, organisational or scientific – respond sensitively to the circumstances and needs of 'localities', communities, places and other cultures? If effective responses, training and 'solutions' are assumed to depend upon formal, 'advanced' expertise, how does the expert come to know about and cope with the endless varieties of events, conditions, communities and places in the real world? How can the techniques we are to give training in, or the technologies to be transferred, be knowingly given? According to communications theory, a 'message' is only as good as the ability to receive and decipher it, let alone respond appropriately.

If it is not apparent that these are questions of foremost importance in the field of public risks, this book has failed in its argument. They are questions posed, however, in the search for answers to a disturbing predicament. They do not necessarily imply there is no place for existing professional and technical knowledge, or a role for government and other centralised organisations. Indeed, the questions arise, for most of us, out of the expectations raised by modern arrangements, not only their failures.

In the dominant view, however, learning from other people's experience and knowledge is almost an admission of defeat. Its definition of an expert is someone who knows better, more rationally and accurately than a 'lay person'. A sad consequence is that so much professional and technical work comes across less as scientific than as *dominating* knowledge and practice. Even at its most generous, it cannot help perpetuating a paternalistic or colonial mentality with respect to the general public and, especially, places and cultures unusually vulnerable to disaster. However well-intentioned, it is predicated on more than the idea of scientific method or specific responsibility and mandate to pursue particular questions. It assumes superior capabilities, and its practitioners are not to be swayed from their path or interfered with in their mission by other practices or ordinary citizens.

There is a dismaying similarity between the approaches to disaster of the dominant view and the one Zygmunt Bauman sees as failing to grasp the roots of the Holocaust. The latter he found to lie not only or mainly in the work of crooks and madmen, or the return to barbarism. Those were necessary agents. 'Madness' may well describe the results, but it was wholly insufficient to *the task*. That required, as we saw, a deadly co-opting of rational, modern instruments. It required all the apparatus of the modern state. More disturbing, however, is the way these proved so very appropriate and effective in carrying out crimes against humanity. Fashionable scientific ideas, the latest engineering and industrial techniques, were handily deployed. Some of the 'best minds' assisted very professionally in race-murder. The way strategic air power developed and deployed the latest technologies and the best-trained forces to destroy cities and defenceless civilians has led others to draw similar conclusions about that (see Snow, 1961; Lifton and Markusen, 1990).

Of course, I do not suggest that the main, contemporary arrangements for responding to disasters are like, or are ever intended to be like, those by which the Nazis destroyed the Jews of Europe or strategic air power wrecked the cities of Japan. However, some of the 'best minds' who *opposed* or who have re-examined those developments believe the great lessons here relate to the dangerous potential of technocratic modernism. They identify such extreme dangers with the limitations of scientific and centralised organisations to avoid disaster and their unique capacities to bring it about. Experts and institutions in this field are often not very vigilant about, if they even recognise, such concerns. It is not uncommon for them to be contemptuous of the value of strong ethical constraints and public accountability.

Suggested reading

Marglin, S. A. and J. Marglin (eds) (1990) *Dominating knowledge: development culture and resistance*. Clarendon Press, Oxford.
Bauman, Z. (1989) *Modernity and the Holocaust*. Cornell University Press, Ithaca, NY.

353

Now, however, let us look at possible alternatives to the dominant view. These are ideas that seem to accord better with evidence of risk and disasters discussed above. They offer ways of overcoming the tendency of all specialised and privileged knowledge to dominate rather than to serve the interests of society at large.

As indicated by the chapter head quotation from Renn, this need not involve rejecting any of the phenomena considered in, or disciplines contributing to, existing work. That would seem to me to throw out the baby with the bath water. At the very least, established expertise is essential to coping with the many risky technologies and arrangements constructed around and dependent upon modern, technocratic management.

Nevertheless, the alternatives suggested are not simply matters of adding information or compromises. As in a new painting, we include observations, subjects, motifs and tools used before and often. That does not mean the work is composed on the same principles, or gives the same result. In fact, we can best present the alternatives as counter-arguments to those of the dominant view, as set out above.

Human ecological alternatives

> Risk, enterprise, progress and modernity are genealogically *interdependent social ideas*.
>
> Colin Gordon (1991, 39, my italics)

A human ecological understanding of risk and disaster may not ignore hazardous agents, or the best understanding of them. Yet it does not look for explanations mainly in terms of them. Rather, our investigations have highlighted:

- the extent to which the degrees and forms of risk, and the occurrence of disasters, are poorly explained by, and not uniquely dependent upon, the nature of the damaging agents that may trigger them. The same damaging processes bring widely varying losses, or none at all. As described throughout this book, damage in disaster reflects, most often, the distribution of human vulnerability, the roles of intervening conditions and response capacities.
- the extent to which risk and damaging events do not depend upon human actions towards or perceptions of hazardous agents, including their nature and frequency, or past experience of them. This is not necessarily a conclusion reached by asking people, although studies of how people perceive hazards and risks often reveal rather hazy, unreliable or stereotyped views. Their exposure to and awareness of particular hazards commonly depend upon other pressures, concerns and goals, including the range and relative severity of all the risks to which they are exposed.

- the extent to which the damaging events, and their causes, internal features and consequences are not explained by conditions and behaviours unique to crises or disasters. How communities respond is best explained from an understanding of pre-existing conditions and values in the society – albeit in efforts to cope with unprecedented devastation and loss. They relate as much to what the permanent arrangements were like as to how much of them are destroyed.
- the capacity to bring about risk reduction and disaster mitigation depends mainly upon improvements in pre-disaster conditions, above all in the security, well-being and rights of the more vulnerable parts of a population.

We may briefly cite the case of famine and the role of agricultural development in food security as providing convincing illustration of these notions. Modern famines have not been simply the result of short-term crises brought on by, say, drought or crop blight. They are not primarily due to a failure to increase food production as fast as persons to be fed. They have not been due to absolute food shortage in the economic sphere to which the starving belong. Rather, *the already hungry* are generally the main casualties of famine. And they starve for the same basic reasons that made them hungry before. Social and economic conditions prevent their gaining access to available food. It has been concluded that in India, at least since 1850, 'famine ceased to be a natural calamity and was transformed into a social problem of poverty and dearth' (Bhatia, 1991, 2).

In many cases, social changes related to, of all things, 'development' have been shown to be integral to the problem of greater famine exposure, rather than its solution. Certain African famines were identified directly as 'the cost of development', the role of droughts being made far worse by 'planning', inappropriate technology and administrative weaknesses (Baker, 1974). Again, famines and related uprooting are usually exaggerated, if not triggered, by social upheavals and armed conflicts that surround struggles to 'modernise'.

Yet famines are disasters typically identified with natural hazards or impersonal forces of sociobiology, notably 'population'. This also accords with the interpretation of hazards in the dominant view. Alternatives, however, direct us to other conditions. In fact, we focus on those that humans, or some of them, have a greater capacity to influence, for better or worse.

To summarise, instead of a separate problem, disaster is seen here as an integral part of everyday or 'ordinary' life. Rather than 'islands' where extreme events occur, the geography of disasters reveals where vulnerable people live. It is, most often, a geography singling out the disadvantaged in society, the poorly governed or abused places in its spatial fabric. Indeed, the earthquakes, mining accidents and armed actions we consider are most often *characteristic* and recurring problems of the places or parts of society they endanger. Rather than highly uncertain or 'unscheduled', they are to be expected there. Large earthquakes in southern California, storm surges in coastal Bangladesh, violent repression in Andean South America or giant oil spills in the English Channel are recurrent, historically well-known and entirely likely events. There is no excuse not to act to reduce their impact now, and to prepare against their certain recurrence.

However, in the modern world, it is not so much the incidence of such events that defines the seriousness of disaster problems. Recent disasters invariably relate to or are brought about by immediately preceding, often drastic changes in society, economy, technology and habitat. If anything, the forms and pace of change in everyday life appear 'abnormal' and unpredictable. But these are orchestrated changes, if not planned and defended as rational or necessary by certain human actors. This applies as much to social and environmental change in Japan and California as in East Africa, Central America or South Asia. Of course, the specific changes or consequences differ substantially in different places.

Suggested reading

Morren, G. (1983) A general approach to the identification of hazards. Chapter 15, 284–297 in Hewitt (ed.) (1983).

Bohle H.-G. (ed.) (1993) *Worlds of pain and hunger: geographical perspectives on disaster vulnerability and food security*. Freiburg Studies in Development Geography no. 5, Verlag Breitenbach, Saarbrücken.

Against grand theories and determinism

As noted earlier, the dominant view goes hand in hand with a preference for explanation by impersonal forces or statistically significant associations. It requires standardised data and procedures. Abstract principles, and global or generalised models and conclusions are favoured. To the extent that societal conditions are invoked, it will be in terms of universal parameters from, say, demographics: measures of economic performance, average life-expectancy, years of education, or distribution of income and of specific technologies and skills. The presence or absence of certain institutions and rights may be added. Such mass measures are generally attached to nation states as the basic units of assessment. They may well be useful in bracketing or informing public awareness and government policy. However, they do not only start from and favour the preoccupations of Western states and powerful institutions. They also tend to be used as evidence for much less rigorous notions.

The measures themselves are, at best, indicators rather than actual sources or causes of problems. Yet the latter tend to be defined in terms of, and even blamed on, the statistics. Thus, the places of danger and misfortune are often labelled and defined through an implied 'excess' or dearth in the relevant parameter – *under-*, *over-*, or *un-* something. Common examples are the 'neo-Malthusian' ones of overpopulation and resource depletion. There are the (liberal) economic identifiers of 'underdevelopment', marginal or peripheral countries to which modernity has yet to diffuse. There are ecological categories of unsustainability, instability and fragility. These are notions arrived at by ignoring the way poverty, overcrowding, social violence, natural disaster, famine and epidemic may reflect deliberate actions, if not necessarily intended effects of human activities. Specifically they ignore the roles of exploitation, unjust laws, the ruthless pursuit of powerful interests, or developments undertaken with little regard to consequences for the general public. Yet these generate or redistribute dangers and undermine existing safety arrangements.

The Cold War and post-1950 regime of 'economic development' made it fashionable to treat profiles of such measures as defining 'worlds': 'First' or 'Third', 'North' or 'South'. Moreover, these have been assumed to represent necessary stages in the development of material life, as opposed to contemporary realms whose condition is related to how they interact in the present, and in a particular configuration of national, regional and global economies. Within such a dominant paradigm, the severity if not the incidence of disasters, among other problems, is strongly associated with the different ideologies or stages of development, as defined by economic and social indicators. Burton *et al.* (1978), for example, described the geography hazards and disasters in terms of a spectrum of states in differing stages of material life, from 'folk' through industrial to 'post-industrial'.

It is not necessary here to go into the ins and outs of these deterministic theories. Rather, the real debate is between any impersonal, deterministic thesis and the notion that societies, to a greater or lesser extent, either make themselves or are forced along certain paths by other, stronger societies.

Are we to view events and changes as, essentially, controlled by impersonal forces, whether of 'nature' or 'nurture', geography or history, and as expressed in mass statistical measures? Or are they substantially conditioned and constrained by persons and institutions? Do human groups shape and perpetuate their own destinies, albeit in accordance with their cultural values and relative powers? Clearly, this would implicate human decisions and priorities directly in the form a society has, and the way risks are allocated. It is to question the idea that humans can, at best, play the hand they have been dealt. Developments, and the situation of different peoples in them, are not explained by the impersonal logic of economic, environmental, technological or biosocial processes. Humans are seen as intelligent actors who shape and modify the world according to shared goals and, especially, the extent of their powers.

Much of the work on hazards and disasters within the dominant view fails to provide a sense of the highly complex social production of material life (Watts 1983, 235). Yet social life and its problems largely reflect human efforts that create and reproduce social and material forms. This is not to say that there are not other constraints. It is to stress the importance of humanly devised arrangements, and the constraints they impose upon those who participate in, and are affected by, those arrangements. Natural environments or human biological inheritance do play a large part. Impersonal, mechanical or statistical models may show that even small tendencies, quantitative balances or types of advantage eventually become dominant, but intelligent, responsive life can offset and alter the meaning or direction of any of these constraints and contexts – if life is possible at all. Rather than inevitability and determinism, there is a sense of intelligent participation in, or 'negotiation' with, nature, technological innovations and other societies. History and geography are restored to a central place in our interpretations. They emerge as theatres of creative human adaptations to habitat and biology, rather than incidental or superficial concerns.

The dominant view does not, of course, deny human innovativeness or the usefulness of ideas, plans and management responses, but it does see the only effective and active possibilities as those imposed upon the public and events. Its faith is put in centralised administration, backed by expert knowledge and technical practice that 'understands' the impersonal forces. The alternatives described also lead us to question that. One doubts that the best or most appropriate responses to dangers are always, or even usually, from and using the powers of the more powerful institutions or 'advanced' sciences. Rather, reasonable and effective response will always incorporate understanding of the conditions of those at risk, or otherwise directly involved. It will require their positive and intelligent participation. People's interpersonal relations, material activities, cultural values and preoccupations must be taken into account for practical as well as ethical reasons. They cannot simply be relegated to the dustbin of history because they do not conform to the latest ideas or fashions in the centres of wealth and power, but their lives and rights may well depend upon the restraint and real motivations of the latter.

Of course, in many political and economic contexts the majority of humans are denied, or prevented from exercising, their capacities. Very often, what appears as a determinism of nature, population or economic attainment turns out to be imposed by human agency. This deliberately suppresses creativity and freedom to respond in accordance with need, if it does not aid in the impoverishment and vulnerability of certain groups. One thinks of how Europeans justified their subordination of native American peoples by seeing the latter as, somehow, prisoners of their environment and race, of heathen and backward ways. This proved to be a self-fulfilling prophecy for subject and impoverished peoples. The analysis generally failed to mention the huge exercise of European arms and punitive or unfavourable trading and employment policies, not to mention enslavement and diseases.

If one accepts the dominant view, then very few human individuals, groups, institutions or nations have the capacity to understand their predicament and to act knowingly to improve it. Risk and disaster become problems *for* society but not *of* it. The alternative interpretations suggested above would, however, situate society and social action at the heart of risk and disaster. The terms in which social action is evaluated are different. The importance of intelligent participation and choices turns less upon an abstract set of options, more upon the available options and power to choose. At least, people would have to have access to the information and resources required to implement a range of adjustments. Instead, we see that more and more of those harmed in disasters have anything but adequate information, resources or options. Often they are second- or third-stage victims of economic disadvantage, uprooting, chaotic urbanisation and political change. For them:

> ... preferences one way or another reveal little or nothing. Much more important than the choices people make are the constraints under which they choose.
>
> S. A. Marglin (1990, 4)

As is increasingly said in critiques of development, real improvements in disaster prevention and mitigation will come about only if they originate in, or are in step with, political improvements and social justice. The more vulnerable only have a real chance to become less vulnerable through greater empowerment. They must have a greater say in matters that affect their lives, and control, especially, over decisions that may endanger them. Those with greater wealth and resources, who say they wish to see risks reduced, should assist this in whatever way they can. In the end, only persons, as opposed to technical abstractions such as population or classes, can improve their situation. And to do so they need a meaningful degree of choice and control over their own destiny.

Any change in risks or disaster response implies, intended or not, social and political change. Therefore, any reliable improvement in public safety comes from social and political improvement. Ours is not merely a field in which humanitarian *assistance* is required. The problems of disaster and its alleviation *are* humanitarian rather than merely technical. They are about decency in public life and social justice before they are about expertise.

Suggested reading

Burton, I. *et al.* (1978) *The environment as hazard.* Oxford University Press, New York.
Cuny, F. C. (1983) *Disasters and development.* Oxford University Press, Oxford.
George, E. (1984) *Ill fares the land: essays on food, hunger and power.* Penguin Books, New York.

Socially constructed risks and paradigms

> Cultural analysis of risk looks behind the perception of physical risks to the social norms or policies that are being attacked or defended.
>
> Steve Rayner (1992, 91)

Suppose we follow the less favoured idea that risk and disasters are primarily a reflection of the socioeconomic order, and of ecological relations of life in particular places. This is not to say that social life lacks rationality, or that natural forces and technological systems have no influence, but risks depend upon the specific constraints within the varied relations between people and places. And those relations are composed and sustained largely by human action in accordance with shared values and goals. Security is reproduced or undermined largely by human activity. Dangers, or their severity, result from the same patterns of living.

In the realm of ideas this leads to a further series of arguments differing from those of the dominant view. Risks are seen to be socially allocated, rather than outcomes of the incidence of natural extremes, or equipment or operator failures. They depend primarily on the social order, rather than climate or, say, weapons potential. They express success or failure in the shared responsibilities and expectations of public life.

Suggested reading

Maskrey, A. (1989) *Disaster mitigation: a community based approach*. Oxfam, Oxford.
Beck, U. (1992) *Risk society: towards a new modernity* (trans. M. Ritter). Sage, London.
Blaikie, P. *et al.* (1994) *At risk: natural hazards, people's vulnerability, and disasters*. Routledge, London.

Such arguments help to define or lead on to a preoccupation with the human ecology of dangers and social geography of endangerment. The risks are seen to reside in the details of human settlements, land uses, practices and ethical restraints. This does include how the most global of influences operate at the local level, but the distinctive social geographies of vulnerability or security attached to each place and society can be known or assessed only by work on the ground, where the risks apply. Cultural and social understanding is as essential as geophysical or technological. To the extent that technical or official actions are required, they need to be informed by the knowledge and cooperation of residents. What we need to know and can do turns upon how individuals and groups 'construct' and value reality. It is not only about the evidence used to assess their performance, or the techniques that may be employed to manipulate aspects of reality.

In such terms, enquiry and interpretation are processes of uncovering, improving or redefining *the nature of our knowledge*, never simply a matter of applying it. That seems a rather persuasive view for our subject. Disaster itself is, after all, a situation in which knowledge and institutions are overwhelmed by circumstances, their practices found wanting. Most of our work is about people and places brought into focus not by research or any prior understanding, but by danger and calamity.

Appendix

Major disasters 1984–1994. A list of some of the most lethal and costly disasters reported around the world. Most of the figures should be regarded as order-of-magnitude estimate only (after Facts of File, Encyclopaedia Britannica Yearbooks and UN/DHA News)

Place	Date	Lives lost	Damage costs	Initiating agent
	1984			
India	03/06	3290	?	dysentery epidemic
India, Bhopal	03/12	7000+		poison gas leak
[Worldwide mid-1984 *c.* 11.5 million registered refugees]				
	1985			
Bangladesh	25/05	*c.* 11 000	(0.25 million cattle lost)	cyclone
Mexico (esp. Mexico City)	19–20/09	7000+	US$4 billion	earthquake
Colombia	15/11	23 000	US$175 million	volcanic eruption
	1986			
USSR, Chernobyl	29/04	30 immediate ? tens of thousands, long-term	US$14 billion	nuclear plant explosion and fire
Cameroon, Lake Nyos	21/08	1700+		volcanic gas eruption
El Salvador	10/10	1000	US$1.5 billion (200 000 homeless)	earthquake
Iran	12	500	US$1.5 billion	floods
	1987			
Ecuador, NE	05–6/03	4000	(110 000 homeless)	earthquake
Bangladesh	08	1600	US$1.6 billion	floods
Philippines	20/12	3000+	?	ferry boat sinking

361

Appendix contd.

Place	Date	Lives lost	Damage costs	Initiating agent
	1988			
UK, North Sea	06/07	?	US$500 million+	oil rig explosion
Gulf of Mexico	10–17/09	1000+	US$10 billion (USA)	Hurricane Gilbert
			US$1 billion (Jamaica)	
			US$880 million (Mexico)	
			(400 000 homeless)	
Bangladesh	08/09	2000	US$1.1 billion	monsoon floods
			(25 million homeless)	
Burma/China	06/11	1000+	500 000 homeless	earthquake
India/Bangladesh	29/11	3000	200 000 homeless	cyclone
USSR, Armenia	07/12	55 000	US$11 billion (250 000 homeless)	earthquake
	1989			
Italy	Winter	–	US$1.5 billion (agriculture)	drought
		–	US$1.3 billion (ski resorts)	
USA, Alaska	24/03		(habitat, wildlife destruction)	oil spill
			c. US$4 billion (clean-up & legal)	(10 million gallons)
Ethiopia	04 (report)	42 000		meningitis epidemic
Caribbean/USA	16–22/09	100+	US$4 billion (Caribbean)	Hurricane Hugo
			US$4.2 billion (USA)	
USA, California	07/10	61	US$7 billion	earthquakes
	1990			
W. Europe	02/01	140	US$4.1 billion (UK)	storms
			$500 million (France)	
USA, Texas, Arkansas	/05	–	$588 million	floods
Iran	21/06	40 000	(500 000 homeless)	earthquake
			(105 000 injured)	

Appendix contd.

Place	Date	Lives lost	Damage costs	Initiating agent
Saudi Arabia, Mecca	02/07	1500	–	stampeded crowd
Philippines, Luzon	16/07	1621	US$500 million	typhoon
Hungary	Summer	–	US$2 billion	drought
	1991			
S. & C. America	(year)	4002	336 560 cases	cholera epidemic
Iraq/Kuwait	01	?	US$500 billion +	war, oil spills & fires
USSR	28/01		US$1 billion	explosion, petrochem.
Bangladesh	30/04	139 000	US$1.4 billion (9 million homeless)	cyclone
Afghanistan	06	5000	–	floods
USA, California	21–22/10	–	$1.2 billion	brush fires
	1992			
Turkey	13/03	4000	180 000 homeless	earthquake
Mexico, Guadalajara	22/04		US$300 million	explosion
USA, Los Angeles	04/5	–	US$775 million	riots
USA	23–26/08	34	US$15–16 billion	Hurricane Andrew
Pakistan	08–18/09	3000	US$1 billion, 3 million evacuated	floods
Indonesia, Flores	12/12	2500	210 000 homeless	earthquake
	1993			
Haiti	16/02	1750	–	ferry sinking
USA, New York	26/02	–	US$591 million	terrorist bombing
India, Nepal, Bangladesh	06–07	3000	US$12.6 billion	floods
USA, mid-West	07–08	50	US$12 billion 100 000 evacuated	floods
China, Qinghai	27/08	1250	US$27 million	dam-break flood
India	30/09	7600	US$80 million	earthquake
USA, S. California	10–11	–	US$950 million	fires

Appendix   contd.

Place	Date	Lives lost	Damage costs	Initiating agent
[Worldwide mid-year 18.2 refugees + 25 million other displaced persons]				
	1994			
USA, Northridge	17/01	56	US$17 billion	earthquake
Mozambique	late/03	240	1.5 million homeless	tropical cyclone
Rwanda	05/07	200 000–500 000 civilians	2 million refugees	civil strife
China	mid/06	1260	US$7.3 billion	cyclone and floods
Nigeria	late/09	40+	400 000 homeless	floods
Finland	28/09	900	–	storm/ferry sinking
China	8/12	300	mostly children	? fire

Bibliography

Abbott, P. L. (1996) *Natural disasters*. Wm. C. Brown, Dubuque, Iowa.

Ahlström, C. (1991) *Casualties of conflict: report for the World Campaign for the Protection of Victims of War*. Dept. of Peace and Conflict Research, Uppsala University, Sweden.

Ahmad, E., J. Dreze, J. Hills and A. Sen (eds) (1991) *Social security in developing countries*. Oxford University Press, Oxford.

Ahmar, T. (1994) AIDS – women and family, *Journal of Women's Studies* (Islamabad, Pakistan), **1** (1), 57–73.

Alabala-Bertrand, J. M. (1993) *Political economy of large natural disasters: with special reference to developing countries*. Clarendon Press, Oxford.

Alexander, D. (1985) Death and injury in earthquakes. *Disasters*, **9** (1), 57–60.

Alexander, D. E. (1989) Urban landslides. *Progress in Physical Geography*, **13**, 157–91.

Alexander, D. (1991) Applied geomorphology and the impact of natural hazards on the built environment. *Natural Hazards*, **4**, 57–80.

Alexander, D. (1993) *Natural disasters*. Chapman and Hall, New York.

Allen, N. J. R. (1987) Impact of Afghan refugees on the vegetation resources of Pakistan's Hindukush-Himalaya. *Mountain Research and Development*, **7**, 200–4.

Allen, N. J. R., G. W. Knapp and C. Stadel (eds) (1988) *Human impact on mountains*. Rowman and Littlefield, Totowa, NJ.

Alvarsson, J. (1989) *Starvation and peace or food and war? Aspects of armed conflict in the Lower Omo valley*. Uppsala Reports in Cultural Anthropology, University of Uppsala, Sweden.

Amnesty International (1975) *Report on torture*. New York.

Amnesty International (1981) *Disappearance: a workbook*. New York.

Amnesty International (1986) *Unlawful killings and torture in the Chittagong Hill Tracts*. September, London.

Amnesty International (1992) *India: torture, rape & deaths in custody*. New York.

Anderson, M. B. and P. J. Woodrow (1989) *Rising from the ashes: development strategies in times of disaster*. Westview Press, Boulder, Colo.

Arendt, H. (1958) *The human condition*. University of Chicago Press, Chicago.

Arney, W. R. (1991) *Experts in the age of systems*. University of New Mexico Press, Albuquerque.

Ashworth, G. J. (1991) *War and the city*. Routledge, London.

Asprey, R. (1975) *War in the shadows: the guerilla in history*, 2 vols, 1st edn. Doubleday, Garden City, New York.

Bachelard, G. (1964) *The poetics of space*, trans. M. Jolas. Orion, New York.
Baker, R. (1974) Famine: the cost of development? *The Ecologist*, **4** (5), 170–5.
Baker, G. W. and D. W. Chapman (eds) (1962) *Man and society in disaster*. Basic Books, New York.
Barkun, M. (1974) *Disaster and the millenium*. Yale University Press, London.
Barry, R. G. (1981) *Mountain weather and climate*. Methuen, London.
Barton, A. H. (1969) *Communities in disaster: a sociological analysis of collective stress situations*. Doubleday, New York.
Bauer, Y. (1989) Jewish resistance and passivity in the face of the Holocaust. chapter 12, 235–51 in Furet, F. (ed.).
Bauman, Z. (1989) *Modernity and the Holocaust*. Cornell University Press, Ithaca, NY.
Beck, V. (1992) *Risk society: towards a new modernity*, trans. M. Ritter. Sage, London.
Benjamin, W. (1968) Franz Kafka: on the tenth anniversary of his death (1934), in *Illuminations: essays and reflections*, H. Arendt (ed.). Harcourt Brace Jovanovich, New York.
Berenbaum, M. (ed.) (1990) *A mosaic of victims: non-Jews persecuted and murdered by the Nazis*. New York University Press, NY.
Berer, M. and R. Sunanda (1990) *Women and AIDS*. Pandora, London.
Berlin, G. L. (1980) *Earthquakes and the urban environment*. CRC Press, Boca Raton, Fla.
Bhatia, B. M. (1991) *Famines in India*, 2nd edn. Konark Publishers, New Delhi.
Bilham, R. (1988) Earthquakes and urban growth. *Nature*, **336**, 625–6.
Blaikie, P., T. Cannon, I. Davis and B. Wisner (1994) *At risk: natural hazards, people's vulnerability, and disasters*. Routledge, London. chapter 5.
Blaikie, P. M. (1985) *The political economy of soil erosion in developing countries*. Longman, New York.
Blong, R. J. (1984) *Volcanic hazards: a sourcebook on the effects of eruptions*. Academic Press, New York.
Bode, B. (1989) *No bells to toll: destruction and creation in the Andes*. Charles Scribner's, New York.
Bodley, J. H. (1975) *Victims of progress*. Cummings Publishing Co., Menlo Park, Calif.
Bohle, H. G. (ed.) (1993a) *Worlds of pain and hunger: geographical perspectives on disaster vulnerability and food security*. Freiburg Studies in Development Geography no. 5, Verlag Breitenbach, Saarbrücken.
Bohle, H. G. (ed.) (1993b) Vulnerability, hunger and famines. *GeoJournal*, **30** (2).
Bommer, J. J. and N. N. Ambraseys (1989) The Spitak, Armenia USSR earthquake of December, 1988. *Journal of Earthquake Engineering and Structural Dynamics*, **18**, 921–5.
Boothby, N. (1987) Children and war. *Cultural Survival Quarterly*, **10** (4), 28–30.
Boulding, K. (1963) The death of the city: a frightened look at postcivilization. 133–45 in Handlin, O. and J. Burchard (eds) *The historian and the city*. MIT Press, Cambridge, Mass.
Bowman, I. (1946) The strategy of territorial decisions. *Foreign Affairs*, **24** (2), 177–94.
Bradley, A. C. (1904) *Shakespearean tragedy: lectures on Hamlet, Othello, King Lear, Macbeth*. Macmillan, London.
Brittain, V. *Wartime chronicle: diary 1939–1965*, Bishop A. and A. Bennett (eds). Gollancz, London.
Brodeur, P. (1989) *Currents of death: power lines, computer terminals, and the attempt to cover up their threat to your health*. Simon and Schuster, New York.
Brogan, D. W. (1963) Implications of modern city growth. 146–64 in Handlin, O. and J. Burchard (eds) *The historian and the city*. MIT Press, Cambridge, Mass.
Brogan, P. (1989) *The fighting never stopped: a comprehensive guide to world conflict since 1945*. Random House, New York.

Brook, D. (1992) Policy in response to geohazards: lessons from the developed world? chapter 18 in McCall *et al.* (eds).

Brook, R. (1990) An introduction to disaster theory for social workers. *Social Work Monographs*, 85, Norwich.

Brookfield, H. (1973) *Interdependent development*. Methuen, London.

Brown, M. (ed.) (1983) *The structure of disadvantage*. Heinemann, London.

Brunswig, H. (1979) *Feuersturm über Hamburg: d. Luftangriffe auf Hamburg im 2. Weltkrieg u. ihre folgen*. Motorbuch-Verlag, Stuttgart.

Bryant, B. and P. Mohai (eds) (1992) *Race and the incidence of environmental hazards*. Westview, Boulder, Colo.

Bunge, W. (1971) *Fitzgerald: geography of a revolution*. Schenkem, Cambridge, Mass.

Bunge, W. (1973) The geography of human survival. *Annals Association American Geographers*, **63**, 275–95.

Burby, R. J. (1991) *Sharing environmental risks: how to control governments' losses in natural disasters*. Westview Press, Boulder, Colo.

Burchell, G., C. Gordon and P. Miller (eds) (1991) *The Foucault effect: studies in governmentality*. Harvester Wheatsheaf, London.

Burton, I. (1962) *Types of agricultural occupance of flood plains in the United States*. Research Paper no. 75, Department of Geography, University of Chicago Press, Chicago.

Burton, I., C. D. Fowle and R. S. McCullogh (eds.) (1982) *Living with risk: environmental risk management in Canada*. Monograph no. 3, Institute of Environmental Studies, University of Toronto.

Burton, I. and R. W. Kates (eds) (1967) *Readings in resource management and conservation*. University of Chicago Press, Chicago.

Burton, I., R. W. Kates and R. E. Snead (1969) *The human ecology of coastal flood hazard in Megalopolis*. Department of Geography, University of Chicago.

Burton, I., R. W. Kates and G. F. White (1978) *The environment as hazard*. Oxford University Press, Oxford.

Calder, A. (1969) *The people's war: Britain, 1939–1945*, 1st American edn. Pantheon Books, New York.

Camus, A. (1948) *The plague*, (trans. S. Gilbert). Modern Library, New York.

Castel, R. (1991) From dangerousness to risk. 281–98 in Burchell *et al.* (eds).

Castells, P. B. (1991) International decade for natural disaster reduction. *UNDRO News*, July/August, 1920.

Castro, E. R. (O. F. Aquino (ed.)) (1991) *Pinatubo: the eruption of the century*. Phoenix Publishing House, Quezon City, Philippines.

Chambers, R. (ed.) (1989) *Vulnerability: how the poor cope*. IDS Bulletin, Special Issue.

Clifford, R. A. (1956) *The Rio Grande flood: a comparative study of border communities in disaster*. National Academy of Sciences, Washington, DC.

Coburn, A., A. Pomonis and S. Sakia (1989) Assessing strategies to reduce casualties in earthquakes, in *International workshop on earthquake injury epidemiology for mitigation and response*. Johns Hopkins University Press, Baltimore, Md.

Coch, N. K. (1995) *Geohazards: natural and human*. Prentice Hall, Englewood Cliffs, NJ.

Cohen, I. and A. Elder (1989) Major cities and disease crises: a comparative perspective. *Social Science History*, **13** (1), 28–63.

Cole, L. A. (1990) *Clouds of secrecy: the army's germ warfare tests over populated areas*. Rowman & Littlefield, Savage, Md.

Collier, B. (1957) *The defence of the United Kingdom*. History of the Second World War, United Kingdom series, HMSO, London.

Collier, R. (1974) *The plague of the Spanish lady: the influenza pandemic of 1918–1919*. Atheneum, New York.

Committee for the Compilation of Materials on Damage Caused by the Atomic Bombs (1981) *Hiroshima and Nagasaki: the physical, medical, and social effects of the atomic bombings*, trans. E. Ishikawa and D. L. Swain. Basic Books, New York.

Conquest, R. (1986) *The harvest of sorrow: Soviet collectivization and the terror-famine.* University of Alberta Press, Edmonton.

Corrigan, S. W. (ed.) (1991) *Readings in aborginal studies I.* Human Services Bearpaw Publishing, Department of Native Studies, Brandon University, Brandon, Manitoba.

Cox, H. (1965) *The secular city: secularization and urbanization in theological perspective.* Macmillan, New York.

Craig, D. and M. Egan (1979) *Extreme situations: literature and crisis from the Great War to the atom bomb.* Macmillan, New York.

Crosby, A. W. (1986) *Ecological imperialism: the biological expansion of Europe, 900– 1900.* Cambridge University Press, Cambridge.

Cuny, F. C. (1983) *Disasters and development.* Oxford University Press, Oxford.

Cuny, F. C. (1994) Cities under siege: problems, priorities and programs. *Disasters,* **18** (2), 152–9.

Currey, B. and G. Hugo (eds) (1984) *Famine as a geographical phenomenon.* Reidel Publishing Co., Boston.

Cutter, S. L. (1993) *Living with risk: the geography of technological hazards.* Edward Arnold, London.

Cutter S. L. (ed.) (1994) *Environmental risk and hazards.* Prentice Hall, Englewood Cliffs, NJ.

Cutter, S. L. and W. D. Solecki (1989) The national pattern of airborne toxic releases. *The Professional Geographer,* **41** (2), 49–161.

Czerniakow, A. (1979) *The Warsaw diary of Adam Czerniakow: prelude to doom.* Hilberg R., S. Staron and J. Kermisz (eds), trans. S. Staron and the staff of Y. Vashem. Stein and Day, New York.

Daniels, G. (1975) The great Tokyo air raid, 9–10 March, 1945. 113–31 in Beasley, W. G. (ed.) *Modern Japan: aspects of history, literature and society.* Berkeley, Calif.

Dardel, E. (1952) *L'homme et la terre: nature de la réalité géographique.* Presses Universitaires de France, Paris.

Davis, I., (ed.) (1981) *Disasters and the small dwelling.* Pergamon Press, Oxford.

Davis, I. (1981) *Disasters and settlements – towards an understanding of the key issues.* 11–23 in Davis, I. (ed.) (1981).

Davis, I. (1984) A critical review of the work method and findings of the Housing and Natural Hazards Group. 200–27 in Miller, K. J. (ed.).

Degg, M. R. (1992) Some implications of the 1985 Mexican earthquake for hazard assessment, in McCall, G. J. H., D. J. C. Laming and S. C. Scott (eds) (1992).

Denniston, D. (1995) *High priorities: conserving mountain ecosystems and cultures.* World Watch Paper 123. Worldwatch Institute, Washington, DC.

Desmond, C. (1983) *Persecution East and West: human rights, political prisoners and amnesty.* Penguin, Markham.

di Castri, F. and G. Glaser (1980) Highlands and islands: ecosystems in danger. *UNESCO Courier,* 6–11 (April).

Dimsdale, J. E. (ed.) (1980) *Survivors, victims and perpetrators: essays on the Nazi Holocaust.* Hemisphere Publishing, New York

Dixon, J. A., L. M. Talbot and G. J-M. LeMoigne (1989) *Dams and the environment: considerations in world bank projects.* The World Bank, Washington, DC.

Dobyn, H. F. (1993) Disease transfer by contact. *Annual Review of Anthropology,* **22,** 273–91.

Doughty, P. L. (1971) From disaster to development. *Americas,* **23** (5), 23–35.

Doughty, P. L. (1986) in Oliver-Smith, A. and A. Hansen (eds) (1986).

Douglas, M. (1985) *Risk acceptability according to the social sciences.* Russell Sage Foundation, New York.

Douglas, M. (1992) *Risk and blame: essays in cultural theory*. Routledge, London.
Douglas, M. and A. Wildavsky (1982) *Risk and culture: as essay on the selection of technical and environmental dangers*. University of California Press, Berkeley.
Drabek, T. E. (1986) *Human system responses to disaster: an inventory of sociological findings*. Springer-Verlag, New York.
Drakakis-Smith, D. (1987) *The Third World city*. Routledge, London.
Dregne, H. E. (1983) *Desertification of arid lands*. Harwood, New York.
Dreze, J. and A. K. Sen (eds) (1990) *The political economy of hunger*, 3 vols. Clarendon Press, Oxford.
Dubhashi, P. R. (1992) Drought and development. *Economic and Political Weekly*, India, **27** (13), A27–A36.
Dudley, N. (1987) *This poisoned Earth: the truth about pesticides*. Piatkus Publishers, London.
Duhl, L. J. (ed.) (1963) *The urban condition*. Basic Books, New York.
Dunin-Wasowicz, K. (1984) Forced labour and sabotage in the Nazi concentration camps. 47–68 in Gutman, Y. and A. Saf (eds) (1985).
Dünninger, J. (1946) Auf dem suche nach dem lebengesetz einer zerstörten stadt, *Franfurter-Hefte*, #5, 51–2.
Dynes, R. R. and E. L. Quarantelli. p.13 in Kreps and Bosworth (1994).
Eagleman, J. R. (1983) *Severe and unusual weather*. Van Nostrand Reinhold, New York.
Earney, F. C. F. *et al.* (1974) Urban snow hazard: Marquette, Michigan. 167–74 in White G. F. (ed.) (1974).
Ebert, C. H. V. (1963) Hamburg's firestorm weather. *National Fire Protection Assoc. Quarterly*, **56** (3).
Ebert, C. H. V. (1993) *Disasters: violence of nature, threats by man*, 2nd edn. Kendall/ Hall, Dubuque, Iowa.
Edoin, H. (1987) *The night Tokyo burned*. St Martin's Press, New York.
Ehrlich, A. H. and J. W. Birks (eds) (1990) *Hidden dangers: environmental consequences of preparing for war*. Sierra Club Books, San Francisco.
Fiber, L. (1993) *'Ich wusste es wird schlimm.' Die Verfolgung der Sinti und Roma in München 1933–1945*. Buchendorfer Verlag, München.
El-Sabah, M. I. and T. S. Murty (eds) (1988) *Natural and man-made hazards*. Proceedings of the International Symposium held at Rimouski, Quebec, Canada, 3–9 August, 1986. D. Reidel, Dordrecht.
Elkins, P. (1992) *A new world order: grassroots movements for global change*. Routledge, London.
Elliott, G. (1973) *Twentieth century book of the dead*. Penguin, London.
Epps, K. (1993) The world's crop of land mines: reaping a deadly harvest. *Ploughshares Monitor*, **XIV** (3), September, 11–14, Waterloo, Ontario.
Epstein, S. S., L. O. Brown and C. Pope (1982) *Hazardous waste in America*. Sierra Club Books, San Francisco.
Ericsen, N. J. (1986) *Creating flood disasters? New Zealand's need for a new approach to urban flood hazard*. National Water and Soil Conservation Authority, Wellington, NZ.
Erikson, K. T. (1976) *Everything in its path: destruction of community in the Buffalo Creek flood*. Simon and Schuster, New York.
Erikson, K. T. (1994) *A new species of trouble: explorations in disaster, trauma, and community*. W. W. Norton, New York.
Evans, R. J. (1987) *Death in Hamburg: society and politics in the cholera years 1830– 1910*. Oxford University Press, Oxford.
Evans, R. (1989) The death industry: world drug economies. *Geographical Magazine*, **LXI**, 5 May, 10–14.
Evans, S. G. (1986) Landslide damming in the Cordillera of western Canada. 93–130 in Schuster, R. L. (ed.).

Ewald, F. (1991) Risk and insurance. 180–201 in Burchell, G. *et al.* (eds) (1991).

Fénelon, F. (1977) *Playing for time*, trans. J. Landry. Atheneum, New York.

Foster, H. D. (1980) *Disaster planning*. Springer-Verlag, New York.

Foster, H. D. (1994) Disease as disaster. *Interdisciplinary Science Reviews*, **19** (3), 237–54.

Frey, R. and P. Safar (eds) (1980) *Types and events of disasters: organization in various disaster situations*. Springer-Verlag, New York.

Fried, M. (1963) Grieving for a lost home. 151–71 in Duhl, L. J. (ed.) *The urban condition*. Basic Books, New York.

Front Line (1943) *Frontline – The Official Story of Civil Defence*. UK Ministry of Information, HMSO, London.

Furet, F. (ed.) (1989) *Unanswered questions: Nazi Germany and the genocide of the Jews*. Schocken Books, New York.

Galtung, J. (1980) *The North–South debate: technology, basic human needs, and the new international order*. Institute for World Order, New York.

Garrett, L. (1994) Human movements and behavioural factors in the emergence of diseases. 312–18 in Wilson, M. E. *et al.* (eds).

Geipel, R. (1982) *Disaster and reconstruction: the Friuli (Italy) earthquakes of 1976*, trans. P. Wagner. Allen and Unwin, London.

Geissler, E. (ed.) (1986) *Biological and toxin weapons today*. Oxford University Press, Oxford, New York.

Geographische Rundschau (1985) *Naturrisiken*. Special issue H.2., 47–92.

George, E. (1984) *Ill fares the land: essays on food, hunger and power*. Penguin Books, New York.

Gertel, J. (1993) Food security within metropolitan Cairo under conditions of structural adjustment. 101–13 in Bohle, H. G. (ed.) (1993a).

Gibson, M. (1991) *Order from chaos: responding to traumatic events*. Venture Press, Birmingham, UK.

Gilbert, M. (1993) *Atlas of the Holocaust*, revised edn. Lester Publishing, Toronto.

Glantz, M. H. (1976) *The politics of natural disaster: the case of the Sahel drought*. Praeger, New York.

Glantz, M. H. (ed.) (1987) *Drought and hunger in Africa: denying famine a future*. Cambridge University Press, Cambridge.

Glass, R. I. *et al.* (1977) Earthquake injuries related to housing in a Guatemalan village. *Science*, **197**, 638–43.

Gleser, G. C., B. L. Green and C. Winget (1981) *Prolonged psychosocial effects of disaster: a study of Buffalo Creek*. Academic Press, New York.

Glickman, T. S. and M. Gough (eds) (1990) *Readings in risk*. Resources for the Future, Washington, DC.

Gluck, S. B. and D. Patai (eds) (1991) *Women's words: the feminist practice of oral history*. Routledge, London.

Goldsmith, E. and N. Hildyard (1984) *The social and environmental effects of large dams*. Wadebridge Ecology Centre, Camelford, Cornwall, UK.

Goodman, D. and M. Redclift (1991) *Refashioning nature: food, ecology and cultures*. Routledge, London.

Gould, P. (1991) *Fire in the rain: the democratic consequences of Chernobyl*. Johns Hopkins University Press, Baltimore, Md.

Grant, G. (1969) *Technology and empire: perspectives on North America*. Anansi, Toronto.

Grosser, G. H., H. Wechsler and M. Greenblatt (eds) (1964) *The threat of impending disaster: contributions to the psychology of stress*. MIT Press, Cambridge, Mass.

Gutman, Y. and A. Saf (eds) (1985) *The Nazi concentration camps: structure and aims, the image of the prisoner, the Jews in the camps*. Proceedings of the Fourth Yad

Vashem International Historical Conference, Jerusalem, January 1980, trans. D. Cohen *et al.* Yad Vashem, Jerusalem.

Hacking, I. (1990) *The taming of chance.* Cambridge University Press, Cambridge.

Hamilton, L. S. (1991) The Philippines storm disaster and logging: the wrong villain? *World Mountain Newsletter,* 4 December, 11.

Hancock, G. (1989) *Lords of poverty.* Mandarin, London.

Handmer, J. and E. Penning-Rowsell (eds) (1990) *Hazards and the communication of risk.* Gower Technical, Brookfield, Vt.

Hansen, A. and A. Oliver-Smith (eds) (1982) *Involuntary migration and resettlement.* Westview, Boulder, Colo.

Hardoy, J. E. and D. Satterthwaite (1989) *Squatter citizen: life in the urban Third World.* Earthscan, London.

Harrell-Bond, B. (1986) *Imposing aid.* Oxford University Press, Oxford.

Harrington, M. (1965) *The accidental century.* Penguin Books, New York.

Harrisson, T. (1976) *Living through the Blitz.* Schocken Books, New York.

Hartmann, B. and J. K. Boyce (1983) *A quiet violence: view from a Bangladesh village.* Zed Books, London.

Hass, J. E., R. W. Kates and M. J. Bowden (eds) (1977) *Reconstruction following disaster.* MIT Press, Mass.

Havens, T. R. H. (1978) *Valley of darkness: the Japanese people and World War Two,* 1st edn. Norton, New York.

Heinemann, M. E. (1986) *Gender and destiny: women writers and the Holocaust.* Greenwood Press, New York.

Herbert, D. T. and R. J. Johnston (eds) (1982) *Geography and the urban environment, Progress in Research and Applications,* vol. 5. John Wiley, New York.

Hewison, R. (1977) *Under siege: literary life in London, 1939–1945,* 1st American edn. Oxford University Press, New York.

Hewitt, K., unpublished data from Public Record Office, London and US Strategic Bombing Survey Archive, National Archives, Washington, DC.

Hewitt, K. (1970) Probabilistic approaches to discrete natural events; a review and theoretical discussion. *Economic Geography,* **46** (Supplement), 332–49.

Hewitt, K. (1976) Earthquake hazards in the mountains. *Natural History,* **85** (5), 30–7.

Hewitt, K. (1983a). The idea of calamity in a technocratic age. 3–32 in Hewitt, K. (ed.) *Interpretations of calamity: from the viewpoint of human ecology.* Allen and Unwin, London.

Hewitt, K. (ed.) (1983b) Place annihilation: area bombing and the fate of urban places. *Annals Association American Geographers,* **73**, 257–84.

Hewitt, K. (1984a) Settlement and change in 'basal zone ecotones': and interpretation of the geography of earthquake risk. 15–41 in Jones, B. G. and M. Tomazevic (eds) (1982).

Hewitt, K. (1984b) Ecotonal settlement and natural hazards in mountain regions: the case of earthquake risks. *Mountain Research and Development,* **4** (1), 31–7.

Hewitt, K. (1987a) The social space of terror: towards a civil interpretation of total war. *Society and Space, Environment and Planning D,* **5**, 445–74.

Hewitt, K. (1987b) Risks and emergencies in Canada: a national overview. *Ontario Geographer,* **29**, 1–36.

Hewitt, K. (1992) Mountain hazards. *GeoJournal,* **27**, 47–60.

Hewitt, K. (1993) Torrential rains in central Karakoram, 9–10 September, 1992: geomorphological impacts and implications for climatic change, *Mountain Research and Development,* 371–5.

Hewitt, K. (1994) When the great planes came and made ashes of our city . . . : towards an oral geography of the disasters of war. *Antipode,* **26** (1), 1–34.

Bibliography

Hewitt, K. and I. Burton (1971) *The hazardousness of a place: a regional ecology of damaging events*. Department of Geography, Research Publication 6, University of Toronto Press.

Hewitt, K. *et al.* (1994) *Mountain hazards in the Americas: a selected bibliography*. Working Paper 3, Cold Regions Research Centre, Wilfrid Laurier University, Waterloo, Ontario.

Hilberg, R. (1985) *The destruction of the European Jews*, 3 vols. Holmes and Meier, New York.

Hilberg, R. (1989) The bureaucracy of annihilation. chapter 7, 119–33 in Furet, F. (ed.) (1989).

Hobart, C. W. (1991) The impact of resource development on the health of native people in the Northwest Territories [Canada]. 83–100 in Corrigan, S. W. (ed.).

Hodgkinson, P. E. and M. Stewart (1991) *Coping with catastrophe: a handbook of disaster management*. Routledge, London.

Hostettler, E. (ed.) (1990) *The Island at war: memories of wartime life in the Isle of Dogs, East London*. Island History Trust, London.

Housner, G. W. (1989) An international decade for natural disaster reduction, 1990–2000. *Natural Hazards*, **2**, 45–75.

Huntington, S. (1968) The Bases of Accommodation. *Foreign Affairs*, **46**, 650.

Huss-Ashmore, R. and S. H. Katz (eds) (1989) *African food systems in crisis Part One: Microperspectives*. Gordon and Breach, New York.

ICIMOD (1993) *Our mountains: the Hindu Kush-Himalayas – a decade of effort towards integrated mountain development, 1983–1993*. Kathmandu, Nepal.

Ienaga, S. (1978) *The Pacific War, 1931–1945: a critical perspective on Japan's role in World War II*, 1st edn. Pantheon Books, New York.

Iklé, F. C. (1958) *The social impact of bomb destruction*, 1st edn. University of Oklahoma Press, Norman.

International Development Centre (Ottawa) (1993) *Reports*, **21** (1), 20–2.

Ives, J. G. and R. Barry (eds) (1974) *Arctic and alpine environments*. Methuen, London.

Ives, J. D. and B. Messerli (1989) *The Himalayan dilemma: reconciling development and conservation*. Routledge, London.

Jacob, G. and A. Kirby (1990) On the road to ruin: the transport of military cargoes. chapter 5, 71–95 in Ehrlich and Birks (eds) (1990).

Janis, I. L. (1951) *Air war and emotional stress: psychological studies of bombing and civilian defense*. Rand Corp., Santa Monica, Calif.

Jansen, R. B. (1980) *Dams and public safety: a water resources technical publication*. US Department of the Interior, Washington, DC.

Jiminez Dias, V. (1992) Landslides and the squatter settlements of Caracas. *Environment and Urbanisation*, **4** (2), 80–9.

Johnson, D. (1988) *The city ablaze: the second Great Fire of London: 29th December, 1940*. William Kimber, London.

Jones, B. G. and M. Tomazevic (eds) (1982) *Social and economic aspects of earthquakes*. Proceedings of the Third International Conference; Bled, Yugoslavia, 1981. Cornell University, Ithaca, NY.

Jones, D. K. C. (1992) Landslide hazard assessment in the context of development. chapter 12, 117–41 in McCall *et al.* (eds).

Jones, D. K. C. (ed.) (1993) Environmental hazards: the challenge of change. *Geography*, 161–98.

Kasperson, R. E. and P. Stallen (eds) (1991) *Communicating risks to the public: international perspectives*. Kluwer Academic Publishers, Boston.

Kates, R. W. (1962) *Hazard and choice perception in flood plain management*. Research Paper no. 78, Department of Geography, University of Chicago Press, Chicago.

Kates, R. W. (1965) *Industrial flood losses: damage estimate in the Lehigh Valley.* Research Paper no. 98, Department of Geography, University of Chicago Press, Chicago.

Kates, R. W. and I. Burton (eds) (1986) *Geography, resources and environment,* vol. II, *Themes from the work of Gilbert F. White.* University of Chicago Press, Chicago.

Kates, R. W., C. Hohenemser and J. X. Kasperson (1985) *Perilous progress: managing the hazards of technology.* Westview Press, Boulder, Colo.

Keefer, D. K. and N. E. Tannaci (1984) *Bibliography on landslides, soil liquefaction, and related ground failure in selected historic earthquakes.* US Geological Survey, Open File Report 81-572, Menlo Park, Calif.

Kenovic, V. (1993) A way of staying alive: an interview with Ademir Kenovic [Bosnian filmmaker] in 'Za Sarajevo', *Lusitania: Journal of Reflection and Oceanography,* **5,** Fall, 14–19.

Kenrick, D. and G. Puxon (1972) *The destiny of Europe's gypsies.* Basic Books, New York.

Kent, G. (1984) *The political economy of hunger: the silent holocaust.* Praeger, New York.

Kent, R. C. (1987) *Anatomy of disaster relief: the international network in action.* Pinter, London.

Kharbanda, O. P. and E. A. Stallworthy (1988) *Safety in the chemical industry: lessons from major disasters.* Heinemann Professional Publishing, London.

Kidron, M. and D. Smith (1983) *The war atlas: armed conflict – armed peace.* Pan Books, London.

Kienholz, H., H. Hafner, G. Schneider and R. Tamrakar (1983) Mountain hazards mapping in Nepal's middle mountains. *Mountain Research and Development,* **3** (3), 195–220.

Killian, L. M. (1956) *A study of response to the Houston, Texas, fireworks explosion.* National Research Council, National Academy of Sciences, Washington, DC.

Kirby, A. (ed.) (1990a) *Nothing to fear: risks and hazards in American society.* University of Arizona Press, Tucson.

Kirby, A. (1990b) Things fall apart: risks and hazards in their social setting. chapter 1, 17–38 in Kirby A. (ed.) (1990a).

Kirby, A. (ed.) (1992) *The Pentagon and the cities.* SAGE, Newbury Park, Calif.

Kishk, M. A. (1993) Rural poverty in Egypt: the case of landless and small farmers' families. 55–70 in Bohle H. G. (ed.) (1993a).

Kogon, E. (1958) *The theory and practice of Hell,* trans. H. Norden. Berkley Publishing, New York.

Kolko, G. (1986) *Vietnam: anatomy of a war 1940–1975.* Allen and Unwin, London.

Konvitz, J. W. (1985) *The urban millennium: the city-building process from the Middle Ages to the present.* Southern Illinois University Press, Carbondale.

Kovach, R. L. (1995) *Earth's fury: an introduction to natural hazards and disasters.* Prentice Hall, Englewood Cliffs, NJ.

Krausnick, H., H. Buchheim, M. Broszat and H-A. Jacobsen (1965) *Anatomy of the SS state,* trans. R. Barry, M. Jackson and D. Long. Walker and Company, New York.

Kreps, G. A. (1984) Sociological inquiry and disaster research. *Annual Review of Sociology,* **10,** 309–30.

Kreps, G. A. (ed.) (1989) *Social structure and disaster.* Associated University Press, Toronto.

Kreps, G. A. and S. L. Bosworth (1994) *Organizing, role enactment, and disaster: a structural theory.* University of Delaware Press, Newark, NJ.

Kreutzmann, H. (1994) Habitat conditions and settlement processes in the Hindukush-Karakoram. *Petermanns Geographische Mitteilungen,* **138** (6), 337–56.

Krider, E. P. (1983) *The thunderstorm in human affairs*. University of Oklahoma Press, Norman.

Krimsky, S. and D. Goulding (eds) (1992) *Social theories of risk*. Praeger, Westport, Conn.

Krimsky, S. and A. Plough (1988) *Environmental hazards: communicating risks as a social process*. Auburn House, Dover, Mass.

Kroll-Smith, J. S. and S. R. Couch (1990) *The real disaster is above ground: a mine fire and social conflict*. University Press of Kentucky, Lexington.

La Farge, H. (ed.) (1946) *Lost treasures of Europe*. Pantheon Books, New York.

Lang, B. (ed.) (1988) *Writing and the Holocaust*. Holmes and Meher, New York.

Langer, L. L. (1991) *Holocaust testimonies: the ruins of memory*. Yale University Press, New Haven, Conn.

La Porte, T. R. and P. M. Consolini (1991) Working in practice but not in theory: theoretical challenges of 'high reliability organisations'. *Journal of Public Administration Research and Theory*, **1** (1), 19–47.

Last, M. (1994) Putting children first. *Disasters*, **18** (3), 192–211.

Lavell, A. (1994a) Comunidades urbanas, vulnerabilidad a desastres y opciones de prevencion y mitigacion. chapter 2, 59–103 in Lavell, A. (ed.) (1994b).

Lavell, A. (ed.) (1994b) *Viviendo en Riesgo: comunidades vulnerables y prevención de desastres en América latina*. La Red, Tercer mundo Editores, Santafé de Bogata, Colombia.

Lerner, S. and T. Schrecker (eds) (1987) Work and environment in a high tech world. *Alternatives*, **14**, 3/4. Waterloo, Ontario.

Levi, P. (1988) *The drowned and the saved*, trans. R. Rosenthal. Michael Joseph, London.

Lewallen, J. (1971) *Ecology of devastation: Indochina*. Penguin Books Inc., Baltimore, Md.

Lewin, R. G. (ed.) (1990) *Witness to the Holocaust: an oral history*. Twayne Publishers, Boston.

Lewis, J. (1987) *Vulnerability and development – and the development of vulnerability: a case for management*. Discussion Paper, Plenum International.

Lifton, R. J. (1967) *Death in life: survivors of Hiroshima*. Random House, New York.

Lifton, R. J. (1970) *History and human survival*. Vintage Books, New York.

Lifton, R. J. (1986) *The Nazi doctors: medical killing and the psychology of genocide*. Basic Books, New York.

Lifton, R. J. and E. Markusen (1990) *The genocidal mentality: Nazi Holocaust and nuclear threat*. Basic Books, New York.

Lintner, B. (1993) A fatal overdose: civilians butchered in fighting between drug gangs [Thai–Burmese border]. *Far Eastern Economic Review*, **156** (22), 26–7.

Little, P. D., M. M. Horowitz and A. E. Nyerges (eds) (1987) *Lands at risk in the Third World: local-level perspectives*. Westview Press, Boulder, Colo.

Livermore, D. M. (1990) Vulnerability to global environment change. 27–45 in Kasperson, R. (ed.) *Understanding global environmental change: the contributions of risk analysis and management*. Clark University, Worcester, Mass.

Livermore, D. M. (1993) Drought impacts in Mexico: climate, agriculture, technology, and land tenure in Sonora and Puebla. *Annals Association American Geographers*, **80** (1), 49–72.

Longmate, N. (1983) *The bombers: the RAF offensive against Germany 1939–1945*. Hutchinson, London.

Lovell, W. G. (1988) Surviving conquest: the Maya of Guatemala in historical perspective. *Latin American Research Review*, **23** (2), 25–57.

McCall, G. J. H., D. J. C. Laming, and S. C. Scott (eds) (1992) *Geohazards: natural and man-made*. Chapman and Hall, New York.

McCoy, A. W. (1972) *The politics of heroin in Southeast Asia.* Harper and Row, New York.

MacIsaac, D. (1976) *Strategic bombing in World War Two: the story of the United States strategic bombing survey.* Garland Publishing Co., New York.

Macksoud, M. S. (1992) Assessing war trauma in children: a case study of Lebanese children. *Journal of Refugee Studies*, **5** (1), 1–15.

McNeil, W. H. (1976) *Plagues and peoples.* Doubleday, New York.

McNicoll, A. (1983) *Drug trafficking: a North–South perspective.* North–South Institute, Ottawa.

Macrae, J. and A. B. Zwi (1993) Food as an instrument of war in contemporary African famines: a review of the evidence. *Disasters*, **16** (4), 299–321.

Madajczk, C. (1984) Concentration camps as a tool of oppression in Nazi-occupied Europe. 47–68 in Gutman, Y. and A. Saf (eds) (1985).

Mann, J. M. *et al.* (eds) (1992) *AIDS in the world.* Harvard University Press, Cambridge, Mass.

Mannion, A. M. and S. R. Bowlby (eds) (1992) *Environmental issues in the 1990s.* John Wiley, New York.

Marglin, F. A. (1990) Smallpox in two systems of knowledge. chapter 4, 102–44 in F. A. and S. A. Marglin (eds).

Marglin, S. A. (1990) Towards the decolonisation of the mind. chapter 1, 1–28 in Marglin and Marglin (eds) (1990).

Marglin, S. A. and F. A. Marglin (eds) (1990) *Dominating knowledge: development, culture and resistance.* Clarendon Press, Oxford.

Marrus, M. R. (1989) *The Holocaust in history.* Penguin Books, New York.

Marske, C. E. (ed.) (1991) *Communities of fate: readings in the social organization of risk.* Saint Louis University, New York.

Maskrey, A. (1989) *Disaster mitigation: a community based approach.* Oxfam, Oxford.

Maskrey, A. (1994) Comunidad y desastres en America Latina: estrategias de intervencio. chapter 1, 25–57 in Lavell, A. (ed.) (1994b).

Mather, J. R. and G. V. Sdasdyk (eds) (1991) *Global change: geographical approaches.* University of Arizona Press, Tucson.

Matousek, J. (1990) The release in war of dangerous forces from chemical facilities. chapter 3, 30–7 in Westing, A. H. (ed.) (1990).

Meyer, E. and E. Poniatowska (1988) Documenting the earthquake of 1985 in Mexico City. *Oral History Review*, **16** (1), 1–31.

Middlebrook, M. and C. Everitt (1987) *The Bomber Command war diaries: an operational reference book, 1939–1945.* Viking, Harmondsworth.

Miller, K. J. (ed.) (1984) *The International Karakoram Project*, vol. 2, Housing and Natural Hazards, 245–358. Cambridge University Press, Cambridge.

Minear, R. H. (ed.) (1990) *Hiroshima: three witnesses.* Princeton University Press, Princeton, NJ.

Mitchell, B. and D. Draper (1983) Ethics in geographical research. *Professional Geographer*, **35** (1), 9–17.

Mitchell, J. K. (1988) Confronting natural disasters: an international decade for natural hazards reduction. *Environment*, **30**, 25–9.

Mitchell, J. K. (1990a) Natural hazard prediction and response in very large cities. *Proceedings: IDNDR Workshop on Prediction and Perception of Hazards.* WARREDOC Centre, Perugia, Italy.

Mitchell, J. K. (1990b) Human dimensions of environmental hazards. 131–75 in Kirby, A. (ed.) (1990a) *Nothing to fear: risks and hazards in American society.* University of Arizona Press, Tucson.

Mitchell, J. K., N. Devine and K. Jagger (1989) A contextual model of natural hazard. *Geographical Review*, **79** (4), 391–409.

Bibliography

Monger, J. W. H. (ed.) (1994) *Geology and geological hazards of the Vancouver region, southwestern British Columbia.* Geological Survey of Canada, Bulletin 481, Ministry of Energy, Mines and Resources, Ottawa.

Morren, G. (1983) A general approach to the identification of hazards. chapter 15, 284–97 in Hewitt, K. (ed.) (1993).

Morrison, C. F., W. R. Schriever and D. E. Kennedy (1960) The collapse of the Listowel Arena. *Canadian Consulting Engineer*, **2**, 36–47.

Mountain Agenda (1992) *The state of the world's mountains – a global report.* Zed Books, London.

Mukul (1992) Threat of mega projects: struggle of two Himalayan villages. *Economic and Political Weekly*, **27** (14), 697–70.

Müller, L. (1968) New considerations on the Vaiont slide. *Rock Mechanics and Engineering Geology*, **6**, 1–91.

Mumford, L. (1963) *Technics and civilization.* Harcourt Brace and World Inc., New York.

Mumford, L. (1967) *The myth of the machine: technics and human development.* Harcourt Brace Jovanovich, New York.

National Academy of Engineering (1986) *Hazards: technology and fairness.* National Academy Press, Washington, DC.

National Research Council (1987) *Confronting natural disasters: an international decade for natural hazards reduction.* National Academy, Washington, DC.

New Internationalist (1983) Dumping: the global trade in dangerous products. NI no. 129, November.

Newburn, T. (1993) *Disaster and after: social work in the aftermath of disaster.* Jessica Kingsley, London.

Nietschmann, B. (1986) Economic development by invasion of indigenous nations: cases from Indonesia and Bangladesh. *Cultural Survival Quarterly*, **10** (2), 2–12.

Nietschmann, B. (1987) Militarization and indigenous peoples: introduction to the Third World War. *Cultural Survival Quarterly*, **11**, 1–16.

Noji, E. K. (1989) The 1988 earthquake in Soviet Armenia: implications for earthquake preparedness. *Disasters*, **13**, 255–62.

Nuñez de la Peña, F. J. and J. Orozco (1988) *El Terremoto: una version corregida.* Iteso, Guadalajara, Mexico.

Office of Technology Assessment (1981) *Assessment of techologies for determining cancer risks from the environment*, Report H-138, Washington, DC.

O'Keefe, P., K. Westgate and B. Wisner (1976) Taking the naturalness out of natural disaster. *Nature*, 260.

Oliver-Smith, A. S. (1986) *The martyred city: death and rebirth in the Andes*, 2nd edn. Waveland Press, Prospect Heights, Ill.

Oliver-Smith, D. (1990) Post-disaster housing reconstruction and social inequality. *Disasters*, **14** (1), 7–19.

Oliver-Smith, D. and A. Hansen (eds) (1986) *Natural disasters and cultural responses*, Studies in Third World Societies, vol. 36. College of William and Mary, Williamsburg, Va.

Openshaw, S., P. Steadman and O. Greene (1983) *Doomsday: Britain after nuclear attack.* Basil Blackwell, Oxford.

O'Riordan, T. (1986) Coping with environmental hazards. vol. II, chapter 10, 212–309 in Kates, R. W. and I. Burton (eds) (1986).

O'Riordan, T. (1990) Hazard and risk in the modern world: political models for programme design. chapter 17, 293–302 in Handmer and Penning-Rowsell (eds) (1990).

Orwell, G. (1946) *Animal Farm.* Harcourt Brace, New York.

Orwell, G. (1949) *Nineteen eighty-four.* Harcourt Brace, New York.

376

Osada, A. (1959) *Children of the A-bomb, the testament of the boys and girls of Hiroshima*, trans. J. Dan and R. Sieben-Morgen. Putnam, New York.

Osgood, J. (1993) From food security to food insecurity: the case of Iraq, 1990–1991. 185–94 in Bohle, H. G. (ed.) (1993b).

Palm, R. (1990) *Natural hazards: an integrative framework for research and planning.* Johns Hopkins University Press, Baltimore, Md.

Palm, R. I. and M. E. Hodgson (1992) *After a California earthquake: attitude and behaviour change.* Geography Research Paper no. 233, University of Chicago, Chicago.

Parasuraman, S. (1995) The impact of the 1993 Latur–Osmanabad (Maharashtra) earthquake on lives, livelihoods and property. *Disasters,* **14** (2), 156–69.

Payne, J. R. and C. R. Phillips (1985) *Petroleum spills in the marine environment: the chemistry and formation of water-in-oil emulsions and tar balls.* Lewis Publishers Inc., Mich.

Pelanda, C. (1982) Disaster and sociosystemic vulnerability. 67–91 in Jones, B. G. and M. Tomazevic (eds) (1982).

Perrow, C. (1984) *Normal accidents: living with high-risk technologies.* Basic Books, New York.

Perrow, C. and M. F. Guillén (1990) *The AIDS disaster: the failures of organizations in New York and the nation.* Yale University Press, New Haven, Conn.

Perry, S. E., E. Silber and D. A. Bloch (1956) *The child and his family in disaster: a study of the 1953 Vicksburg tornado.* Publ. no. 394. National Academy of Sciences, National Research Council, Washington, DC.

Petak, W. J. and A. A. Atkinson (1982) *Natural hazard risk assessment and public policy: anticipating the unexpected.* Springer-Verlag, New York.

Petts, G. E. (1984) *Impounded rivers: perspectives for ecological management.* John Wiley and Sons, Toronto.

Pflaker, G. *et al.* (1971) Geological aspects of the May 31, 1970, Peru earthquake. *Bulletin Seismological Society of America,* **61** (3), 543–57.

Phillips, D. R. and Y. Verhasselt (eds) (1994) *Health and development.* Routledge, London.

Pickering, K. T. and L. A. Owen (1994) *An introduction to global environmental issues.* Routledge, London.

Pingel, F. (1984) The concentration camps as part of the Nazi system. 3–17 in Gutman, Y. and A. Saf (eds) (1985).

Platt, R. H. (1986) Floods and man: a geographer's agenda. vol. II, chapter 2, 28–68 in Kates, R. W. and I. Burton (eds) (1986).

Preble, E. A. (1990) Impact of HIV/AIDS on African children. *Social Science Medicine,* **31** (6), 671–80.

Prete, R. A. and A. H. Ion (eds) (1984) *Armies of occupation.* Wilfrid Laurier University Press, Waterloo, Ontario.

Price, L. W. (1981) *Mountains and man: a study of process and environment.* University of California Press, Berkeley.

Prince, S. H. (1920) *Catastrophe and social change.* Columbia University Press, New York.

Procacci, G. (1991) Social economy and the government of poverty. chapter 7 in Burchell *et al.* (eds) (1991).

Project Ploughshares (1991) *Armed conflicts report, 1991.* Project Ploughshares and Institute of Peace and Conflict Studies, Waterloo, Ontario.

Project Ploughshares (1995) *Armed conflicts report, 1995.* Project Ploughshares and Institute of Peace and Conflict Studies, Waterloo, Ontario.

Purdey, M. (1994) Anecdote and orthodoxy: degenerative nervous diseases and chemical pollution. *The Ecologist,* **24** (3), 100–4.

Quarantelli, E. L. (ed.) (1978) *Disasters: theory and research.* Sage, London.

Quarantelli, E. L. (1982) What is disaster? An agent specific or an all-disaster spectrum approach to socio-behavioural aspects of earthquakes? 453–78 in Jones B. G. and M. Tomazevic (eds) (1982).

Ramana Dhara, V. and D. Kriebel (1993) An exposure–response method for assessing the long term health effects of the Bhopal gas disaster. *Disasters*, **17** (4), 281–90.

Rao, M. (1992) Of cholera and the post-modern world. *Economic and Political Weekly*, India, **27** (31), 1792–6.

Raphael, B. (1986) *When disaster strikes: how individuals and communities cope with catastrophe*. Basic Books, New York.

Reed, S. A. (1906) *The San Francisco conflagration of April, 1906: special report*. National Board of Fire Underwriters, New York.

Reich, M. R. (1991) *Toxic politics: responding to chemical disasters*. Cornell University Press, Ithaca, NY.

Reiter, L. (1990) *Earthquake hazard analysis: issues and insights*. Columbia University Press, New York.

Reitlinger, G. (1971) *The Final Solution*. Sphere Books, London.

Riebsame, W. E., S. A. Chagnon Jr and T. R. Karl (1991) *Drought and natural resources management in the United States: impacts and implications of the 1987–89 drought*. Westview Press, Boulder, Colo.

Rittner, C. (1990) Forward: the triumph of memory. xi–xv in Berenbaum, M. (ed.) (1990).

Robinson, R., H. F. Franco, R. M. Casterejon and H. R. Bernard (1986) It shook again – the Mexico City earthquake of 1985. 81–122 in Oliver-Smith, A. (ed.) (1986).

Rogers, G. C. (1980) A documentation of slope failure during the British Columbia earthquake of 23 June, 1946. *Canadian Geotechnical Journal*, **17**, 122–7.

Romme, W. H. and D. G. Despain (1989) The Yellowstone fires. *Scientific American*, **261**, 37–46.

Rooney, J. F. (1967) The urban snow hazard in the United States: an appraisal of disruption. *Geographical Review*, **57**, 538–59.

Rousset, D. (1947) *L'univers concentrationnaire (The other kingdom)*, trans. R. Guthrie. Boyer and Hitchcock, New York.

Ruel, S. (1993) The scourge of land mines, *UN/DHA News*, Sept./Dec., 10–12.

Rumpf, H. (1963) *The bombing of Germany*, trans. E. H. Fitzgerald, 1st edn. Holt, Rinehart and Winston, New York.

Ryan, C. (1993) Crime, violence, terrorism and tourism: an accidental or intrinsic relationship? *Tourism Management*, **14** (3), 173–82.

Saarinen, T. F. (1966) *Perception of the drought hazard in the Great Plains*. Research Paper no. 106, Department of Geography, University of Chicago.

Sachs, W. (1990) Development. 26–37 in *The development dictionary*, Sachs, W. (ed.) Zed Books, London.

Sagan, S. D. (1993) *The limits of safety: organizations, accidents, and nuclear weapons*. Princeton University Press, Princeton, NJ.

Sakowska, R. *et al.* (1988) *Warszawskie Getto (Warsaw Ghetto) 1943–1988*. Wydawnictwo Interpress, Warszawa.

Salgado Andrade, E. (1988) Epilogue: one year later. *Oral History Review*, **16** (1), 21–31.

Saylor, C. F. (ed.) (1993) *Children and disasters*. Plenum Press, New York.

Scarpacci, J. L. and L. J. Frazier (1993) State terror: ideology, protest and the gendering of landscapes. *Progress in Human Geography*, **17** (1), 1–21.

Scarry, E. (1985) *The body in pain*. Oxford University Press, New York.

Schimper, A. F. (1903) (1964) *Plant-geography upon a physiological basis*, trans. W. P. Fisher. J. Cramer, Weinheim.

Schuster, R. L. (ed.) (1986) *Landslide dams: processes, risk, and mitigation*. Geotechnical Special Publication no. 3. American Society of Civil Engineers, New York.

Seaman, J., S. Leivesey and C. Hogg (1984) *Epidemiology of natural disasters*. Karger, Basle.

Seamon, D. (1979) *A geography of the lifeworld: movement, rest, and encounter*. St Martin's Press, New York.

Seipel, M. M. O. (1994) Disability: an emerging global challenge. *International Social Work*, **37**, 1965–78.

Seley, J. E. and J. Wolpert (1982) Negotiating urban risk. vol. 5, ch. 8, 279–305 in Herbert, D. T. and R. J. Johnston (eds).

Sen, A. (1981) *Poverty and famines: an essay on entitlement and deprivation*. Clarendon Press, Oxford.

Sewell, W. R. D. (1963) *Water management and floods in the Fraser River basin*. Research Paper no. 115, Department of Geography, University of Chicago.

Shah, B. V. (1983) Is the environment becoming more hazardous? – A global survey 1947 to 1980. *Disasters*, **7** (3), 202–9.

Shaw, J. A. and J. J. Harris (1994) Children of war and children at war: child victims of terrorism in Mozambique. chapter 10, 287–305 in Ursano, R. J. *et al.* (eds) (1994).

Sheets, P. D. and D. K. Grayson (eds) (1979) *Volcanic activity and human ecology*. Academic Press, New York.

Sherry, M. S. (1987) *The rise of American air power: the creation of Armageddon*. Yale University Press, New Haven, Conn.

Shkilnyk, A. M. (1985) *A poison stronger than love: the destruction of an Ojibwa community*. Yale University Press, New Haven, Conn.

Shrader-Frechette, K. S. (1991) *Risk and rationality: philosophical foundations for populist reforms*. University of California Press, Berkeley.

Shrader-Frechette, K. S. (1993) *Burying uncertainty: risk and the case against geological disposal of nuclear waste*. University of California Press, Berkeley.

Shrivastava, P. (1989) *Bhopal: anatomy of a crisis*. Balliger, Cambridge, Mass.

Shroder, J. F., Jr (1989) Hazards of the Himalaya. *American Scientist*, **77**, 565–73.

Simpson, R. H. and H. Riehl (1981) *The hurricane and its impact*. Louisiana State University, Baton Rouge.

Simpson-Lewis, W., N. McKechnic and V. Neimanis (eds) (1983) *Stress on land in Canada*. Environment Canada, Ottawa.

SIPRI (Stockholm International Peace Research Institute) (1971) *The problem of chemical and biological warfare*, 5 vols. Almqvist and Wiksell, Stockholm.

SIPRI (1975) *Incendiary weapons*. MIT Press, Cambridge, Mass.

SIPRI (1993) *World armaments and disarmament, SIPRI yearbook*. Taylor and Francis, London.

Sivard, R. L. (ed.) (1993) *World military and social expenditures 1992*. World Priorities, Inc., Washington, DC.

Slovic, P., B. Fischhoff and S. Lichtenstein (1979) Rating the risks. *Environment*, **21** (3), 14–20, 36–9.

Small, M. and J. D. Singer (1982) *Resort to arms: international and civil wars, 1816–1980*. Sage, Beverly Hills.

Smith, K. (1992) *Environmental hazards: assessing risk and reducing disaster*. Routledge, London.

Solomatine, N. (1988) International solidarity for Armenian Quake Victims. *UNDRO News*, Nov./Dec., 4–7.

Solway, L. (1994) Urban developments and megacities: vulnerability to natural disasters. *Disaster Management*, **6** (1) (special issue: New York City and Urban Emergency Management, T. Horlick-Jones (ed.)) 160–9.

Sorokin, P. A. (1941) *The crisis of our age*. Dutton and Co., New York.

379

Sorokin, P. A. (1942) *Man and society in calamity: the effects of war, revolution, famine and pestilence upon the human mind, behaviour, social organisation and cultural life.* E. P. Dutton, New York.

Starosolszky, O. and O. M. Melder (eds) (1989) *Hydrology of disasters: proceedings of the Technical Conference in Geneva, November 1988.* World Meteorological Organisation. James and James, London.

Starr, C. (1969) Social benefit versus technological risk. *Science*, **165**, 1332–38.

Stephens, S. (1987) Chernobyl fallout: a hard rain for the Sami. *Cultural Survival Quarterly*, **11** (2), 66–71.

Stern, J. (1947) *The hidden damage.* Harcourt and Brace, New York.

Stonehouse, B. (ed.) (1986) *Arctic air pollution.* Cambridge University Press, Cambridge.

Susman, P., P. O'Keefe and B. Wisner (1983) Global disasters, a radical approach. 263–83 in Hewitt, K. (ed.) (1983).

Sylves, R. T. and W. L. Waugh (eds) (1990) *Cities and disasters.* Charles C. Thomas, Springfield, Ill.

Tachakra, S. S. (1989) The Bhopal disaster: lessons for developing countries. *Environmental Conservation*, **16** (1), 65–6.

Tarr, J. A. and G. Dupuy (eds) (1988) *Technology and the rise of the networked city in Europe and America.* Temple University Press, Philadelphia, Pa.

Taylor, D. (1994) *My children, my gold: meetings with women of the fourth world.* Virago, London.

Thayer, T. C. (1985) *War without fronts: the American experience in Vietnam.* Westview, Boulder, Colo.

Thrift, N. and D. Forbes (1986) *The price of war: urbanisation in Vietnam 1954–1985.* Allen and Unwin, London.

Timmerman, P. (1981) *Vulnerability, resilience, and the collapse of society: a review of models and possible climatic applications.* Environmental monograph 1, Institute of Environmental Studies, University of Toronto.

Tobias, M. (ed.) (1986) *Mountain people.* University of Oklahoma Press, Norman.

Torry, W. I. (1978) Natural disasters, social structure and changes in traditional societies. *Journal of Asian and African Studies*, **13**, 167–83.

Torry, W. I. (1980) Urban earthquake hazard in developing countries: squatter settlements and the outlook for Turkey. *Urban Ecology*, **4**, 317–27.

Torry, W. I. (1986) Morality and harm: Hindu peasant adjustments to famines. *Social Science Information*, 12.

Trunk, I. (1979) *Jewish responses to Nazi persecution: collective and individual behaviour in extremis.* Stein and Day, New York.

Tsuchiya, Y. and Y. Kawata (1988) Historical changes of storm surge disasters in Osaka. 279–303 in El-Sabh, M. I. and T. S. Murty (eds).

Tuan, Y. (1979) *Landscapes of fear.* Pantheon Books, New York.

Tufnell, L. (1984) *Glacier hazards.* Longman, London.

Turner, B. A. (1978) *Man-made disasters.* Wykeham Publications, London.

Turner, T. E. (ed.) (1990) *Love and money: grandmother's strategies with spouses and bosses.* Women's Studies, University of Massachusetts, Amherst.

US Food Safety Council (1980) UA proposed system for food safety assessment. Final Report, Scientific Committee, Washington, DC.

Uman, M. A. (1984a) *Lightning.* Dover, New York.

UN Department of Humanitarian Affairs (UN/DHA) (1993) 1993 in review. *DHA News*, special edn. no. 7, Jan./Feb.

UN/DHA (1994a) Dignity and sorrow in troubled states. *DHA News*, **8**, March/April.

UN/DHA (1994b) *Disasters around the world – a global and regional overview.* World Conference on Natural Disaster Reduction, Yokohama, Japan, 23–27 May, Information Paper no. 4, Geneva.

UN/DHA (1994c) World conference on natural disaster reduction, Yokohama: a muted warning. *DHA News*, 9–10 (May/Aug.), 33–4.

UN Disaster Relief Organisation (UNDRO) (1990) World launches international decade for natural disaster reduction. *UNDRO News*, special issue, Jan./Feb. United Nations Disaster Relief Organisation, Geneva.

United Nations Environmental Programme (1994) Desertification. *Our Planet*, 6 (5), 2–33.

United States Geological Survey (1989). Lessons learned from the Lomo Prieta, California earthquake of October 17, 1989. Circular no. 1053.

Ursano, R. J., B. G. McCaughey and C. S. Fullerton (eds) (1994) *Individual and community response to trauma and disaster: the structure of human chaos.* Cambridge University Press, Cambridge.

US National Academy of Sciences (1984) *Hurricane Alicia, Galveston and Houston, Texas, August 17–18, 1983.* Committee on Natural Disasters, National Research Council, National Academy Press, Washington, DC.

US OFDA (1988) *Disaster history: significant data on major disasters worldwide, 1900–present.* Office of US Foreign Disaster Assistance, Agency for International Development, Washington, DC.

Vaughan, M. (1987) *The story of an African famine: gender and famine in twentieth-century Malawi.* Cambridge University Press, Cambridge.

Waddell, E. (1977) The hazards of scientism: a review article. *Human Ecology*, 5 (1), 67–76.

Waddell, E. (1983) Coping with frosts, governments and disaster experts: some reflections based on a New Guinea experience and a perusal of the relevant literature. chapter 2, 33–43 in Hewitt (ed.) (1983).

Wagner, P. (1960) *The human use of the Earth.* Free Press, New York.

Walford, C. (1879) *Famines of the World, past and present,* (reissued 1980). Burt Franklin, New York.

Walmsley, D. F. (1988) *Urban living: the individual in the city.* Longman, London.

Ward, R. C. (1978) *Floods: a geographical perspective.* Wiley, New York.

Waterstone, M. (ed.) (1992) *Risk and society: the interaction of science, technology, and public policy.* Kluwer Academic Publishers, Boston.

Watts, M. J. (1983a) *Silent violence: food, famine, and peasantry in northern Nigeria.* University of California Press, Berkeley.

Watts, M. J. (1983b) On the poverty of theory: natural hazards research in context. 231–62 in Hewitt, K. (ed.) (1983)

Watts, M. J. and H. G. Bohle (1993) The space of vulnerability: the causal structure of hunger and famine. *Progress in Human Geography*, 17 (1), 43–67.

Webster, C. and N. Frankland (1961) *The strategic air offensive against Germany 1939–1945.* HMSO, London.

Weisberg, B. (1970) *Ecocide in Indochina: the ecology of war.* Harper and Row, New York.

Weisner, L. A. (1988) *Victims and survivors: displaced persons and other war victims in Viet-Nam, 1954–1975.* Greenwood Press, New York.

Westermeyer, J. (1982) *Poppies, pipes and people: opium and its use in Laos.* University of California Press, Berkeley.

Westing, A. H. (ed.) (1984a) *Explosive remnants of war: mitigating the environmental effects.* Taylor & Francis, London.

Westing, A. H. (ed.) (1984b) *Environmental warfare: a technical, legal, and policy appraisal.* Prepared by Stockholm International Peace Research Institute. Taylor & Francis, London.

Westing, A. H. (ed.) (1984c) *Herbicides in war: the longterm ecological and human consequences.* Prepared by Stockholm International Peace Research Institute. Taylor & Francis, London.

Westing, A. H. (ed.) (1990) *Environmental hazards of war: releasing dangerous forces in an industrialized world.* Sage Publications, London.

White, G. F. (1945) *Human adjustment to floods: a geographical approach to the flood problem in the United States.* University of Chicago.

White, G. F. (1964) The choice of use in resource management, *Natural Resources Journal*, 1, 254.

White, G. F. (1972) Geography and public policy, in Kates and Burton (eds) (1986).

White, G. F. (1974) *Natural hazards: local, national, global.* Oxford University Press, Toronto.

White, G. F. (1991) Greenhouse gases, Nile snails and human choice. 276–305 in Jessor, R. (ed.) *Perspectives on behavioural science.* Westview Press, Boulder, Colo.

White, G. F. *et al.* (1961) *Papers on flood problems.* Research Paper no. 70. Department of Geography, University of Chicago.

Whyte, A. V. and I. Burton (1980) *Environmental risk assessment* (SCOPE: 15). Wiley, New York.

Wijkman, A. and L. Timberlake (1986) *Natural disasters: acts of God or acts of man?* Earthscan, London.

Wildavsky, A. B. (1988) *Searching for safety.* Transaction Books, New Brunswick, NJ.

Wilson, M. E., R. Levins, A. Spielman and I. Eckardt (eds) (1994) Diseases in evolution: global changes and emergence of infectious diseases. *Annals, New York Academy of Sciences*, 740, 15 December.

Wilson, R. (1979) Analyzing the daily risks of life. *Technology Review*, 81 (4), 41–6.

Winchester, P. (1992) *Power, choice and vulnerability: a case study of disaster mismanagement in South India, 1977–1988.* James and James Science Publishers, London.

Wisner, B. (1988) *Power and need in Africa.* Earthscan, London.

Wisner, B. (1993) Disaster vulnerability, geographical scale and existential reality. 13–52 in Bohle, H. G. (ed.) (1993a) *Worlds of pain and hunger: geographical perspectives on disaster vulnerability and food security.* Freiburg Studies in Development Geography, 5 Breitenbach, Fort Lauderdale, Fla.

Wolfenstein, M. (1957) *Disaster, a psychological essay.* Free Press, Glencoe, Ill.

Wolman, M. G. and J. P. Miller (1960) Magnitude and frequency of geomorphic processes. *Journal of Geology*, 68, 54–74.

Wood, H. O. and F. Neumann (1931) The modified Mercalli intensity scale. *Bulletin, Seismological Society of America*, 21, 277–83.

Woolf, V. (1980) *The letters of Virginia Woolf*, volume VI, 1936–1941. Harcourt Brace Jovanovich, New York.

World Health Organisation (1994) *World health statistics annual.* WHO, Geneva.

Zeigler, D. J., J. H. Johnson and S. D. Brunn (1983) *Technological hazards.* Association of American Geographers, Washington, DC.

Zelinski, W. and L. A. Kosinski (1991) *The emergency evacuation of cities: a cross-national historical and geographical study.* Rowman and Littlefield, Totowa, NJ.

Ziegler, P. (1969) *The Black Death.* Harper and Row, New York.

Zolberg, A. R., A. Suhrke and S. Aguayo (1989) *Escape from violence: conflict and the refugee crisis in the developing world.* Oxford University Press, Oxford.

Zwi, A. and A. Ugalde (1991) Political violence in the Third World: a public health issue. *Health Policy and Planning*, 6 (3), 203–17.

Index

Note: *Where not explicit, most items should be taken to imply discussion and illustration in relation to risk, hazards, vulnerability, disasters, etc.; for example,* **agriculture** *or* **aviation** *(risks/ hazards/disasters);* **children** *(at risk/in disasters and vulnerability); or* **debris flow** (hazards/ disasters). Geographical names are referenced where there is actual data from and discussion of the place concerned, or have been mentioned several times.

Aberfan 34, 97, 103, 104, 105
active perspective 30, 169–94, 278–80
'Acts of God' 22, 186
adjustments 29–30, 172–88, 341, 343
    alternative, range of 30, 172–6, 342–6
    choice of 30, 172, 173, 193–4
Afghanistan 124, 135, 209, 212, 243, 244,
    248–9, 251, 271, 350, 363
Africa 112, 134, 155, 159, 160, 162, 240, 247,
    267, 284, 285, 355
agriculture 59, 67–8, 69, 87, 204–5, 249, 251
AIDS/HIV 56, 70, 159–64, 168, 275
    cities and 284–7
    war and 162–3
airborne chemical hazards 95, 98, 127
air raids 42, 43–4, 46, 47, 49, 118–20, 145,
    247, 277, 296–320, 330
Alabala-Bertrand, J.M. 183, 184, 186, 188,
    194
Alaska 209, 211
Alexander, David 89, 202, 218
Algeria 212, 214
Amnesty International 139–40, 190, 191, 248
Andes 234, 235, 238, 241, 247, 251, 355
Angola 124, 130, 134, 137, 162
anti-personnel weapons 119–22
Arendt, Hannah 193
Argentina 241, 245, 270
armed violence (see violence hazards, wars,
    weapons hazards)
Armenia 201, 204, 211, 214, 220, 229, 362

Asia 59, 135, 160, 222, 240, 247, 250, 251,
    267, 271, 285
Aswan Dam, Egypt 157
atmospheric hazards 56, 63–5
Auschwitz (Auschwitz-Birkenau Death
    Camp) 34, 127, 324–5, 326, 331–2,
    334–5, 336, 337, 338, 340
Australia 267
avalanches (snow) 233, 235, 236, 239–42, 243,
    257, 260
aviation 7, 64, 72–3, 90, 287

Bachelard, Gaston 45
Bangladesh 135, 154, 247, 248, 271, 355,
    361–3
Barkun, Michael 1
barrier hazards 72, 74, 76, 81
Barton, Allan, H. 37, 39
'basal zone' environments (mountain) 225–30
Bauman, Zygmunt 348, 351
Berlin 123, 300, 303, 310–11
Bhopal, India 34, 38, 58, 96, 101, 127, 144,
    287–8, 361
'Big Fire' raid, Tokyo 305
biocides 9
biological hazards 55, 56, 67–70, 78
Birmingham, UK 123, 303, 306–8, 316
'Black Death' 39, 69, 139
Blaikie, Piers 39, 168, 177, 223, 360
'Blitz', the 301, 302
blizzards 64, 74–7, 199
Bode, Barbara 51–2

Index

Bohle, Hans-Georg 152, 168, 187, 278, 356
Bolivia 251, 270, 284
Bradley, A.C. 139
Brazil 267, 270, 283
British Columbia, Canada 262–5
bubonic plague 39, 69
Buffalo Creek disaster, USA 43, 96, 243, 245
built environment 67, 202–7, 266–8, 272–3, 280
Bunge, William 317
Burma (Myanmar) 271, 284, 362
Burton, I 13, 39, 74, 96, 154, 172, 173, 194, 233, 270, 357, 359

California 202, 209, 210, 213, 214, 215, 222, 224, 225, 227, 229, 355, 362, 363
Cambodia 124, 135, 192, 271
Camus, Albert 46
Canada 6, 64, 66, 69, 70, 75–6, 77, 125, 160, 240, 252, 262–5, 267, 269, 274
Canetti, Elias 11
captive populations 330–6, 338, 342–6
carbon monoxide poisoning 77, 243
'carceral geography' 330–6
Caribbean Sea/countries 269–70, 285, 362
casualties 131, 155
    in disasters 59–61, 64–5, 97–9, 218–19, 220, 241, 243, 245, 248, 269–72, 282, 288, 292
    of war 112, 118, 120, 131, 133, 140, 197, 198–200, 204, 297–8, 301, 302–16, 321–4, 333, 334, 336
catastrophes 38, 95–9, 198–202
CBT (chemical, biological and toxin) weapons 119–22, 127
chemical hazards 92–3, 95–6, 105, 288
Chernobyl 34, 38, 45, 52, 98, 103, 109–10, 127, 288
children 98, 116, 133, 145, 148, 159, 177, 199, 210, 218, 249, 250
    war and 124, 132, 133, 298, 301, 307, 314–6, 324
Chile 199, 209, 212, 225, 227, 241, 267, 270
China 66, 71, 136, 145, 198, 201, 208, 210, 212, 222, 241, 246, 250, 255, 267, 284, 362, 364
cholera 69, 70
chronic hazards 4–5, 6, 8, 96, 187
civilians/civil society 276, 278–80, 290, 294, 329
    in war 112–16, 128–39, 149–50, 276–7, 289–93, 294, 296–320
civil defence 115, 174–5, 183, 298, 301, 313
civil strife 111, 123, 272
climate (see atmospheric hazards)
Clydebank, Scotland 314
cocaine 251, 252
cold, cold waves/spells 64, 211
Cold War 13, 112, 244, 357
Colombia 270, 282
complex disasters 8, 71, 272

compound hazards 57, 70–1, 74, 100, 243, 253
concentration camps 127, 155, 324, 328, 333–6
conflagration 313
conflicts (see violence, war hazards)
context (see also intervening conditions) 28–9, 71, 98–105, 170, 171, 186–8, 193–4, 232, 234, 236
coping 29–30, 36, 167, 180, 183–4, 337, 343–6
Coventry, UK 301, 303
crop failure 69, 86, 87
cultural annihilation 53
Cutter, Susan 13, 94–5, 96, 233, 288
cyclones (Indian Ocean) 64–5, 233

damage 21, 31–4, 64, 70, 81, 206, 269–72
    forms of 31, 77–8, 84, 129
    primary, secondary, tertiary 31–4, 37, 71, 76–7, 81, 207–12, 220
dams, artificial 96, 157, 242–3, 244
dams, natural 237, 238, 241, 242, 258, 260
dam-break floods 97, 110, 116–17, 201, 210, 212, 226
    natural 241, 243, 257
    artificial 97–8, 243, 245–6
Dardel, Eric 40, 41, 45, 54, 71, 232, 320
Darwin, Charles 48
Darmstadt, Germany 33, 120, 305, 312, 317, 318
Davis, Ian 217, 255
'death marches' 321, 328, 336
death tolls (see casualties)
debris flows 56, 68, 238, 241, 257, 260
debris torrents 235, 238, 263
defencelessness 27–8, 149, 151, 154, 161, 164, 165, 324, 340
deforestation 165, 215, 256
'demographic violence' 129, 133, 328
deprivation hazards 78, 81, 86, 128–9
desertification 165
'destructive labour' 321, 328, 334
development (economic) 3, 27–8, 29, 187, 226, 230, 254, 255–65, 355, 357, 359
disadvantage 27, 30, 146–8, 277, 316, 317, 355
disasters 7, 9, 11, 35–9, 64–5, 66, 68, 69, 73, 75–6, 82–4, 92, 97–9, 170, 355–6, 361–4
    defined 6, 8, 35, 38
    geography of 10, 35, 40–1, 97, 171, 203, 219, 351, 355
    phases of 36–7, 171, 183–4, 185
'disaster raids' 300, 303, 304, 307, 311, 314, 318, 320
disaster reduction 10, 13–15, 59, 170, 171, 176–7, 188, 206, 278
disease hazards (see also epidemics, pandemics) 6, 26, 56, 67–70, 157–64, 168, 211, 366

displacement of populations 12, 45, 53, 124, 133–8, 166, 248–50, 272, 279, 289–93, 296–320, 326, 328, 330
displaced persons 154, 329
dispossession 165–6
domestic space 46–8, 145, 152, 156–7, 161, 202–6, 215, 218, 294, 298, 305, 308–10, 312, 314, 319, 337
'dominant view' 351–4, 355, 356, 357, 358
'dominating knowledge' 352–4
Douglas, Mary 28, 191
drought hazards 59, 65–7, 86–8, 164–5, 273, 355
and famine 58, 86, 152
drug hazards/traffic 161, 163, 251–3, 284, 290
dubious weapons 116–22, 125–8, 244
defined 116
forms of 119–20
dwelling 45

earthquake hazards/disasters 2, 31, 48, 52, 67, 68, 151, 173, 197–231, 237–42, 254, 260, 272, 280–1, 283
geography of 203, 225–30
Ebola 283
ecocide (ecological devastation) 53, 129, 138
'ecologies of disaster' 270
ecotone, 'ecotonal risks' 229–30
Ecuador 214, 244, 270, 361
Egypt 267
electricity 105
Elliot, Gil 21, 34, 114, 116, 118, 131, 140
'emergency communities' 37, 174
emergency measures/actions 31, 183–4, 337
endangerment, forms of 71, 77–8, 141, 153, 166–7
entitlements 152, 155–6, 166, 187, 317
environmental determinism 13, 229
environmental warfare/weapons 116–17
epidemics 7, 56, 67–70, 146, 158, 159–64, 273, 283–7, 333
Erikson, Kai 43, 44, 46, 48, 52, 110
ethics 9, 139, 191–3
Ethiopia 130, 134, 362
'ethnic cleansing' 129, 133, 135, 250
Europe 136, 155, 159, 160, 167, 240, 247, 251–2, 267, 284, 285, 289, 303, 321-48
evacuation 173, 175, 180, 288, 290, 314
events (see also extreme events) 21, 31, 34
experience (extreme) 42–8
exposure (and vulnerability) 27, 144–5, 146, 165
explosions (industrial) 92, 95–6
extreme events 5, 237–8
Exxon Valdez (oil spill) 98, 101, 110

fallout, nuclear 44, 93, 98, 250
famine 29, 36, 58–9, 86, 87, 130, 133, 152, 155–6, 187, 358
filiariasis 157

fire 208, 226
forest 69, 79, 145, 237–9, 273
grassland 145
mass 102, 118, 145, 301, 312
urban 31, 102, 118, 272–3, 287, 301, 306, 311–12
firestorm 118, 120, 199, 303, 307, 312–3
First World War 121, 136
flood hazards/disasters 59, 65–7, 80–5, 212, 237–42, 243, 245, 254, 258–60, 263, 272, 280
adjustments to 172, 173, 176
geography of 85, 87, 164–5
flood plain 81, 85, 172–3
fog hazards 72–4
food security/shortages 155–6, 355
'food as a weapon' 128–9, 133, 166
France 245
Frankfurt-am-Main 46, 120, 296, 318
Friuli, Italy 202, 206, 218, 228

gender (see also women) 132, 148, 218, 261, 276, 285, 340
genocide 129, 135, 138, 248–50, 321–48
defined 53
geography 12, 40, 116, 205, 217, 219–20, 233, 252, 325
of disasters 40–1, 171, 203, 219, 269–74, 292, 326, 351
of risk 13, 40–1, 71, 103–4, 109–10, 164, 272, 300, 338
of vulnerability 27–8, 63, 65, 146, 157, 164–7, 177, 264, 275
'geography that matters' 12, 38, 41–6, 320, 326
geographicalness of risks 12, 40–54
geographical disasters 8, 12, 36, 53–4, 133–8, 319–20, 326
'geographic shock' 42–6
geographical violence (see also geographical disasters) 129, 133–8
geological hazards 56, 67, 207–8
classes of 56, 67
Germany 118, 120, 130, 245, 267, 297, 303–5, 310–13, 314–15, 321–8
glacier hazards 237, 258–60
global change 9–10, 99–100, 165, 187
globalisation 9, 107, 161, 283
Grassy Narrows, Ontario 44, 107
Greece 211, 215
Guatemala 244, 248
Guatemala City 200, 204, 206, 217, 218, 228, 229, 272

hail hazards, hailstorms 64
Halifax Harbour explosion 92, 115
Hamburg 120, 123, 127, 303, 307, 312–16, 318
Hanoi, Vietnam 289–93

Index

Hardoy, Jorge E. and Satterthwaite, D. 266, 277–8
Harrisson, Tom 43, 140, 300
hazardous waste 99
hazards 8, 13, 144, 307
  defined 25, 55
  classes of 24, 56–7, 328
  mapping 40–1, 242
hazards perspective 25, 55, 152, 164, 173
hazards paradigm 55, 58, 100, 142, 156, 158, 163, 167, 184, 193, 205, 242, 294, 340, 351
heatwave 64
hepatitis 283
heroin 251, 252
high-risk technologies 103, 147, 185
Hilberg, Raul 322, 327, 333, 344
Himalayas 234, 238, 241, 244, 248, 254, 260
Hiroshima 43, 44, 49–50, 58, 130, 288, 294, 296, 306, 316
Hitler, Adolf 133, 327, 330, 343
HIV/AIDS (see AIDS/HIV)
Ho Chi Minh City, Vietnam (formerly Saigon) 289–93
The Holocaust 45, 50, 54, 127, 167, 321–48
  'processes of' 328–36
  and vulnerability 330, 336–41, 354–6
homeless/'dehoused' 81–4, 123, 198–201, 202, 206, 292, 305–14
human ecology 22, 24, 29, 36, 53, 71, 72, 103, 143, 205–6, 354–9
humanitarian issues/responses 37, 111, 114, 184, 190, 351
human rights 3, 27–8, 139, 187
hurricanes (Atlantic Ocean) 64–5, 71, 272
hydrological hazards 56, 65–7, 365

ice (glaze, freezing rain) storms 77
icebergs 1–2
ideas, perspective of 349–60
incendiary weapons 117–20, 301, 307
India 109, 136, 198, 201, 205, 206, 209, 212, 218–19, 250, 267, 271, 284, 287, 355
indigenous peoples 135, 140, 146, 158, 166, 167, 235, 247–8, 262
Indonesia 109, 211, 245, 247, 267, 363
Indus River 241, 250, 255–61
industrial hazards/disasters 98–100, 127, 273, 274, 287, 311
infestations 67–70
influenza 69
infrastructure (modern) 29, 41
inner city disasters 300, 306, 310, 311, 312, 316–18
insurance 22, 23, 142, 173, 185–6
intensity (of impacts) 79
intercontinental ballistic missiles (ICBMs) 97, 100, 106
international aid/relief 14

International Decade for Natural Disaster Reduction (IDNDR) 13–15, 59
International Federation of Red Cross and Red Crescent Societies (IFRCRCS) 119, 160, 171, 174
intervening conditions/variables 28–9, 31, 86, 208, 214–15, 221–2, 223, 224–5, 233, 354
  defined 28
'invisible contamination' 44–5
Iran 137, 199–201, 209, 212, 225, 247, 249, 361
Iraq 124, 137, 247, 350, 363
irrigation systems 106, 157, 204, 257
Italy 6, 198, 211, 214, 225, 243, 245–6
Ives, J.D. 236, 237

Japan 6, 130, 145, 199, 209, 210, 211, 215, 222, 225, 230, 248, 267, 280–2, 297, 303–6, 313–15, 316

Karakoram Himalaya 238, 239, 241, 250, 255–61
Kates, Robert W. 13, 110, 172, 194
Kenya 113, 134, 162
Kirby, Andrew 13, 140, 350
Kobe, Japan 2, 47, 202, 210, 222, 227, 230, 232, 280
Köln, Germany 299, 300, 311–12, 316–17

lahars (volcanic mudflows) 237, 238, 282
landslide hazards 198–201, 208–12, 226, 228, 237–42, 257, 260, 263–5
  catastrophic landslides 31, 51, 272, 273, 283
Lassa fever 273, 283
Lavell, Alan 153
Latin America 112–13, 160, 166, 222, 240, 247, 251, 255, 267, 269–70, 283, 284, 285, 363
Lebanon 132, 137
Leningrad (St Petersburg) 131
Levi, Primo 341, 344
'lifelines' 41, 204
Lifton, Robert J. 44, 49, 127, 184, 296, 340, 345, 348, 350, 353
lightning hazards 64, 77, 237
Lima, Peru 151, 153, 227, 270, 277–8
liquefaction of soil 208, 209, 214–15, 216, 225, 228
Livermore, Diane M. 27, 147
Liverpool 303, 316
'living with risk' 96–9
Lomo Prieta (California) earthquake 190, 194, 106
London (UK) 74, 118, 123, 300, 301, 303, 305–6, 314, 316, 317
loneliness 43–4, 45
loss bearing 173, 175, 185, 337
loss of community 51–2
loss sharing 175, 183, 185

Love Canal, New York 34, 44

magnitude and frequency of hazardous events 79
malaria 157, 283
Malawi 134, 162
Managua, Nicaragua 200, 214, 217
marginalisation 59, 154
Marglin, S.A. 353, 359
Maskrey, Andrew 151, 153, 194, 277–8, 360
mass movements (see landslide hazards)
Mass Observation (UK) 296, 308, 310
media, the 7, 109, 278
Mediterranean Sea/lands 225, 227, 229, 239, 240, 285
megacities 222, 268
meteorite hazards 79
methyl icocyanate (MIC) 98, 127, 144, 287–8
Mexico 6, 244, 247, 267, 363
Mexico City 98, 201, 204, 209, 210, 221–3, 228, 229, 232, 273
   earthquake disaster, 1985 45
mining 217, 222–3, 226
'mines plague' (unexploded munitions) 124, 248
misrule 31, 165, 276
   'regions of misrule' 165–6
Mitchell, Bruce 192–3
Mitchell, James Kenneth 25, 39, 268
mitigation 170, 177–82, 191
modernity 3, 9–10, 94, 96, 99, 106, 293, 347, 351
Modified Mercalli Intensity Scale (earthquake) 206–7
Mongolia 209
monsoons 239, 245
Morocco 209, 214
Morren, George 356
mountain environments 205, 225–30, 232–65
Mozambique 124, 130, 134, 137, 162, 364
mudflows 38, 49, 237–8
Mumford, Lewis 101, 108, 110, 289

Nagasaki 34, 273, 316
napalm 117–20
National Socialism (Nazism) 122–3, 133, 155, 167, 192, 321, 324, 326–7, 330–6, 339–40, 346
natural hazards/disasters 7, 14–15, 55–89, 142, 173, 237–42, 257–61, 262, 266, 269–72, 273, 280–3
   defined 59
   classes 26, 56, 63–5
New York 284, 286, 289
New Zealand 212, 267
Nicaragua 212
Nietschmann, Bernard 140, 248
Nepal 135, 214, 239, 241, 260, 267, 271, 363
non-governmental organisations (NGOs) 15, 180, 184, 190–1

novel risks 4, 9, 16, 92
nuclear 9
   hazards/safety 107, 127
   power plants 97–8, 125
   weapons/war 122, 125–6, 250
'nuclear winter' 138

oil spills 93, 97–8
Oliver-Smith, Anthony 51, 355
omnicide 129, 138
Ontario, Canada 74, 269, 273, 274
O'Riordan, Timothy 141, 151–2
organisations 100–106, 171, 184, 188–91
Osaka, Japan 280–2
ozone depletion 99

Pakistan 86, 88, 135, 161, 199, 204, 211, 215, 218, 249–51, 254, 255–61, 267, 363
Palm, Risa 55, 176, 194
pandemic 56, 139
Pelanda, Carlo 141, 204
Perrow, Charles 163, 185, 284, 286
Peru 200, 208, 211, 241, 244–5, 251, 267, 270
pests 56, 67–70
Philippines 209, 247, 254, 361, 363
photochemical smog 73
place 34, 43, 164–5, 319–20, 352
   attachment to, sense of, 40, 42, 49–53, 320
   defined 320
place annihilation 53, 129, 133, 296–320
plague 56, 69
Plymouth, UK 303, 308
Poland 322–4, 332–6
Pompeii and Herculaneum 282
poverty 3, 147–8, 153–4, 166, 167, 217, 218, 231, 277, 316–7, 355
powerlessness 27–8, 141, 151–3, 162
Prince, S.H. 92, 115
Project Ploughshares 113, 139, 186, 249
public health 70, 140, 157, 158, 276, 279
public policy 13, 92, 169, 188, 193, 294
public safety 4, 10, 171, 179, 181, 188, 192, 278, 294

Quarantelli, E.L. 38, 170
Quetta, Pakistan 179

radiation fog 72
radiation hazards 44–5, 57, 250
rate of onset (of destructive process) 79
reconstruction after disaster 31, 37, 181, 185
Red Cross, Red Crescent Societies 180, 191, 192
refugees 130, 133–8, 249–51, 254, 326, 343
Reitlinger, Gerald 127, 324
release of dangerous forces 116–18, 125, 140
relief measures 32, 37
relief (topographical) 49–50, 205–6, 213–14, 235
resilience 27, 48–9

Index

Richter magnitude (earthquake scale) 213–14
risk 11, 261
  acceptable/unacceptable 9, 359
  assessment 22–3
  defined 22
  elements of 6–13, 22–3
  geography of 224–30, 232–6
  routine 38
  voluntary/involuntary 104, 154, 182
'risk society' (U. Beck) 360
road accidents 6
rockslides, catastrophic 238, 243, 257
Romany (and Sinti) peoples 155, 321, 323
routine risks 4, 96, 187
Russian Republic 267
Rwanda 113, 134, 162, 364

Salang Tunnel, Afganistan 243
San Francisco 31, 34, 58, 208, 216, 227, 272
Santa Ana winds/fires 239
Sarajevo (Bosnia) 40
Scarry, Elaine 131
schistosomiasis (bilharzia) 157
Schrader-Frechette, K.S. 39
sea ice 46, 56
Second World War 118, 120, 122–3, 130, 133,
  136, 145, 167, 244, 297–318, 324, 330
Second Indo-China War (Vietnam War) 120,
  135–8, 289–93
security 143, 276, 279, 294, 348, 359–60
  geography of 48–9
  sense of, loss of 42, 47, 52
  national 247
  social 153
seismic hazards (see earthquakes)
Sen, Amartya 155–6, 187
sex workers (and AIDS) 160–1, 162, 284–7
'slum raids' 316–17
shipping (see transportation)
slope stability 49–50, 187–8, 206–9, 235, 243
smallpox 119, 146
Smith, Keith 87, 89, 233
smog (see photochemical smog)
snow hazards 74–7, 173, 273
snowdrifts 76
social construction of risk/disaster 22, 71, 93,
  106, 115, 147, 219–20, 359
social geography 30, 37, 157, 164, 177, 205,
  217, 275, 278, 300, 305–18, 329, 360
social hazards/violence 12, 26, 57, 111–40,
  244–53, 296–320, 367
social justice 180, 188, 191, 194, 347, 359
Somalia 115, 124, 130, 134, 137, 244, 248,
  350
Sorokin, Pitirim, A. 115
South Africa 214, 267
spatial dimensions 78–9, 85, 86
spatial disorganisation 10, 35, 41
special forces 26
storm surges 52, 260, 262

storms 240, 245, 258, 260–1, 272
Sudan 130, 134, 248
survivors 183–4, 210, 220, 325, 339
'susceptible' soils 209, 214–15, 216, 225, 226,
  228
sustainability 143, 179–80, 188, 233
Switzerland 241

Tangshan, China 38, 201, 222
Tanzania 134, 162
technocracy, technocratic approach 70, 152,
  172, 192, 352–4
technological hazards 7, 91–109, 127, 142,
  190, 242–4, 262, 269–72, 273, 274, 287–8
  defined 91
  classes 26, 57, 99–100
temporal dimensions 36–7, 79
terror, terrorism 116–23, 125, 165–6, 251,
  276, 289
Thailand 135, 254, 271, 284
Three Mile Island 103
thresholds 25, 85, 269
  defined 80
  role of 80
Tibet 218
Titanic 1, 92, 97, 102
Tokyo 31, 199, 208, 273, 305–6, 308, 313–15,
  317
tornadoes 42, 64, 233, 280
Toronto 269, 274
torture 123, 140
tourism 236–7, 239, 242, 243, 245–6, 251,
  260, 261–2
toxic waste 94, 125
toxin weapons (see CBT weapons)
traffic accidents (see road accidents)
transportation (see also aviation, road, rail,
  shipping) 93–4, 237, 242–3, 257, 273,
  275, 280, 387
tropical cyclones (see cyclones, hurricanes,
  typhoons)
tsunamis 56, 198–201, 208, 211, 226, 227, 280
tuberculosis 70, 159
Turkey 199, 209, 212, 224, 244, 247, 250, 251
Turner, Barry A. 36–7, 92, 104–7, 110
typhus 69
typhoons (Pacific Ocean) 64–5, 71, 280–2

Uganda 113, 134, 162, 267, 284, 350
Ukraine 133
'unnatural hazards' 68, 197–231
United Kingdom (Britain) 2, 74, 91–92,
  103–4, 122, 134, 166, 245, 267, 297,
  300–3,
  305–10, 362
United Nations 13–15, 59, 180, 192, 233
United Nations Department of Humanitarian
  Affairs (DHA) 111, 124, 139
United States of America (USA) 2, 6, 74–7,
  109, 122, 125, 159, 160, 161, 172, 233,

240, 245–6, 250, 251–2, 255, 267, 280, 283, 284, 285, 286, 332, 362, 363, 364
uprooting (see displacement)
urban-industrial societies 9, 92, 107–8
urban risks 63, 72–3, 74, 79, 108, 118, 150, 158, 206, 222–4, 230, 251, 266–95, 296–320
urbanisation 59, 157, 230–1, 263, 266–9, 272, 279, 285, 286, 293
  forced 290–2
USSR (former) 125, 126, 133, 136, 250, 363

Vaiont Dam disaster 31, 97, 243
Venezuela 209, 214, 252, 270
Vietnam 120, 135–8, 166, 284–93, 271, 350
Vietnam War (see Second Indo-China War)
violence (see also social hazards, war) 8, 21, 170, 244–53, 276
  defined 114
  hazards 116–22, 129, 131–2, 244–53
visibility hazards 72, 74, 76, 77
volcanic hazards/disasters 67, 77, 237–42, 272, 281
  classes 56
vulnerability 26–8, 31, 63, 66–7, 80, 84, 99, 102, 121, 141–68, 177–83, 215–18, 220–3, 231, 253–4, 255, 266–8, 276–8, 279, 298, 305–18, 330, 336–40, 354–6, 359, 360
  age and 143, 148, 159
  defined 27, 141–2
  enforced 154, 165–7, 187
  classes 27, 144–9, 153, 165
  gender and 48, 145, 159, 161, 168
  occupation and 143, 153
'vulnerability paradigm' 167
vulnerability perspective 28, 70, 141–3, 163, 167, 272
'vulnerability syndrome' 148

Wagner, Philip 89, 105, 214

warning 36, 172, 180, 298–9
war hazards/disasters 7, 8, 111–40, 163, 166, 174–5, 244–53, 277, 284, 289–93, 296–320
  defined 114
  classes 26
'war of extermination' 129, 138–9
war on culture/symbols 129, 134–5
Warsaw ghetto 326, 333, 344
watershed management (and hazards) 49, 153, 158
Waterstone, Marvin 39
Watts, Michael, J. 13, 152, 168, 187, 357
weather modification 97, 118, 153, 158
weapons (see anti-personnel, dubious, CBT, incendiary, nuclear and war hazards)
White, Gilbert, F. 13, 30, 89, 169, 172, 193, 194
wind storms 64, 273
Wisner, Ben 141, 148, 156–7
Wolfenstein, Martha 42, 44, 45
women 99, 130, 148, 218, 249, 275, 278, 284–6, 289, 324, 337, 338, 339–40
  AIDS and 159, 161–2
  casualties 131, 159
  vulnerability of 48, 145, 148, 168, 324, 337, 338, 339–40, 342
  war and 130, 131, 133, 145, 301, 307, 314–16, 317
Würzburg (Germany) 49, 120

yellow fever 56, 69
Yoko Ota (Hiroshima survivor/author) 49–50
Yugoslavia (former) 123, 135, 209, 211, 214, 244, 248
Yungay, Peru 51–2, 200, 208, 221

Ziegler, Donald, M. 13, 25, 107, 110
Zimbabwe 113, 134, 138, 162